S0-AKY-020

WAVE AND STABILITY IN FLUIDS

WAVE AND STABILITY IN FLUIDS

D. Y. Hsieh
The Hong Kong University of Science and Technology

S. P. Ho
Hong Kong Polytechnic

Published by

World Scientific Publishing Co. Pte. Ltd.

P O Box 128, Farrer Road, Singapore 9128

USA office: Suite 1B, 1060 Main Street, River Edge, NJ 07661

UK office: 73 Lynton Mead, Totteridge, London N20 8DH

Library of Congress Cataloging-in-Publication Data

Hsieh, D. Y. (Din-Yu), 1933–
Wave and stability in fluids / D.Y. Hsieh, S.P. Ho.
p. cm.
Includes bibliographical references and index.
ISBN 9810218702
1. Fluid dynamics. 2. Waves. 3. Stability. I. Ho, S. P.
II. Title.
QC151.H75 1994
532'.059--dc20 94-30338
CIP

Printed in Singapore by Uto-Print

PREFACE

The book is based on the courses given by the first author (DYH) over the years at Brown University, Wuhan University and the Hong Kong University of Science and Technology. It is designed for students in applied mathematics and engineering, who have already been exposed to basic fluid dynamics. The main purpose of the book is to provide relatively easy access to the understanding of the essence of the mechanisms involved in waves and stabilities in fluid. The contents of the book is inevitably influenced by the interests and biases of the authors.

Except for Sections 4.1, 4.2 and 4.3, the writing of the book is essentially carried out by the first author. He has to bear almost all the responsibility of the shortcomings of the book.

The book is largely self-contained. References are given only for supplementation of missing details which cannot be found readily in familiar books such as that listed at the end of the book.

We would like to acknowledge the contributions of graphs in Chapter 6 to our colleague, Dr. Xiao-Ping Wang, and to Mr. Chow Chak On, a second year undergraduate student at HKUST.

Our special thanks to Miss Grace Yeung who typed and produced this camera-ready manuscript, and ingeniously helped us to solve many problem relating to the production of the book.

D.Y. HSIEH
S.P. HO

Hong Kong
June 1994

TABLE OF CONTENTS

CHAPTER 1

Preliminaries on Waves

What is wave? It is a disturbance that propagates from one place to another. For the simplest case of linear, nondissipative, and nondispersive plane waves, the shape of the disturbance remains unchanged while propagating with constant speed over large distances. The spatial dimension of a wave may be characterized by the wavelength of its dominant sinusoidal mode, while its undulating nature by the corresponding frequency. The spatial scale of the waves can range from that of a γ-ray which is electromagnetic wave of subatomic dimension to that of density waves in galaxies of the order of tens of thousands of light years. Indeed, wave phenomena are present everywhere in the universe.

In daily life encounters, both sound and light are linear and nondispersive waves. Light is the electromagnetic wave with wavelengths in a definite spectral range, and electromagnetic waves in vacuum is intrinsically linear and nondispersive. On the other hand, sound wave is linear only when the amplitude is small.

Sound waves originate from compressions and rarefactions of the media through which they travel. Compressibility of the medium is the essential element for its occurrence. Another important type of the waves in fluids, the water waves, has a different origin. Water waves arise from the presence of a free surface, or an interface separating air from water, even though water may be considered as incompressible. An interface between media of different densities may be considered as a special case of density stratification. Indeed, waves can be generated in a stratified fluid due to density stratification.

Wave patterns are in general regular and bounded. However for some physical systems there may be certain states in which disturbances tend to grow out of control. These states are then unstable. The instability is usually characterized by unbounded growth or noise. The origin of the instability may be the breakdown of the mathematical model describing the physical system. A purpose of studying stability problems is to understand the mechanism of instability and attempt to find meaning of the subsequent development of the real physical system.

1.1 Governing Equations of the Fluid Motion

We shall deal only with the Newtonian fluids. The basic governing equations for Newtonian fluids are the equation of continuity, the equation of motion and the equation of energy together with thermodynamic relations and constitutive relations that characterize the material properties of the medium. These equations and relations are described in the following.

Equation of Continuity

Let ρ be the density of the fluid, and $\mathbf{v}$ the velocity of the fluid element, then the equation of continuity is

$$\frac{\partial \rho}{\partial t} + \nabla \cdot (\rho \mathbf{v}) = 0. \tag{1.1.1}$$

In tensorial notation, (1.1.1) becomes

$$\frac{\partial \rho}{\partial t} + \frac{\partial}{\partial x_i}(\rho v_i) = 0 , \tag{1.1.2}$$

where the summation convention of repeated index i has been adopted. Thus $\frac{\partial}{\partial x_i}(\rho v_i)$ means $\sum_{i=1}^{3} \frac{\partial}{\partial x_i}(\rho v_i)$. This summation convention will be adopted throughout the book unless otherwise noted. Equation (1.1.1) is the mathematical statement of the conservation of mass.

Equation of Motion

Let σ_{ik} be the component of the rate of stress tensor, and **b** the external body force, then the equation of motion can be expressed in the following tensorial form :

$$\rho\left(\frac{\partial v_i}{\partial t}+v_k\frac{\partial v_i}{\partial x_k}\right)=\frac{\partial \sigma_{ik}}{\partial x_k}+\rho b_i, \qquad i=1,2,3. \tag{1.1.3}$$

For Newtonian fluids, we have the following constitutive relation relating the rate of stress to fluid velocity and the pressure p :

$$\sigma_{ik}=-p\delta_{ik}+\sigma_{ik}', \tag{1.1.4}$$

$$\sigma_{ik}'=\mu\left(\frac{\partial v_i}{\partial x_k}+\frac{\partial v_k}{\partial x_i}-\frac{2}{3}\delta_{ik}\frac{\partial v_l}{\partial x_l}\right)+\mu'\,\delta_{ik}\frac{\partial v_l}{\partial x_l}, \tag{1.1.5}$$

where μ is the shear viscosity coefficient, μ' the bulk viscosity coefficient and δ_{ik} is the Kronecker delta.

While Equation (1.1.3) is quite general, Equations (1.1.4) and (1.1.5) define the Newtonian fluid. For Newtonian fluid with constant viscosities, Equation (1.1.3) can then be expressed in the following vectorial form :

$$\frac{\partial \mathbf{v}}{\partial t}+(\mathbf{v}\cdot\nabla)\mathbf{v} = -\frac{1}{\rho}\nabla p+\upsilon\left[\frac{4}{3}\nabla(\nabla\cdot\mathbf{v})-\nabla\times(\nabla\times\mathbf{v})\right]$$

$$+\ \upsilon'\,\nabla(\nabla\cdot\mathbf{v})+\mathbf{b}, \tag{1.1.6}$$

where $\upsilon=\dfrac{\mu}{\rho}$, and $\upsilon'=\dfrac{\mu'}{\rho}$, are the kinematic viscosity coefficients.

Equation (1.1.6) is known as the Navier-Stokes equation.

Equation of State and Thermodynamic Relations

The pressure of the fluid element p is related to the density ρ and the entropy s. This relationship is called the equation of state which can be expressed as :

$$p = p\,(\rho, s). \tag{1.1.7}$$

In many instances, the entropy s can be assumed to be constant throughout the fluid, then we have

$$p = p\,(\rho). \tag{1.1.8}$$

Such flow is called the barotropical flow. For ideal fluids, i.e. if there is no dissipation due to viscosity and thermal conduction, the flow is barotropic if the entropy is constant throughout the fluid at some instant.

For barotropic flows, Equations (1.1.1), (1.1.6) and (1.1.8) form a closed system. Otherwise, we need another equation to determine the entropy.

Let U be the internal energy of the fluid element, H the enthaply, T, the temperature, c_v, the specific heat at constant volume, and c_p the specific heat at constant pressure. According to the second law of thermodynamics, we have :

$$Tds = dU - \frac{p}{\rho^2}d\rho = c_v dT - \frac{p}{\rho^2}d\rho\ , \tag{1.1.9}$$

or

$$Tds = dH - \frac{1}{\rho}dp = c_p dT - \frac{1}{\rho}dp\ . \tag{1.1.10}$$

For processes that the density change is small, we have approximately

$$Tds = c_v dT \ , \tag{1.1.11}$$

while for processes that the pressure variation is small, we have approximately :

$$Tds = c_p dT \ . \tag{1.1.12}$$

It is often argued that since for most cases the fluid is free to expand, the pressure tends to remain constant. Thus (1.1.12) is commonly adopted. As for water, which can be considered as an incompressible fluid, the value of $\left(\dfrac{c_p - c_v}{c_p}\right)$ is very small, being less than 10^{-2} at room temperature. Therefore, there is not much difference between equations (1.1.11) and (1.1.12).

Equation of Energy

The equation of energy can be expressed as follows :

$$\rho c_v \left[\frac{\partial T}{\partial t} + (\mathbf{v} \cdot \nabla)T\right] = -p(\nabla \cdot \mathbf{v}) + \sigma'_{ik}\frac{\partial v_i}{\partial x_k} + \nabla \cdot (\kappa \nabla T) \ , \tag{1.1.13}$$

where κ is the thermal conductivity. The left hand side of (1.1.13) represents the rate of increase of the internal energy, since $dU = c_V \, dT$ according to (1.1.9). The terms on the right hand side represent successively the rate of work done by the pressure, the dissipation due to viscous stress, and the heat inflow due to thermal conduction.

Using (1.1.1) and (1.1.9), Equation (1.1.13) can also be put in the form :

$$\rho T\left[\frac{\partial s}{\partial t}+(\mathbf{v}\cdot\nabla)s\right]=\sigma_{ik}'\frac{\partial v_i}{\partial x_k}+\nabla\cdot(\kappa\nabla T)\ , \tag{1.1.14}$$

which expresses explicitly the role played by viscous dissipation and thermal conduction on the production of entropy. The term $\sigma_{ik}'\frac{\partial v_i}{\partial x_k}$ can be rewritten as :

$$\sigma_{ik}'\frac{\partial v_i}{\partial x_k}=\frac{\mu}{2}\left(\frac{\partial v_i}{\partial x_k}+\frac{\partial v_k}{\partial x_i}-\frac{2}{3}\delta_{ik}\frac{\partial v_l}{\partial x_l}\right)^2+\mu'\left(\frac{\partial v_l}{\partial x_l}\right)^2 . \tag{1.1.15}$$

If the viscous dissipation is neglected, and the approximation (1.1.12) is adopted, then (1.1.14) for constant κ becomes :

$$\frac{\partial T}{\partial t}+(\mathbf{v}\cdot\nabla)T=D_T\nabla^2 T\ , \tag{1.1.16}$$

where

$$D_T=\frac{\kappa}{\rho c_p}\ , \tag{1.1.17}$$

is known as the thermal diffusivity. Equation (1.1.16) is the well-known equation for heat transfer.

Eulerian and Lagrangian Coordinates

The coordinate system ($\mathbf{x}$, t) which we have used to describe the flow variables is known as the Eulerian coordinate system. The meaning of $\mathbf{v}(\mathbf{x}, t)$ is the velocity of a fluid particle which at time t is located at the position $\mathbf{x}$ in the space. This description is an instantaneous picture taken by a camera fixed in space. It does not track the history or future development of the fluid particle in question.

Another description is to follow a definite fluid particle which is assigned a label ξ. ξ is of the same dimension as **x**, since fluid particles filled the whole space. We shall use τ to denote the time variable. τ and t are the same. But τ is an independent variable with repect to ξ, while t is the independent variable with respect to **x**. This coordinate system (ξ, τ) is known as the Lagrangian coordinate system.

We may express the relationship between Eulerian and Lagrangian coordinate systems by :

$$\mathbf{x} = \mathbf{x}(\xi, \tau)\,, \tag{1.1.18}$$

$$t = \tau\,. \tag{1.1.19}$$

The inverse relationship is

$$\xi = \xi(\mathbf{x}, t)\,, \tag{1.1.20}$$

$$\tau = t. \tag{1.1.21}$$

Since **v** is the rate of change of position **x** for the same fluid particle, it is by definition that

$$\mathbf{v} = \frac{\partial \mathbf{x}}{\partial \tau}\,. \tag{1.1.22}$$

Equation (1.1.22) connects the Eulerian and Lagrangian coordinate systems.

The operator $\frac{\partial}{\partial \tau}$ is the time differentiation following the fluid particle. Using (1.1.22) we have

$$\frac{\partial}{\partial \tau} = \frac{\partial t}{\partial \tau}\frac{\partial}{\partial t} + \frac{\partial x_i}{\partial \tau}\frac{\partial}{\partial x_i} = \left(\frac{\partial}{\partial t} + \mathbf{v}\cdot\nabla\right). \qquad (1.1.23)$$

We may note that in writing down the basic equations, we have employed the notion of the time differentiation following the fluid particle.

1.2 The Wave Equation

We begin our discussion of waves with small amplitudes. Consider first the fluid in an equilibrium state, i.e., $p = p_0$, $\rho = \rho_0$, $s = s_0$ all constants, and $\mathbf{v} = \mathbf{v}_0 = 0$. Take the case that there is no external force, i.e., $\mathbf{b} = 0$. Fluids in such an equilibrium state clearly satisfy the governing equations discussed in the previous section.

Now let the density, pressure and other physical quantities of the fluid be of the form :

$$\rho = \rho_0 + \rho' , \quad p = p_0 + p' , \quad \cdots ,$$

and substitute into (1.1.1), (1.1.6) and (1.1.14). Since we are interested in the small amplitude perturbations of the equilibrium system, we shall retain only the first order terms of the perturbed quantities ρ', p' etc. Thus we obtain a set of linearized equations. For simplicity of notation, we shall write ρ for ρ', p for p', etc., i.e. drop all the superscripts in the linearized equations. Then, Equations (1.1.1), (1.1.6) and (1.1.14) become

$$\frac{\partial \rho}{\partial t} + \rho_0 (\nabla \cdot \mathbf{v}) = 0 \ , \tag{1.2.1}$$

$$\frac{\partial v}{\partial t} = -\frac{1}{\rho_0} \nabla p + \upsilon \nabla^2 \mathbf{v} + \left(\upsilon' + \frac{\upsilon}{3} \right) \nabla (\nabla \cdot \mathbf{v}) \ , \tag{1.2.2}$$

$$\frac{\partial s}{\partial t} = \frac{\kappa}{\rho_0 T_0} \nabla^2 T \ . \tag{1.2.3}$$

For barotropical flows, p depends only on ρ, then we can eliminate $\mathbf{v}$ from (1.2.1) and (1.2.2), and obtain

$$\frac{\partial^2 \rho}{\partial t^2} = c^2 \nabla^2 \rho + \overline{\upsilon} \nabla^2 \frac{\partial \rho}{\partial t} \ , \tag{1.2.4}$$

where

$$c^2 = \left(\frac{dp}{d\rho} \right)_0 \ , \tag{1.2.5}$$

$$\overline{\upsilon} = \frac{4}{3} \upsilon + \upsilon' \ . \tag{1.2.6}$$

If $\overline{\upsilon}$ may be neglected, Equation (1.2.4) is reduced to the well known wave equation :

$$\frac{\partial^2 \rho}{\partial t^2} = c^2 \nabla^2 \rho \ , \tag{1.2.7}$$

where c is the velocity of propagation of the waves, and is the sound speed for this case.

In Equation (1.2.4), the term with $\overline{\upsilon}$ represents the viscous damping. In general the equation of state is $p = p(\rho, s)$, then there will be damping due also to thermal conduction.

Under atmospheric pressure, sound speed in dry air at 15°C is about 340.6 m/s. In the audible frequency range, sound propagation is essentially adiabatic. Thus $c^2 = \left(\frac{\partial p}{\partial \rho}\right)_s$. The sound speed in the water at 20°C is about 1484 m/s, which is much higher than the sound speed in air.

1.3 Simple Solutions : Sinusoidal Waves

Let us consider the wave equation we derived in the previous section :

$$\frac{\partial^2 \rho}{\partial t^2} - c^2 \nabla^2 \rho = 0 . \tag{1.3.1}$$

The pressure perturbation also satisfies the same equation :

$$\frac{\partial^2 p}{\partial t^2} - c^2 \nabla^2 p = 0 . \tag{1.3.2}$$

For inviscid fluids and when the flow is irrotational, we may introduce the velocity potential φ and express $\mathbf{v}$ in terms of φ :

$$\mathbf{v} = \nabla \varphi . \tag{1.3.3}$$

From (1.2.2), we thus obtain

$$p = -\rho_0 \frac{\partial \varphi}{\partial t} . \tag{1.3.4}$$

Hence, φ also satisfies the wave equation :

$$\frac{\partial^2 \varphi}{\partial t^2} - c^2 \nabla^2 \varphi = 0 . \tag{1.3.5}$$

Let us discuss first the propagation of waves in one spatial dimension, i.e. consider first the case that $\varphi = \varphi(x,t)$. Then (1.3.5) becomes

$$\frac{\partial^2 \varphi}{\partial t^2} - c^2 \frac{\partial^2 \varphi}{\partial x^2} = 0 . \tag{1.3.6}$$

If we change the variables (x, t) to (ζ, η) with

$$\zeta = x - ct , \quad \eta = x + ct , \tag{1.3.7}$$

then Equation (1.3.6) becomes

$$\frac{\partial^2 \varphi}{\partial \zeta \partial \eta} = 0 . \tag{1.3.8}$$

Therefore we obtain

$$\varphi = f(\zeta) + g(\eta) , \tag{1.3.9}$$

or

$$\varphi = f(x - ct) + g(x + ct) , \tag{1.3.10}$$

where f and g are arbitrary functions.

$f(x - ct)$ represents a wave travelling in the positive x-direction, while $g(x + ct)$ a wave progressing in the negative x-direction, both with the same propagation speed c.

From the perspective of 3-dimensional waves, these solutions represent the plane waves, since the wavefronts are planes $x =$ constant. Now

$$\mathbf{v} = \nabla \varphi = \frac{\partial \varphi}{\partial x} \mathbf{e}_x , \tag{1.3.11}$$

where $\mathbf{e}_x$ is the unit vector in x-direction. Therefore the fluid velocity is parallel to the direction of propagation. Consequently, it is a longitudinal wave. It is worth noting that the sound wave, which is a wave of compression and rarefaction, is a longitudinal wave.

The general wave profile $\varphi(x,t)$ can be expressed in terms of Fourier integrals, i.e.,

$$\varphi(\mathbf{x},t) = \int_{-\infty}^{\infty} \varphi_\omega(\mathbf{x}) e^{-i\omega t} d\omega \ . \qquad (1.3.12)$$

In this representation, $\varphi(\mathbf{x},t)$ is considered to be a superposition of monochromatic waves with different ω. Since the wave equation is linear, the principle of superposition is valid, i.e., superposition of any solutions is also a solution of the equation. Therefore much can be learned from the discussion of monochromatic waves. Let us now consider a monochromatic wave.

$$\varphi(\mathbf{x},t) = \varphi(\mathbf{x}) e^{-i\omega t} \ . \qquad (1.3.13)$$

For 1-dimensional monochromatic waves, $\varphi(x)$ satisfies the following equation :

$$\frac{d^2\varphi}{dx^2} + k^2\varphi = 0 \ , \qquad (1.3.14)$$

where

$$k = \frac{\omega}{c} \ . \qquad (1.3.15)$$

(1.3.14) has solutions e^{ikx} and e^{-ikx} . Therefore $\varphi(\mathrm{x},t)$ is in general of the form :

$$\varphi(x,t) = Ae^{i(kx-\omega t)} + Be^{-i(kx+\omega t)} \ . \qquad (1.3.16)$$

$\varphi(x,t)$ is known as sinusoidal wave, since it is expressed in terms of sinusoidal functions. In this expression, ω is called frequency, and k wavenumber. The expression $\lambda = \dfrac{2\pi}{k}$ is called wavelength, and $\tau = \dfrac{2\pi}{\omega}$ is called period. The exponent $\Phi = (kx - \omega t)$ or $(kx + \omega t)$ is called phase. Thus the wave speed $c = \dfrac{\omega}{k} = \left(\dfrac{dx}{dt}\right)_{\Phi}$ is also called the phase velocity, since it is the velocity for constant phase. The quantities A and B in (1.3.16) are known as amplitudes.

Consider the sinusoidal waves moving in the positive x direction only, i.e. take $B = 0$ in (1.3.16). Then a general wave profile can be constructed from the monochromatic waves by superposition :

$$\varphi(\mathrm{x},t) = \int A(\omega) e^{i(kx-\omega t)} d\omega \,. \qquad (1.3.17)$$

For the present case, since the phase velocity $c = \dfrac{\omega}{k}$ is independent of ω, we can rewrite (1.3.17) as :

$$\varphi(x,t) = \int A(\omega) e^{i\frac{\omega}{c}(x-ct)} d\omega = \varphi(x-ct) \,. \quad (1.3.18)$$

Therefore the wave propagates with c without change of the original shape. Those waves, for which the phase velocity is independent of frequency or wavelength, are known as non-dispersive waves, and the sound wave is non-dispersive. If the phase velocity varies with frequency then the waves are dispersive. For dispersive waves, it is evident from (1.3.18) that $\varphi(x,t)$ cannot be put in the form of $\varphi(x-ct)$ since c varies with ω. The original shape of the wave profile cannot be maintained for dispersive waves.

The 3-dimensional monochromatic plane waves are of the form :

$$\varphi(\mathbf{x},t) = Ae^{i(\mathbf{k}\cdot\mathbf{x}-\omega t)} + Be^{-i(\mathbf{k}\cdot\mathbf{x}+\omega t)} , \qquad (1.3.19)$$

where

$$|\mathbf{k}| = \frac{\omega}{c} .$$

(1.3.19) represents plane waves propagating in directions parallel to **k**, the wavenumber vector. A general monochromatic wave is superposition of monochromatic plane waves propagating in various directions.

1.4 Fundamental Solutions of Wave Equation

It is illuminating to discuss the fundamental solutions of the wave equation. The fundamental solution of the 3-dimensional wave equation in the infinite space-time satisfies the following equation :

$$\frac{\partial^2 \varphi}{\partial t^2} - c^2\nabla^2\varphi = \delta(x-x')\delta(y-y')\delta(z-z')\delta(t-t') , \quad (1.4.1)$$

where $\delta(x)$ is the Dirac δ-function. The solution of (1.4.1) represents the disturbance caused by impulse struck at the point (x', y', z') at time t'. There is no loss of generality to set $x'=y'=z'=t'=0$, since it is equivalent to change the variables from $(\mathbf{x}, t)$ to $(\mathbf{x} - \mathbf{x}', t - t')$. Therefore, let us consider

$$\frac{\partial^2 \varphi}{\partial t^2} - c^2\nabla^2\varphi = \delta(x)\delta(y)\delta(z)\delta(t) . \qquad (1.4.2)$$

Introduce the 4-fold Fourier transform $\Phi(\mathbf{k},\omega)$ of $\varphi(\mathbf{x},t)$ by :

$$\Phi(\mathbf{k},\omega) = \frac{1}{(2\pi)^2}\iiiint_{-\infty}^{\infty} dx\, dy\, dz\, dt\, \varphi(\mathbf{x},t)e^{-i(\mathbf{k}\cdot\mathbf{x}-\omega t)} , \quad (1.4.3)$$

and

$$\varphi(\mathbf{x},\, t) = \frac{1}{(2\pi)^2}\iiiint_{-\infty}^{\infty} dk_x\, dk_y\, dk_z\, d\omega\, \Phi(\mathbf{k},\omega)\, e^{i(\mathbf{k}\cdot\mathbf{x}-\omega t)} . \quad (1.4.4)$$

The transformed equation of (1.4.2) is

$$(\omega^2 - k^2c^2)\Phi = -\frac{1}{(2\pi)^2} ,$$

or

$$\Phi(\mathbf{k},\omega) = -\frac{1}{(2\pi)^2(\omega^2 - c^2k^2)} , \quad (1.4.5)$$

where $k = |\mathbf{k}|$.

Hence

$$\varphi(\mathbf{x},t) = -\frac{1}{(2\pi)^4}\iiiint_{-\infty}^{\infty} dk_x\, dk_y\, dk_z\, d\omega\, \frac{e^{i(\mathbf{k}\cdot\mathbf{x}-\omega t)}}{\omega^2 - c^2k^2} . \quad (1.4.6)$$

Introduce the spherical polar coordinates (k,θ,ψ) in $\mathbf{k}$-space. Let the polar axis be directed along the vector $\mathbf{x}$. Then

$$\iiint_{-\infty}^{\infty} dk_x\, dk_y\, dk_z \frac{e^{i\mathbf{k}\cdot\mathbf{x}}}{\omega^2 - c^2k^2} = 2\pi\int_0^{\infty} dk \int_{-1}^{1} d(\cos\theta)\frac{k^2 e^{ikr\cos\theta}}{\omega^2 - k^2c^2}$$

$$= \frac{2\pi}{ir}\int_{-\infty}^{\infty} dk \frac{ke^{ikr}}{\omega^2 - c^2k^2} ,$$

where $r = |\mathbf{x}|$.

Thus

$$\varphi(x,t) = -\frac{1}{(2\pi)^3 ir}\int_{-\infty}^{\infty} dk \int_{-\infty}^{\infty} d\omega \frac{ke^{i(kr-\omega t)}}{\omega^2 - c^2k^2} \,. \quad (1.4.7)$$

The integral over ω can only be performed with appropriate interpretation, since there are singularities along the path of integration at $\omega = \pm ck$.

The difficulty can be overcome if we appeal to the causality principle that there should be no disturbance before the impulse struck at $t = 0$; i.e., $\varphi(\mathbf{x}, t) = 0$, for $t < 0$. Then the path of the integration in the ω-plane should be the one that passes above the singularities, i.e., the path C as shown in Fig. 1-1.

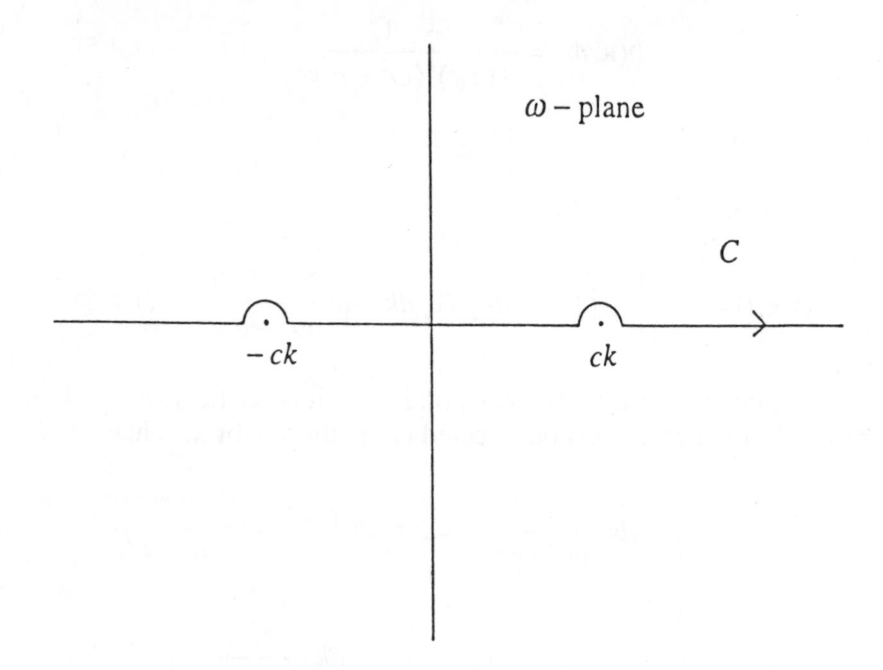

Fig. 1-1 The ω-plane showing the path of integration above the singularities of the integrand in (1.4.7).

It should be pointed out that the difficulty is overcome by appealing to the physical principle of causality. Mathematically, since (1.4.2) is symmetric with respect to time reversal, the issue cannot be resolved. An alternate way is to consider the more realistic case with viscous damping, i.e., going back to Equation (1.2.4). Then the demoninator of the integrand in (1.4.7) becomes $(\omega^2 - c^2k^2 + i\overline{\upsilon}\omega k^2)$. For $\overline{\upsilon}$ infinitesimally small, the singularities are located at

$$\omega = \pm ck - \frac{i\overline{\upsilon}k^2}{2} . \tag{1.4.8}$$

Therefore the integration along real ω can be performed and the result is the same as that of deforming the contour above the singularities. The inclusion of dissipation destroys the symmetry of time reversal, and it defines unambiguously the direction of future.

With these preliminaries, the integration of (1.4.7) can be carried out by deforming the contour C, and we obtain :

$$\varphi(\mathbf{x},t) = 0, \quad \text{for } t < 0 ,$$

and

$$\varphi(\mathbf{x},t) = \frac{1}{4\pi cr}\delta(r - ct), \quad \text{for } t > 0.$$

Or, going back to Equation (1.4.1), we obtain the fundamental solution :

$$\varphi(\mathbf{x},t;\mathbf{x}',t') = \begin{cases} 0, & t < t' \\ \dfrac{1}{4\pi c|\mathbf{x}-\mathbf{x}'|}\delta[|\mathbf{x}-\mathbf{x}'| - c(t-t')], & t > t' \end{cases} . \tag{1.4.9}$$

The general inhomogeneous wave equation

$$\frac{\partial^2 \varphi}{\partial t^2} - c^2\, \nabla^2 \varphi = f(\mathbf{x},t)\ , \qquad (1.4.10)$$

can then be solved with the aid of the fundamental solution (1.4.9), and we obtain

$$\varphi(\mathbf{x},t) = \frac{1}{4\pi c}\int\!\!\int\!\!\int_{-\infty}^{\infty} d^3\mathbf{x}' \int_{-\infty}^{\infty} dt'\, \frac{f(\mathbf{x}',t')\,\delta[|\mathbf{x}-\mathbf{x}'|-c(t-t')]}{|\mathbf{x}-\mathbf{x}'|}$$

$$= \frac{1}{4\pi c^2}\int\!\!\int\!\!\int_{-\infty}^{\infty} d^3\mathbf{x}'\, \frac{f(\mathbf{x}',t-\frac{1}{c}|\mathbf{x}-\mathbf{x}'|)}{|\mathbf{x}-\mathbf{x}'|}\ . \qquad (1.4.11)$$

This is called the retarded potential, since the influence of the source f is evaluated at a previous time $(t-\frac{1}{c}|\mathbf{x}-\mathbf{x}'|)$, a time retarded by an amount $\frac{1}{c}|\mathbf{x}-\mathbf{x}'|$ which is required for the wave to travel with velocity c from the source point $\mathbf{x}'$ to the field point $\mathbf{x}$.

It is quite simple to obtain from the fundamental solution of the 3-dimensional wave equation the fundamental solutions of the 2-dimensional and 1-dimensional wave equations. We shall denote them $G_3(x-x',\, y-y',\, z-z',\, t-t')$, $G_2(x-x',\, y-y',\, t-t')$ and $G_1(x-x',t-t')$ respectively. Thus we have :

$$\frac{\partial^2 G_3}{\partial t^2} - c^2\left[\frac{\partial^2 G_3}{\partial x^2}+\frac{\partial^2 G_3}{\partial y^2}+\frac{\partial^2 G_3}{\partial z^2}\right]$$

$$= \delta(x-x')\,\delta(y-y')\,\delta(z-z')\,\delta(t-t')\ , \qquad (1.4.12)$$

$$\frac{\partial^2 G_2}{\partial t^2} - c^2\left[\frac{\partial^2 G_2}{\partial x^2}+\frac{\partial^2 G_2}{\partial y^2}\right] = \delta(x-x')\,\delta(y-y')\,\delta(t-t') \qquad (1.4.13)$$

and

$$\frac{\partial^2 G_1}{\partial t^2} - c^2 \frac{\partial^2 G_1}{\partial x^2} = \delta(x - x')\, \delta(t - t') . \qquad (1.4.14)$$

Then it is clear that

$$G_2(x - x', y - y', t - t')$$

$$= \int_{-\infty}^{\infty} G_3(x - x', y - y', z - z', t - t')\, dz , \qquad (1.4.15)$$

and

$$G_1(x - x', t - t') = \int_{-\infty}^{\infty} G_2(x - x', y - y', t - t')\, dy . \quad (1.4.16)$$

Now G_3 is the $\varphi(\mathbf{x}, t; \mathbf{x}', t')$ in (1.4.9). By carrying out the integrations, we thus obtain,

$$G_2(x - x', y - y', t - t')$$

$$= \begin{cases} 0 , & (t - t') < \frac{1}{c}|\mathbf{x} - \mathbf{x}'| , \\ \frac{1}{2\pi c[c^2(t - t')^2 - |\mathbf{x} - \mathbf{x}'|^2]^{1/2}} , & (t - t') > \frac{1}{c}|\mathbf{x} - \mathbf{x}'| , \end{cases} \qquad (1.4.17)$$

where

$$|\mathbf{x} - \mathbf{x}'|^2 = (x - x')^2 + (y - y')^2 ,$$

and

$$G_1(x - x', t - t') = \begin{cases} 0, & (t - t') < \frac{1}{c}\,|x - x'|, \\ \frac{1}{2c}, & (t - t') > \frac{1}{c}\,|x - x'|. \end{cases} \tag{1.4.18}$$

CHAPTER 2

Sound Waves

In the narrow sense, sound wave refers to the oscillatory motion of small amplitude in a compressible fluid. However, we shall also touch upon finite amplitude wave in a compressible fluid. We shall also discuss the effect of dissipation on the propagation of waves in a compressible fluid. The emphasis is to bring forth the essential features displayed by these mechanisms.

2.1 Waves in Compressible Fluids

The governing equations for fluid motion have been described in Section 1.1. To summarize, Equations (1.1.1), (1.1.6), (1.1.7), (1.1.12) and (1.1.13) constitute the basic system governing the motion of a compressible, viscous and thermally conducting fluid. When the viscosity and themal conductivity can be neglected as a first approximation, the basic system is reduced to :

$$\frac{\partial \rho}{\partial t} + \nabla \cdot (\rho \mathbf{v}) = 0 \;, \tag{2.1.1}$$

$$\frac{\partial \mathbf{v}}{\partial t} + (\mathbf{v} \cdot \nabla)\mathbf{v} = -\frac{c^2}{\rho}\nabla \rho + \mathbf{b} \;, \tag{2.1.2}$$

where

$$c^2 = \left(\frac{dp}{d\rho}\right) \tag{2.1.3}$$

is the sound speed.

The study of sound waves, both linear and nonlinear, deals mainly with the Equations (2.1.1) and (2.1.2).

For discussion on generation of sound, instead of (2.1.1), we often deal with the following equation :

$$\frac{\partial \rho}{\partial t} + \nabla \cdot (\rho \mathbf{v}) = q \ , \tag{2.1.4}$$

where q represents a source term. The Equation (2.1.2) should then more appropriately be in the conservation form :

$$\frac{\partial}{\partial t}(\rho v_i) + \frac{\partial}{\partial x_j}(\rho v_i v_j) = -c^2 \frac{\partial \rho}{\partial x_i} + \rho b_i \ , \quad i = 1,2,3 \ . \tag{2.1.5}$$

Differentiating (2.1.4) with respect to t and differentiating (2.1.5) with respect to x_i and then substract, we obtain

$$\frac{\partial^2 \rho}{\partial t^2} - \frac{\partial}{\partial x_i}\left(c^2 \frac{\partial \rho}{\partial x_i} \right) = \frac{\partial q}{\partial t} - \frac{\partial}{\partial x_i}(\rho b_i) + \frac{\partial^2}{\partial x_i \partial x_j}(\rho v_i v_j) \ . \tag{2.1.6}$$

Or

$$\frac{\partial^2 \rho}{\partial t^2} - c_0^2 \frac{\partial^2 \rho}{\partial x_i^2} = \frac{\partial q}{\partial t} - \frac{\partial}{\partial x_i}(\rho b_i) + \frac{\partial^2}{\partial x_i \partial x_j} T_{ij} \ , \tag{2.1.7}$$

where

$$c_0^2 = \left(\frac{dp}{d\rho} \right)_0 \ , \tag{2.1.8}$$

$$T_{ij} = \rho v_i v_j + [(p - p_0) - (\rho - \rho_0) c_0^2] \delta_{ij} \ , \tag{2.1.9}$$

and p_0, ρ_0, c_0 are all constants referring to the quantities in equilibrium condition.

When the source term q and the external force are absent, and when we are dealing with disturbance of small amplitudes, then Equation (2.1.7) become the familiar wave equation :

$$\frac{\partial^2 \rho}{\partial t^2} - c_0^2 \nabla^2 \rho = 0. \qquad (2.1.10)$$

In the following, when we discuss small amplitude wave, we shall neglect the subscript 0 in c_0, and just use c for the constant sound speed.

2.2 Multi-pole Radiation

The wave equation (2.1.10) or equivalently (1.3.6), in terms of the spherical polar coordinates (r, θ, ψ), has the following form for the velocity potential φ :

$$\frac{1}{r^2}\frac{\partial}{\partial r}\left(r^2 \frac{\partial \varphi}{\partial r}\right) + \frac{1}{r^2 \sin\theta}\frac{\partial}{\partial \theta}\left(\sin\theta \frac{\partial \varphi}{\partial \theta}\right)$$

$$+ \frac{1}{r^2 \sin^2\theta}\frac{\partial^2 \varphi}{\partial \psi^2} - \frac{1}{c^2}\frac{\partial^2 \varphi}{\partial t^2} = 0 . \qquad (2.2.1)$$

If the problem possesses the spherical symmetry, then $\varphi = \varphi(r,t)$, and (2.2.1) can be written as

$$\frac{\partial^2}{\partial r^2}(r\varphi) - \frac{1}{c^2}\frac{\partial^2}{\partial t^2}(r\varphi) = 0 . \qquad (2.2.2)$$

The general solution of (2.2.2) is

$$\varphi = \frac{1}{r} f(r - ct) + \frac{1}{r} g(r + ct) . \qquad (2.2.3)$$

The first term on the right hand side of (2.2.3) represents the outgoing wave, while the second term represents the incoming wave. In contrast with the plane waves represented by (1.3.10), the amplitudes of the outgoing and incoming waves diminishes as the distance from the origin, r, increases.

If the problem does not possess spherical symmetry, let us consider a particular Fourier mode, and let

$$\varphi(\mathbf{x},t) = \varphi(\mathbf{x})e^{-i\omega t} . \tag{2.2.4}$$

Then $\varphi(\mathbf{x})$ satisfies the following Helmholtz equation :

$$\nabla^2\varphi + k^2\varphi = 0 , \tag{2.2.5}$$

where $k = \dfrac{\omega}{c}$.

The general solution of (2.2.5), using the method of separation of variables, may be expressed as follows :

$$\varphi(\mathbf{x},t) = \sum_{l,m} \begin{Bmatrix} j_l(kr) \\ n_l(kr) \end{Bmatrix} \begin{Bmatrix} P_l^m(\cos\theta) \\ Q_l^m(\cos\theta) \end{Bmatrix} \begin{Bmatrix} e^{im\psi} \\ e^{-im\psi} \end{Bmatrix} e^{-i\omega t} , \tag{2.2.6}$$

where j_l and n_l are the spherical Bessel functions, P_l^m and Q_l^m are the associated Legendre functions. Some of the properties of these functions are described in Appendix 1. We know that $Q_l^m(z)$ is singular at $z = \pm 1$; and $P_l^m(z)$ is also singular at $z = -1$, if l is not an integer. Therefore, if the problem under consideration covers the whole region of (θ, ψ), a regular solution of φ cannot contain Q_l^m. Moreover, l and m need to be integers with $l \geq m$. Then the general solution (2.2.6) becomes :

$$\varphi(\mathbf{x},t) = \sum_{l=0}^{\infty} \sum_{m=-l}^{l} \left[C_{lm} j_l(kr) + D_{lm} n_l(kr) \right] Y_{lm}(\theta,\psi)\, e^{-i\omega t} , \tag{2.2.7}$$

where

$$Y_{lm}(\theta, \psi)$$

$$= \left[\frac{(2l+1)}{4\pi} \frac{(l-|m|)!}{(l+|m|)!} \right]^{1/2} P_l^{|m|}(\cos\theta)\, e^{im\psi} \times \begin{cases} (-1)^m, & (m \geq 0)\,, \\ 1\,, & (m < 0)\,. \end{cases} \tag{2.2.8}$$

Y_{lm} is known as the spherical harmonics which satisfies the following orthonormal relation :

$$\int d\Omega\, Y^*_{lm}(\Omega)\, Y_{l'm'}(\Omega) = \delta_{ll'} \delta_{mm'}\,, \tag{2.2.9}$$

where Ω is the solid angle and $d\Omega = \sin\theta\, d\theta\, d\psi$.

The solution (2.2.7) can also be expressed in terms of the spherical Hankel function :

$$\varphi(\mathbf{x}, t) = \sum_{l=0}^{\infty} \sum_{m=-1}^{l} \left[A_{lm} h_l^{(1)}(kr) + B_{lm} h_l^{(2)}(kr) \right] Y_{lm}(\theta, \psi)\, e^{-i\omega t}\,. \tag{2.2.10}$$

Making use of the asymptotic expressions of the spherical Hankel functions, we obtain, for large r, that

$$\varphi(\mathbf{x}, t) \approx \sum_{l,m} \left[\frac{(-i)^{l+1}}{kr} A_{lm} e^{i(kr-\omega t)} + \frac{i^{l+1}}{kr} B_{lm} e^{-i(kr+\omega t)} \right] Y_{lm}(\theta, \psi)\,. \tag{2.2.11}$$

It is evident that the terms on the right hand side of (2.2.11) are divided into two groups. One group represents the outgoing waves, while the other the incoming waves. The variations along the radial direction are either $\frac{1}{r} e^{i(kr-\omega t)}$ or $\frac{1}{r} e^{-i(kr+\omega t)}$. But the angular variations are determined by Y_{lm}, depending on the values of l and m. These are called multi-pole radiations. In particular, we have the following designations :

(i) l=0 : Monopole radiation.

Because $Y_{00} = \left(\dfrac{1}{4\pi}\right)^{1/2}$, monopole radiation is spherically symmetric.

(ii) l=1 : Dipole radiation.

The angular variations are given by

$$Y_{10} = \left(\frac{3}{4\pi}\right)^{1/2} \cos\theta\,, \quad Y_{1,\pm1} = \mp\left(\frac{3}{8\pi}\right)^{1/2} \sin\theta\, e^{\pm i\psi}\,.$$

(iii) l=2 : Quadrupole radiation.

The angular variations are given by

$$Y_{20} = \left(\frac{5}{16\pi}\right)^{1/2} \left(3\cos^2\theta - 1\right),$$

$$Y_{2,\pm1} = \mp\left(\frac{15}{8\pi}\right)^{1/2} \sin\theta\cos\theta\, e^{\pm i\psi}\,,$$

$$Y_{2,\pm2} = \left(\frac{15}{32\pi}\right)^{1/2} \sin^2\theta\, e^{\pm 2i\psi}\,.$$

It may be noted that, among those spherical Bessel functions, only $j_l(kr)$ is regular at r=0. Thus if the problem has no singularity at r=0, then only the terms associated with j_l can be present in (2.2.7). Therefore, only the standing waves are present in the neighborhood of r=0, i.e. the amplitudes of the outgoing and incoming waves are the same there.

2.3 Generation of Sound

A moving body in the fluid will cause disturbances in the fluid, therefore the sound wave. It is very difficult to find the solution of this general boundary value problem. In this section, we shall discuss only the basic physical ideas and make some approximate estimates.

We shall consider the case that the motion of the body can be largely characterized by a characteristic frequency ω. Thus a characteristic wavelength $\lambda = \dfrac{2\pi c}{\omega}$ can be obtained. Let the linear dimension of the body be L. We shall discuss the cases when $\lambda << L$ and $\lambda >> L$.

The case $\lambda << L$.

For this case, the surface of the body can be considered to compose of many elemental parts, each element small enough to be considered as plane. However the linear dimension of each "plane" is still large compared with λ. Then plane waves are emitted by the motion of these planes. Only the movements normal to the planes will cause fluid motions. Let u_n be the normal component of the body velocity on its surface. These then will be the fluid velocities of the plane waves on the surface of the body. Thus the sound field is composed of many plane waves in all directions issuing from the body surface with varying amplitudes.

We can estimate the intensity of sound generated in the following way. Intensity of sound I is defined as the time rate of acoustic energy radiated. In the plane wave, we have $p = \rho_0 cv$. Now the energy flux in the plane wave is pv, since this is the work done by the fluid element per unit area and unit time, and fluid velocity is in the direction of propagation. Therefore the energy flux is $\rho_0 cv^2$. Bearing in mind the oscillatory nature of the sound, we obtain the mean energy flux as $\rho_0 c\overline{v^2}$. Now $v = u_n$ on the body.

Let the surface of the body be denoted by S, we thus obtain the intensity of sound generated by the moving body :

$$I = \rho_0 c \int_S \overline{u_n^2}\, dS \,. \qquad (2.3.1)$$

It is worth noting that for this case, I is independent of frequency ω.

The case $\lambda >> L$.

For monochromatic waves, the sound field is given by (2.2.10) which becomes (2.2.11) as $r \to \infty$. When we are dealing with the problem of sound generation, it is evident that there are only the outgoing waves and no incoming waves. Thus the sound field is for this case :

$$\varphi = \sum_{l,m} A_{lm} h_l^{(1)}(kr) Y_{lm}(\theta, \psi) e^{-i\omega t} \,. \qquad (2.3.2)$$

Now $\lambda >> L$ means $kL << 1$. Let us consider a sphere with radius r_1, which just barely contains the whole body. Therefore $kr_1 << 1$ also. From (A1.3) and (A1.7), we obtain that

$$h_l^{(1)}(kr_1) \approx (-i) \frac{(2l-1)!!}{(kr_1)^{l+1}} \,. \qquad (2.3.3)$$

Consider the sound field φ at $r = r_1$, it is clear that unless there is peculiar angular distribution of the sound field, we can obtain the following estimate :

$$A_{lm} = 0((kL)^{l+1}) \,. \qquad (2.3.4)$$

Now in the far field for large r, we have

$$\varphi \approx \sum_{l,m} (-i)^{l+1} \frac{A_{lm}}{kr} Y_{lm}(\theta, \psi)\, e^{i(kr - \omega t)} \,. \qquad (2.3.5)$$

Therefore, the smaller the value of l, the more important is the contribution from A_{lm}. If $A_{00} \neq 0$, then the term associated with A_{00} is the most important one.

A_{00} is closely related to mass flux I_m, i.e. the total mass flow out of a controlling surface surrounding the body per unit time. Let us take the controlling surface be the sphere of radius r. Then we have

$$I_m = \rho_0 \int \frac{\partial \varphi}{\partial r} r^2 d\Omega$$

$$= \frac{\rho_0 A_{00}}{(4\pi)^{1/2}} e^{-i\omega t} \int \frac{d}{dr}\left[h_0^{(1)}(kr)\right] r^2 d\Omega \ , \qquad (2.3.6)$$

since

$$\int Y_{lm}(\theta, \psi) d\Omega = 0 \ , \quad for\, l \neq 0 \ .$$

Now, we have from (A1.6), that

$$\frac{d}{dr}\left[h_0^{(1)}(kr)\right] = -\frac{e^{ikr}}{ikr^2}[1 - ikr] \ .$$

Therefore at $r = r_1$, where $kr_1 << 1$, we obtain

$$I_m \approx \frac{(4\pi)^{1/2} i\rho_0 A_{00}}{k} e^{-i\omega t} \ . \qquad (2.3.7)$$

Thus the mass flux is independent of r near the body.

Suppose the volume of the body $V(t)$ is oscillating with frequency, i.e.

$$V(t) = \overline{V} + V_0 e^{-i\omega t} \ . \qquad (2.3.8)$$

The mass flux induced in the fluid is evidently given by

$$I_m = \rho_0 \frac{dV}{dt} = -i\omega\rho_0 V_0 e^{-i\omega t} \ . \tag{2.3.9}$$

Comparing (2.3.9) and (2.3.7), we thus obtain

$$A_{00} = -\frac{\omega k V_0}{(4\pi)^{1/2}} \ . \tag{2.3.10}$$

The expression (2.3.5) is thus dominated by the leading term :

$$\varphi \approx \frac{i\omega V_0}{4\pi r} e^{-i\omega\left(t-\frac{r}{c}\right)} \ . \tag{2.3.11}$$

The above analysis can be generalized to general non-monochromatic waves. The main idea is that the leading term associated with A_{00} is the spherically symmetric outgoing wave. For the general spherically symmetric wave, we have

$$\varphi \approx \frac{1}{r} f\left(t - \frac{r}{c}\right) . \tag{2.3.12}$$

Since we are still dealing with the case that the representative wavelength λ is much larger than the body length L, $f\left(t-\frac{r}{c}\right)$ is practically just $f(t)$ near the body. In other words, it is practically the potential flow near the body, i.e.,

$$\varphi \approx \frac{1}{r} f(t) , \qquad \text{for } r \text{ small.} \tag{2.3.13}$$

The mass flux I_m calculated according to (2.3.6) is thus

$$I_m = -4\pi\rho_0 f(t) \ . \tag{2.3.14}$$

Now, we have (2.3.9) :

$$I_m = \rho_0 \frac{dV}{dt} \equiv \rho_0 \dot{V}(t) , \tag{2.3.15}$$

where we have used the notation $\dot{V}$ to denote the total derivative. Compare (2.3.14) and (2.3.15), we obtain

$$f(t) = -\frac{1}{4\pi} \dot{V}(t) . \tag{2.3.16}$$

Therefore, we obtain from (2.3.12), that the sound field is given by:

$$\varphi = -\frac{1}{4\pi r} \dot{V}\left(t - \frac{r}{c}\right) . \tag{2.3.17}$$

The intensity of sound generated by the body is given by

$$I = \int p\mathbf{v} \cdot \mathbf{n} dS , \tag{2.3.18}$$

where $\mathbf{n}$ is the outward unit normal to a controlling surface S. We shall take the controlling surface to be a large sphere with radius r. Since $p = -\rho_0 \frac{\partial \varphi}{\partial t}$, we thus have

$$I = -\rho_0 \int \frac{\partial \varphi}{\partial t} \frac{\partial \varphi}{\partial r} r^2 d\Omega . \tag{2.3.19}$$

Using (2.3.17), and let $r \to \infty$, we thus obtain

$$I = \frac{\rho_0}{4\pi c} \ddot{V}^2 \left(t - \frac{r}{c}\right) . \tag{2.3.20}$$

Therefore the average power output is

$$\langle I \rangle = \frac{\rho_0}{4\pi c} \langle \ddot{V}^2 \rangle . \tag{2.3.21}$$

For $V(t)$ given by (2.3.8), we obtain

$$\langle I \rangle \propto \omega^4 . \tag{2.3.22}$$

The power output is thus proportional to the fourth power of frequency.

If the volume of the body does not change, then there is no mass flux, and $A_{00} = 0$ in the expression of (2.3.2). The leading terms will be those associated with A_{1m}, and the most important radiation is the dipole radiation. For the dipole radiation, the sound field may be represented by

$$\varphi = \nabla \cdot \left[\frac{\mathbf{A}\left(t - \frac{r}{c}\right)}{r} \right] . \tag{2.3.23}$$

It is evident that (2.3.23) is a solution of the wave equation, since each component of the term inside the square bracket satisfies the wave equation and the differential operators commute with each other.

Near the body, since $\lambda >> L$, just like the previous case, we have practically the dipole potential flow :

$$\varphi = \nabla \cdot \left[\frac{\mathbf{A}(t)}{r} \right] = \mathbf{A}(t) \cdot \nabla \left(\frac{1}{r} \right) , \quad \text{for } r \text{ small} . \tag{2.3.24}$$

For large r, we have approximately,

$$\varphi \approx - \frac{\dot{\mathbf{A}}\left(t - \frac{r}{c}\right) \cdot \mathbf{e}_r}{cr} , \quad r \to \infty . \tag{2.3.25}$$

The corresponding expression for the monochromatic case of (2.3.25) is

$$\varphi \approx -\sum_m \frac{A_{1m}}{kr} Y_{1m}(\theta, \psi) e^{i(kr-\omega t)} \ . \qquad (2.3.26)$$

We can again compute the power output from (2.3.19). Since for $r \to \infty$,

$$\frac{\partial \varphi}{\partial t} \approx -\frac{\ddot{\mathbf{A}} \cdot \mathbf{e}_r}{cr} ,$$

and

$$\frac{\partial \varphi}{\partial r} \approx \frac{\ddot{\mathbf{A}} \cdot \mathbf{e}_r}{c^2 r} ,$$

we obtain

$$I = \frac{\rho_0}{c^3} \int \left(\ddot{\mathbf{A}} \cdot \mathbf{e}_r\right)^2 d\Omega = \frac{4\pi\rho_0}{3c^3} \left(\ddot{\mathbf{A}}\right)^2 . \qquad (2.3.27)$$

$\mathbf{A}(t)$ can be determined from the motion and shape of the body. It is by no means trivial for the general case. We may mention just the simple case of a sphere with radius R moving with velocity $\mathbf{u}$. For this case, it can be shown that

$$\mathbf{A} = \frac{1}{2} R^3 \mathbf{u} \ . \qquad (2.3.28)$$

Although this is a simple case, it is still useful to have some feeling of how the dimension and velocity of the body come into play. Then the power output for this case is

$$I = \frac{\pi\rho_0}{3c^3} R^6 (\ddot{\mathbf{u}})^2 \ . \qquad (2.3.29)$$

If the body is executing an oscillatory motion such that

$$\mathbf{u} \approx -i\omega \mathbf{a} e^{-i\omega t} ,$$

then we have

$$\langle I \rangle \propto \omega^6 . \tag{2.3.30}$$

In contrast with the case of volume oscillation, now the power output is proportional to the sixth power of frequency.

2.4 Scattering of Sound

The problem of scattering of sound may be stated as follows. There is a scattering region D. D may be a material body, or may be represented by certain potential or force. The velocity potential φ satisfies the wave equation :

$$\nabla^2 \varphi - \frac{1}{c^2}\frac{\partial^2 \varphi}{\partial t^2} = 0 , \qquad \text{outside } D . \tag{2.4.1}$$

There is the incident wave $\varphi_{\mathrm{i}}(\mathbf{x},t)$, which is a prescribed solution of wave equation (2.4.1) for the whole region if the scattering region D is absent. Now because of the presence of the scattering region D, the solution of (2.4.1) is modified to become

$$\varphi = \varphi_{\mathrm{i}}(\mathbf{x},t) + \varphi_{\mathrm{s}}(\mathbf{x},t) , \tag{2.4.2}$$

where $\varphi_{\mathrm{s}}(\mathbf{x},t)$ is the scattered wave. The problem of scattering is to find the solution of (2.4.1) based on

(i) Given incident wave $\varphi_{\mathrm{i}}(\mathbf{x},t)$;

(ii) Boundary conditions at the boundary of the scattering region D,

(iii) The scattered wave $\varphi_{\mathrm{s}}(\mathbf{x},t)$ can only be outgoing waves.

The requirement (iii) is natural, since the scattered wave can only originate from the scattering region.

Since both φ and φ_i are solutions of (2.4.1), φ_s is also a solution of (2.4.1). Thus

$$\nabla^2 \varphi_s - \frac{1}{c^2}\frac{\partial^2 \varphi_s}{\partial t^2} = 0 , \quad \text{outside } D . \tag{2.4.3}$$

Let us consider the case of monochromatic wave and also let the incident wave be plane wave. The general case of non-monochromatic wave can be obtained by superposition of monochromatic waves. Also the general incident wave can be constructed from plane waves propagating in various directions. In practice, such superposition may be quite cumbersome. However the study of this representative case can help us to understand the essential features of the scattering problem. Thus we have

$$\varphi_i(\mathbf{x},t) = e^{ikz-i\omega t} , \tag{2.4.4}$$

where we have chosen propagation direction to be $+z$ direction, and the amplitude to be 1 for simplicity.

The energy flux in φ_i is given by

$$J_i = pv = -\rho_0\left(\frac{\partial \varphi_i}{\partial t}\right)\left(\frac{\partial \varphi_i}{\partial z}\right)$$

$$= \rho_0 \omega k \varphi_i^2 . \tag{2.4.5}$$

It should be remarked that although we have been using complex representations such as $e^{ikz-i\omega t}$, we really mean the real part of such expression, i.e. $\cos(kz-\omega t)$. The usage of complex representation is for its convenience. So long as the operations are linear, we can take the real part at any stage. However, when we are dealing with nonlinear quantities such as J_i, then we have to take real part first before we perform the nonlinear operation. Since we really have $\varphi_i = \cos(kz-\omega t)$, therefore the average of the energy flux is

$$\langle J_i \rangle = \frac{1}{2}\rho_0 \omega k \,. \qquad (2.4.6)$$

Now the scattered wave, being an outgoing wave like (2.3.2), can thus be expressed as

$$\varphi_s = \sum_{l,m} S_{lm} h_l^{(1)}(kr) Y_{lm}(\theta, \psi) e^{-i\omega t} \,, \qquad (2.4.7)$$

where S_{lm} is the scattering amplitude of the (l,m)-wave.

Far away from the scattering region, as $r \to \infty$, we have

$$\varphi_s \approx \sum_{l,m} (-i)^{l+1} \frac{S_{lm}}{kr} Y_{lm}(\theta, \psi) e^{i(kr-\omega t)} \,. \qquad (2.4.8)$$

The intensity of the scattered wave I_s , as (2.3.18) and (2.3.19), is given by

$$I_s = \int p\mathbf{v} \cdot \mathbf{n} dS = -\rho_0 \int \frac{\partial \varphi}{\partial t} \frac{\partial \varphi}{\partial r} r^2 d\Omega \,. \qquad (2.4.9)$$

Substitute (2.4.8) into (2.4.9) and use the orthonormal relation for Y_{lm} (2.2.9), we thus obtain the average intensity :

$$\langle I_s \rangle = \frac{\rho_0 \omega}{2k} \sum_{l,m} |S_{lm}|^2 \,. \qquad (2.4.10)$$

An important quantity in scattering studies is the scattering cross section σ_s. It is defined as follows :

$$\sigma_s = \frac{\langle I_s \rangle}{\langle J_i \rangle} . \qquad (2.4.11)$$

It is clear σ_s has the dimension of the cross sectional area. For a beam of particles bounced off a body according to the law of classical mechanics, the maximum cross sectional area of the body normal to the direction of the beam is σ_s. This is the physical meaning associated with σ_s. From (2.4.6), (2.4.10) and (2.4.11), we thus obtain

$$\sigma_s = \frac{1}{k^2}\sum_{l=0}^{\infty}\sum_{m=-l}^{l}|S_{lm}|^2 . \tag{2.4.12}$$

From the properties of Legendre polynomials and the spherical Bessel functions, we know that the quantity e^{ikz} can be expand in terms of Legendre polynomials P_l :

$$\begin{aligned} e^{ikz} &= e^{ikr\cos\theta} \\ &= \sum_{l=0}^{\infty} i^l(2l+1)j_l(kr)P_l(\cos\theta) \\ &= \sum_{l=0}^{\infty}\left(\frac{2l+1}{2}\right)i^l[h_l^{(1)}(kr)+h_l^{(2)}(kr)]P_l(\cos\theta) . \end{aligned} \tag{2.4.13}$$

From (2.4.13), we see clearly that the incident wave φ_{i} actually consists of both the incoming and the outgoing waves of the same intensity. Let us denote the intensity of the incoming waves I_i. The incoming waves are the terms associated with $h_l^{(2)}(kr)$. Using the asymptotic expression of $h_l^{(2)}(kr)$ for large r [see (A1.9)], and carrying out similar computations for I_s, we obtain

$$\langle I_i\rangle = \frac{\rho_0\omega\pi}{2k}\sum_{l=0}^{\infty}(2l+1) . \tag{2.4.14}$$

As the total sound field φ consists of just φ_i and φ_s, and φ_s has outgoing waves only, $< I_i >$ is in fact the average intensity of the incoming waves of the total sound field φ. On the other hand, the intensity of the outgoing waves of the total sound field φ has contributions from both φ_i and φ_s. Denote φ_o the total outgoing wave field, we thus have

$$\varphi_o = e^{-i\omega t}\sum_{l=0}^{\infty}\left\{\left[S_l + \frac{i^l(2l+1)}{2}\right]P_l(\cos\theta) + \sum_{\substack{m=-l\\ m\neq 0}}^{l} S_{lm}Y_l(\theta,\psi)\right\}h_l^{(1)}(kr), \tag{2.4.15}$$

where

$$S_l = \left(\frac{2l+1}{4\pi}\right)^{1/2} S_{l0}\,. \tag{2.4.16}$$

We can thus compute the intensity of the total outgoing waves I_0 from φ_0 by using the similar method for I_i, and obtain

$$\langle I_o\rangle = \frac{\rho_0\omega}{2k}\left\{\sum_{l=0}^{\infty}\left[(2l+1)\pi\left|1+\frac{2(-i)^l}{2l+1}S_l\right|^2\right] + \sum_{l=0}^{\infty}\sum_{|m|=1}^{l}|S_{lm}|^2\right\}. \tag{2.4.17}$$

The difference between $\langle I_i\rangle$ and $\langle I_0\rangle$, $[\langle I_i\rangle - \langle I_0\rangle]$, is the intensity of energy loss to the scattering region D. The loss is due to some absorbing mechanism which may be present in D. This energy loss leads to the definition of another important quantity σ_a, the absorption cross section :

$$\sigma_a = \frac{\langle I_i\rangle - \langle I_0\rangle}{\langle J_i\rangle}. \tag{2.4.18}$$

It is worth noting that while $\langle I_i\rangle$ and $\langle I_0\rangle$, as represented by (2.4.14) and (2.4.17) respectively, are both divergent, $[\langle I_i\rangle - \langle I_0\rangle]$

is finite. From (2.4.18), using (2.4.6), (2.4.14) and (2.4.17), we obtain

$$\sigma_a = \frac{1}{k^2}\sum_{l=0}^{\infty}\left\{(2l+1)\pi\left[1-\left|1+\frac{2(-i)^l}{2l+1}S_l\right|^2\right]-\sum_{|m|=1}^{l}|S_{lm}|^2\right\}. \quad (2.4.19)$$

Therefore, once S_{lm} is found, both the scattering cross section σ_s and the absorption cross section σ_a can be readily found. However, it is not easy to find S_{lm} for the general case.

Scattering by a Spherical Bubble in a Liquid

Let us illustrate the previous general discussion by the example of scattering of sound wave by a spherical gas bubble with radius a in a liquid. We shall consider the case that the gas is barotropic for simplicity. Furthermore let us assume that there is only the bulk viscosity, and the shear viscosity coefficient is zero. These assumptions are not quite realistic but are convenient here for illustration.

Let p_g, ρ_g and μ'_g be the gas pressure, density and bulk viscosity coefficient inside the bubble respectively. Then p_g satisfies an equation similar to (1.2.4) :

$$\frac{\partial^2 p_g}{\partial t^2} = c_g^2\nabla^2 p_g + \upsilon\nabla^2\frac{\partial p_g}{\partial t}, \qquad r \le a\,, \qquad (2.4.20)$$

where

$$c_g^2 = \left(\frac{dp_g}{d\rho_g}\right)_0, \qquad \upsilon = \frac{\mu'_g}{\rho_{go}}\,, \qquad (2.4.21)$$

and the subscript 0 denotes the equilibrium value.

For monochromatic waves with $p_g(\mathbf{x},t) = p_g(\mathbf{x})e^{-i\omega t}$, then (2.4.20) becomes

$$\nabla^2 p_g + k_i^2 p_g = 0, \quad r \le a \,, \tag{2.4.22}$$

where

$$k_i^2 = \frac{k_g^2}{1 - i\left(\dfrac{k_g}{k_\upsilon}\right)^2} \,,$$

$$k_g^2 = \frac{\omega^2}{c_g^2}, \quad k_\upsilon^2 = \frac{\omega}{\upsilon} \,. \tag{2.4.23}$$

Since the bubble is spherical, therefore the problem of scattering of plane wave by the bubble possesses the axial symmetry. With the requirement that the solution is regular at r=0, the solution of (2.4.22) is

$$p_g = \sum_{l=0}^{\infty} A_l j_l(k_i r) P_l(\cos\theta), \quad r \le a \,. \tag{2.4.24}$$

From (1.2.1) and (1.2.2), we obtain

$$-i\omega v_r = -\frac{1}{\rho_{g0}}\frac{\partial p_g}{\partial r} + \frac{i\upsilon\omega}{\rho_{g0}c_g^2}\frac{\partial p_g}{\partial r}, \quad r \le a \,,$$

or

$$v_r = -\frac{i\omega}{\rho_{g0}c_g^2}\left(\frac{1}{k_g^2} - \frac{i}{k_\upsilon^2}\right)\frac{\partial p_g}{\partial r}, \quad r \le a \,. \tag{2.4.25}$$

The viscous stress in the radial direction is given by

$$\sigma_{rr} = p_g - \mu'_g \nabla \cdot \mathbf{v}$$

$$= p_g + \frac{\upsilon}{c_g^2}\frac{\partial p_g}{\partial t}, \quad r \le a , \tag{2.4.26}$$

after using again (1.2.1).

The flow field outside the bubble φ is given by (2.4.4) for φ_i and (2.4.7) for φ_s; and we can compute p and v_r for $r \ge a$ accordingly. Neglecting the effect of surface tension, the interfacial conditions at the bubble wall $r=a$, are given by the continuity of velocity and stress in the radial direction, i.e.

$$p(r \ge a) = \sigma_{rr}(\, r \le a), \quad r = a \ ,$$

and

$$v_r(r \ge a) = v_r(\, r \le a), \quad r = a \ .$$

Using the orthonormality of the spherical harmonics or just Legendre polynomials, since the problem is axially symmetric, the scattering coefficients S_l which is related to S_{l0} by (2.4.16) can be readily computed. Straightforward calculations lead to :

$$S_0 = -\frac{ix}{\left(1 - \dfrac{m}{x}\right) + i\left(x + \dfrac{mn}{x}\right)}, \tag{2.4.27}$$

and

$$S_l = \frac{-i^l(2l+1)(1-lm_l)}{\left\{(1-lm_l) - \dfrac{iR_l^2}{(2l+1)x^{2l+1}}\left[1+(l+1)m_l\right]\right\}}, \quad l \ge 1 \tag{2.4.28}$$

where

$$x = ka, \qquad m = \left(\frac{3}{ka}\right)\left(\frac{\rho_{g0}}{\rho_0}\right)\left(\frac{c_g}{c}\right)^2,$$

$$m_l = \frac{\rho_{g0}}{l\rho_0}, \quad n = \left(\frac{k_g}{k_\upsilon}\right)^2, \qquad R_l = (2l+1)!!\,. \tag{2.4.29}$$

For small x, i.e. when the wavelength is large compared with the radius of the bubble, the major contribution to scattering is due to the monopole term S_0, since $|S_l| \propto x^{2l+1}$ for small x. Therefore it is meaningful to single out only the monopole term in the cross sections. Using (2.4.12) and (2.4.19) the monopole terms for the scattering and absorption cross sections are found to be respectively :

$$\sigma_s^{(0)} = \frac{4\pi}{k^2}\frac{x^2}{\left(1-\dfrac{m}{x}\right)^2+\left(x+\dfrac{mn}{x}\right)^2}, \tag{2.4.30}$$

and

$$\sigma_a^{(0)} = \frac{4\pi}{k^2}\frac{mn}{\left(1-\dfrac{m}{x}\right)^2+\left(x+\dfrac{mn}{x}\right)^2}\,. \tag{2.4.31}$$

It is worth noting that the absorption cross section vanishes if $\mu'_g = 0$. Also contribution to the absorption cross section comes only from the monopole term, since dissipation by the bulk viscosity can only be caused by volume changes of the bubble.

Scattering by a Small Body

If the scattering body dimension is small compared with the incident wavelength, approximate general expression can be obtained for scattering sound field. From the example of scattering by the spherical bubble, we may appreciate that the major contribution to the scattering sound field will come from the monopole term, and the second most important contribution from the dipole term. Therefore, following the similar arguments that leading to (2.3.17) and (2.3.25), we can write approximately,

$$\varphi_s = -\frac{\dot{V}_s\left(t-\frac{r}{c}\right)}{4\pi r} - \frac{\dot{\mathbf{A}}_s\left(t-\frac{r}{c}\right)\cdot\mathbf{r}}{cr^2}, \qquad (2.4.32)$$

where V_s and $\mathbf{A}_s$ are the effective volume and dipole strength for the scattering field.

To find $\dot{V}_s$, we observe first that $\dot{V}_s = 0$, if the body is absent. Let V be the volume of the body. If the body is absent, then V is occupied by the external fluid itself. The sound field will produce density variation in V, which is practically uniform spatially since the body is small. Therefore it is equivalent to the existence of sink of strength $\left(V\frac{\dot{\rho}}{\rho_0}\right)$. But now there is this body with volume V, thus the "sink" has been removed. Or an equivalent source of the strength $\left(V\frac{\dot{\rho}}{\rho}\right)$ is present to cancel that "sink". This source is just $\dot{V}_s$. Let the amplitude of the incident wave be characterized by the fluid velocity in the sound field v. Then $\frac{\dot{\rho}}{\rho_0} = \frac{\dot{v}}{c}$, and we have

$$\dot{V}_s = V\frac{\dot{v}}{c}. \qquad (2.4.33)$$

If the scattering body is a rigid body with volume V_0, then

$$V = V_0. \qquad (2.4.34)$$

If the scattering body is a fluid of density ρ_i with equilibrium volume V_0, then we have

$$V = V_0\left(1 - \frac{\dot{\rho}_i \rho_0}{\rho_{i0}\dot{\rho}}\right), \tag{2.4.35}$$

since the fluid body is compressible and hence the effective volume will be diminished by its compressibility.

$$\dot{\rho}_i = \left(\frac{d\rho_i}{dp_i}\right)_0 \dot{p}_i = \left(\frac{d\rho_i}{dp_i}\right)_0 \dot{p} = \left(\frac{d\rho_i}{dp_i}\right)_0 \left(\frac{dp}{d\rho}\right)_0 \dot{\rho}$$

$$= \frac{c^2}{c_i^2}\dot{\rho}, \tag{2.4.36}$$

since pressures are the same inside and outside the body. Therefore we obtain

$$\dot{V}_s = \frac{V_0 \dot{v}}{c}\left(1 - \frac{\rho_0}{\rho_{i0}}\frac{c^2}{c_i^2}\right). \tag{2.4.37}$$

For the scattering of sound by a gaseous body in a liquid, the term $\left(\frac{\rho_0}{\rho_{i0}}\frac{c^2}{c_i^2}\right)$ is much larger than 1. Therefore we should have

$$\dot{V}_s \approx -\frac{V_0 \dot{v}}{c}\left(\frac{\rho_0}{\rho_{i0}}\frac{c^2}{c_i^2}\right), \tag{2.4.38}$$

which is consistent with (2.4.27) when $ka << 1$.

To find $\dot{\mathbf{A}}_s$, we may note that the quantity $\mathbf{A}$ is associated with the linear motion of a body in potential flow. From the theory of potential flow, a body with volume V_0 moving with velocity $\mathbf{u}$ in a fluid with density ρ_0, the components of the vector $\mathbf{A}$ can be expressed as

$$A_i = \frac{1}{4\pi\rho_0} m_{ik} u_k + \frac{V_0}{4\pi} u_i , \qquad i = 1,2,3, \quad (2.4.39)$$

where m_{ik} is the induced mass tensor of the body. If the body is absent, then the fluid inside V_0 is moving with velocity **v** of the incident sound wave. The presence of the body thus leads to expression of $\mathbf{A}_s$ with **u** replaced by (-**v**) in **A**. Therefore the components of $\mathbf{A}_s$ are

$$A_{si} = -\frac{1}{4\pi\rho_0} m_{ik} v_k - \frac{V_0}{4\pi} v_i , \qquad i = 1,\ 2,\ 3\ . \quad (2.4.40)$$

The induced mass tensor of a body of arbitrary shape is difficult to find. For a sphere of radius a, it is found that

$$m_{ik} = \frac{2}{3}\pi\rho_0 a^3 \delta_{ik} = \frac{1}{2}\rho_0 V_0 \delta_{ik}\ . \qquad (2.4.41)$$

With $\dot{V}_s$ and $\mathbf{A}_s$ thus obtained, the scattering cross section can be readily computed.

2.5 Nonlinear Acoustics

Acoustic waves usually refer to the small amplitude waves in compressible fluids or solids. It is thus usually a linear phenomenon. But the equations governing the fluid motion are nonlinear. If the amplitude of the waves is not very small, and second order terms are included in the discussion, then we are dealing with nonlinear acoustics. Highly nonlinear waves in fluids are usually no longer referred to as acoustics. In the following, we shall discuss briefly some topics in nonlinear acoustics.

Quadrupole Radiation

The motion of fluid can be put in the compact form (2.1.7) :

$$\frac{\partial^2 \rho}{\partial t^2} - c_0^2 \frac{\partial^2 \rho}{\partial x_i^2} = \frac{\partial q}{\partial t} - \frac{\partial}{\partial x_i}(\rho b_i) + \frac{\partial^2}{\partial x_i \partial x_j} T_{ij} , \quad (2.5.1)$$

where

$$T_{ij} = \rho v_i v_j + \left[(p - p_0) - (\rho - \rho_0) c_0^2 \right] \delta_{ij} . \quad (2.5.2)$$

In (2.5.1), q represents a source term. If there is no source, no external force **b**, and only linear terms are retained, then (2.5.1) reduces to the familiar wave equation. As it stands now, (2.5.1) is not only nonlinear, but also not complete, since the right hand side contains other unknowns **v** besides ρ. Still, we may treat the right hand side of (2.5.1) as known quantities, and use the formula for the retarded potential (1.4.11) to obtain an expression of ρ in terms of those quantities on the right hand side. Now since the wave equation is linear, the contributions from various inhomogeneous terms can be superposed. Moreover, if (1.4.11) is the solution for the inhomogeneous term f, the solution for inhomogeneous $\frac{\partial f}{\partial t}$ or $\frac{\partial f}{\partial x_i}$ will simply be the derivative of the solution (1.4.11) with respect to t or x_i respectively, except for an additive term served as integration constant. Therefore

$$c_0^2(\rho - \rho_0) = \frac{\partial}{\partial t} \iiint_D d^3\mathbf{y} \frac{q\left(\mathbf{y},\ t - \frac{1}{c}|\mathbf{x}-\mathbf{y}|\right)}{4\pi |\mathbf{x}-\mathbf{y}|}$$

$$- \frac{\partial}{\partial x_i} \iiint_D d^3\mathbf{y} \frac{(\rho b_i)\left(\mathbf{y},\ t - \frac{1}{c}|\mathbf{x}-\mathbf{y}|\right)}{4\pi |\mathbf{x}-\mathbf{y}|}$$

$$+ \frac{\partial}{\partial x_i \partial x_j} \iiint_D d^3\mathbf{y} \frac{T_{ij}\left(\mathbf{y},\ t - \frac{1}{c}|\mathbf{x}-\mathbf{y}|\right)}{4\pi |\mathbf{x}-\mathbf{y}|} , \quad (2.5.3)$$

where D is the region for nonvanishing q, b_i and T_{ij} .

Now if D is a compact region, and since $|\mathbf{x}|$ is usually very large for radiation problems, therefore we have $|\mathbf{x}-\mathbf{y}| \approx |\mathbf{x}| = r$ inside the integrals over D. Therefore we can rewrite (2.5.3) approximately as

$$c_0^2(\rho-\rho_0) = \frac{\partial}{\partial t}\left[\frac{1}{4\pi r}\iiint_D g\left(\mathbf{y},\ t-\frac{r}{c}\right)d^3\mathbf{y}\right]$$

$$-\frac{\partial}{\partial x_i}\left[\frac{1}{4\pi r}\iiint_D (\rho b_i)\left(\mathbf{y},\ t-\frac{r}{c}\right)d^3\mathbf{y}\right]$$

$$+\frac{\partial^2}{\partial x_i \partial x_j}\left[\frac{1}{4\pi r}\iiint_D T_{ij}\left(\mathbf{y},\ t-\frac{r}{c}\right)d^3\mathbf{y}\right]. \quad (2.5.4)$$

The first term on the right hand side of (2.5.4) represents the monopole radiation due to material source q, the second term represents the dipole radiation due to external force or momentum source, and the third term represents the quadrupole radiation.

If there is no external material body in the fluid, hence q and b_i are zero, then the dominant mode of radiation is the quadrupole radiation due to T_{ij}.

The source of the quadrupole radiation T_{ij} originates entirely from the fluid motion. In acoustics, the terms $(\rho-\rho_0)$, $(p-p_0)$ and v_i are all small quantities of the order of some small parameter ε. Since $\left(\dfrac{dp}{d\rho}\right)_0 = c_0^2$, it is clear from (2.5.2), that $T_{ij} = 0(\varepsilon^2)$. Therefore, the amplitude of the fluid motion needs to be substantial in the region D to generate sound of sufficient intensity. Moreover the generated radiation is the quadrupole radiation.

Return to (2.5.4), set $q = b_i = 0$. Again making use of the fact that r is very large, we obtain the quadrupole radiation field :

$$c^2(\rho - \rho_0) = \frac{x_i x_j}{4\pi c^2 r^3} \iiint_D d^3\mathbf{y}\, \ddot{T}_{ij}\left(\mathbf{y},\ t - \frac{r}{c}\right). \quad (2.5.5)$$

Take the noise generated by a jet with velocity U. $T_{ij} \propto U^2$. The characteristic frequency of jet usually also increases like U. Thus $\ddot{T}_{ij} \propto U^4$. As the sound intensity I is proportional to $(\rho - \rho_0)^2$, therefore the intensity of jet noise I is proportional to U^8:

$$I \propto U^8 . \quad (2.5.6)$$

Scattering of Sound by Sound

When there is no source, nor external force, (2.5.1) can be written as

$$\mathbf{L}\rho = -\frac{1}{c_0^2}\frac{\partial^2}{\partial x_i \partial x_j} T_{ij} , \quad (2.5.7)$$

where we have used the operator $\mathbf{L}$ to denote

$$\mathbf{L} = \nabla^2 - \frac{1}{c_0^2}\frac{\partial^2}{\partial t^2} . \quad (2.5.8)$$

If the computation is carried only to the first order small quantities, then (2.5.7) becomes approximately

$$\mathbf{L}\rho_1 = 0 , \quad (2.5.9)$$

which leads to the usual solutions of sound waves. At this level, since the wave equation (2.5.9) is linear, there is no interaction between one beam of sound wave with another. However (2.5.9) is not exact. Therefore, when the amplitudes of the sound waves are not small, there will be interactions between different beams of sound waves, and one beam of sound may be scattered by another beam of sound. The amplitudes of the scattered sound field will be second order small quantities. Write

$$\rho = \rho_1 + \rho_s, \quad p = p_1 + p_s\,, \tag{2.5.10}$$

then, up to the second order of small quantities, we have from (2.5.7) and (2.5.2)

$$\mathbf{L}\rho_s = -\frac{1}{c_0^2}\frac{\partial^2}{\partial x_i \partial x_j}\left[\rho_0 v_i v_j + \frac{1}{2}\left(\frac{d^2 p}{d\rho^2}\right)_0 \rho_1^2 \delta_{ij}\right], \tag{2.5.11}$$

where ρ_1 and $\mathbf{v}$ are all solutions of first order wave equation (2.5.9). The first order solutions are the principal waves. In the following, we shall describe the main features of the problem. We shall refer the details to works by Westervelt (1957, 1963).

Since

$$\nabla \cdot \mathbf{v} = -\rho_0 \frac{\partial \rho_1}{\partial t}\,, \quad \frac{\partial \mathbf{v}}{\partial t} = -\frac{c_0^2}{\rho_0}\nabla \rho_1\,, \tag{2.5.12}$$

and thus $\nabla \times \mathbf{v} = 0$, (2.5.11) can be written as

$$\mathbf{L}\rho_s = -\frac{\rho_0}{c_0^2}\left[(\nabla \cdot \mathbf{v})^2 + \mathbf{v}\cdot(\nabla^2 \mathbf{v}) + \nabla^2\left(\frac{v^2}{2}\right)\right]$$

$$-\frac{\alpha}{2\rho_0}\nabla^2\left(\rho_1^2\right), \tag{2.5.13}$$

where

$$\alpha = \frac{\rho_0}{c_0^2}\left(\frac{d^2 p}{d\rho^2}\right)_0 . \qquad (2.5.14)$$

Using (2.5.12), (2.5.13) can, after some calculation, be put in the form :

$$\mathbf{L}\rho_s = \frac{1}{c_0^2}\left[\mathbf{L}E - \nabla^2(2T + \alpha V)\right], \qquad (2.5.15)$$

where

$$T = \frac{1}{2}\rho_0 v^2, \quad V = \frac{1}{2}\frac{c_0^2}{\rho_0}\rho_1^2, \quad E = T + V , \quad (2.5.16)$$

T is the kinetic energy density, V the potential energy density and E the total energy density.

Take the case that there are two principal waves with sound fields given by $(\rho_a, \mathbf{v}_a)$ and $(\rho_b, \mathbf{v}_b)$ respectively. Then

$$T = \frac{1}{2}\rho_0 v^2 = \frac{1}{2}\rho_0(\mathbf{v}_a + \mathbf{v}_b)^2$$

$$= \frac{1}{2}\rho_0\left[v_a^2 + v_b^2 + 2\mathbf{v}_a \cdot \mathbf{v}_b\right]. \qquad (2.5.17)$$

Now in (2.5.17), the terms v_a^2 and v_b^2 represent effect of self-interaction. It is a nonlinear effect. But it has nothing to do with the scattering of sound beam a by the sound beam b. Thus in calculating ρ_s from (2.5.15), we need only retain in T and V the mutually interacting terms :

$$T_{ab} = \rho_0 \mathbf{v}_a \cdot \mathbf{v}_b \quad \text{and} \quad V_{ab} = \frac{c_0^2}{\rho_0}\rho_a \rho_b . \quad (2.5.18)$$

Thus (2.5.15) becomes

$$\mathbf{L}\rho_s = \frac{1}{c_0^2}\left[\mathbf{L}E_{ab} - \nabla^2(2T_{ab} + \alpha V_{ab})\right] . \qquad (2.5.19)$$

Now let the angle between the two principal beams be θ, i.e., $\mathbf{v}_a \cdot \mathbf{v}_b = v_a v_b \cos\theta$. Consider the case that these two principal beams are both monochromatic with frequencies ω_a and ω_b respectively. Then

$$T_{ab} = V_{ab}\cos\theta \ . \qquad (2.5.20)$$

Now Let

$$W_{ab} = \frac{c_0^2}{\rho_0 \omega_a \omega_b}\frac{\partial \rho_a}{\partial t}\frac{\partial \rho_b}{\partial t} , \qquad (2.5.21)$$

then after somewhat lengthy calculation, (2.5.19) becomes

$$\mathbf{L}\rho_s = \mathbf{L}\psi , \qquad (2.5.22)$$

where

$$\psi = \frac{E_{ab}}{c_0^2} - \frac{(2\cos\theta + \alpha)}{2\omega_a \omega_b(\cos\theta - 1)}\nabla^2 W_{ab} . \qquad (2.5.23)$$

If $\cos\theta \neq 1$, i.e., if these two principal beams are not parallel to each other, then the solution of (2.5.22) is

$$\rho_s = \psi . \qquad (2.5.24)$$

But $\psi \neq 0$ only in the interaction region of these two principal beams. $\psi = 0$ elsewhere. Therefore, the scattered field is zero from the viewpoint of scattering, since $\rho_s = 0$ outside the region of interaction.

When $\cos\theta = 1$, ψ is singular, then there is scattering. (2.5.19) can be rewritten as

$$\mathbf{L}\rho_s = -\rho_0 \frac{\partial q}{\partial t}, \tag{2.5.25}$$

where

$$q = \frac{1}{\rho_0^2 c_0^4}\left(1 + \frac{\alpha}{2}\right)\frac{\partial}{\partial t}\left(p_1^2\right), \tag{2.5.26}$$

and p_1 is the pressure due to those two parallel principal beams.

Consider the case that the amplitude of these two parallel principal beams are the same. We may express p_1 by

$$p_1 = P_0\, e^{-\beta x}\left[\cos(\omega_a t - k_a x) + \cos(\omega_b t - k_b x)\right] \tag{2.5.27}$$

where β is some damping constant due to dissipation mechanisms.

The practically important case is when $\omega_a \approx \omega_b$. Write $\omega_s = |\omega_a - \omega_b|$, and let $k_s = \dfrac{\omega_s}{c_0}$, then we can compute the interacting part of q, q_{ab}, from (2.5.26). Now q_{ab} consists of two parts, one part vary with $e^{i2(k_a x - \omega_a t)}$, and the other part vary with $e^{i(k_s x - \omega_s t)}$. At large distance from the interacting region, the part with difference frequency ω_s will be dominant, since the high frequency part with frequency $2\omega_a$ will be damped out. Therefore the remaining part of q_{ab}, expressed in complex form is

$$q = \frac{-i\omega_s}{\rho_0^2 c_0^4}\left(1 + \frac{\alpha}{2}\right) P_0^2\, e^{-i\omega_s t + i(k_s + 2i\beta)x}. \tag{2.5.28}$$

Then p_s may be obtained by solving (2.5.25) using the retarded potential solution (1.4.11).

Acoustic Streaming

In linear theory, for sinusoidal plane waves, the fluid velocity averaged over time will vanish. However, if the amplitude is not very small, the nonlinear contribution to the average velocity can be nonzero, and that is called acoustic streaming.

Let us start with (1.1.2)

$$\frac{\partial \rho}{\partial t}+\frac{\partial}{\partial x_i}(\rho v_i)=0\ , \tag{2.5.29}$$

and write (1.1.6) in the following form :

$$\frac{\partial}{\partial t}(\rho v_i)+\frac{\partial}{\partial x_j}(\rho v_i v_j)+\frac{\partial p}{\partial x_i}$$

$$=\mu\nabla^2 v_i+\left(\mu'+\frac{\mu}{3}\right)\frac{\partial}{\partial x_i}\left(\frac{\partial v_j}{\partial x_j}\right),\quad i=1,\ 2,\ 3\ . \tag{2.5.30}$$

Introduce the small parameter ε, and let

$$\rho=\rho_0+\varepsilon\rho_1+\varepsilon^2\rho_2+\mathrm{L}\ ,$$

$$p=p_0+\varepsilon p_1+\varepsilon^2 p_2+\mathrm{L}\ ,$$

$$v_i=\varepsilon v_{1i}+\varepsilon^2 v_{2i}+\mathrm{L}\ .$$

If p_0 and ρ_0 are constant, then the equations for first and second order of ε are as follows :

$0(\varepsilon)$:

$$\frac{\partial \rho_1}{\partial t} + \rho_0 \frac{\partial v_{1i}}{\partial x_i} = 0 \,, \tag{2.5.31}$$

$$\rho_0 \frac{\partial v_{1i}}{\partial t} + \frac{\partial p_1}{\partial x_i} = \mu \nabla^2 v_{1i} + \left(\mu' + \frac{\mu}{3} \right) \frac{\partial}{\partial x_i} \left(\frac{\partial v_{1j}}{\partial x_j} \right), \quad i = 1,2,3 \tag{2.5.32}$$

$0(\varepsilon^2)$:

$$\frac{\partial \rho_2}{\partial t} + \frac{\partial}{\partial x_i} (\rho_0 v_{2i} + \rho_1 v_{1i}) = 0 \,, \tag{2.5.33}$$

$$\begin{aligned} &\frac{\partial}{\partial t}(\rho_0 v_{2i} + \rho_1 v_{1i}) + \rho_0 \frac{\partial}{\partial x_i}(v_{1i} v_{1j}) + \frac{\partial p_2}{\partial x_i} \\ &= \mu \nabla^2 v_{2i} + \left(\mu' + \frac{\mu}{3} \right) \frac{\partial}{\partial x_i} \left(\frac{\partial v_{2j}}{\partial x_i} \right), \qquad i = 1,2,3 \,. \end{aligned} \tag{2.5.34}$$

From (2.5.31) and (2.5.32), we can obtain the linear wave solutions. They will be damped waves. We may consider a sound column moving in the $+x$ direction of the following form :

$$\mathbf{v}_1 = U(r) e^{-\alpha x} \sin(kx - \omega t) \cdot \mathbf{e}_x \,, \tag{2.5.35}$$

which can be considered as known.

Let us introduce the time average of a quantity f by :

$$\langle f \rangle = \lim_{T \to \infty} \frac{1}{T} \int_0^T f \, dt \,. \tag{2.5.36}$$

Then we obtain from (2.5.31) that

$$\frac{\partial}{\partial x_i} \langle v_{1i} \rangle = 0 \,. \tag{2.5.37}$$

Also we obtain from (2.5.33) that

$$\rho_0 \frac{\partial}{\partial x_i}\langle v_{2i}\rangle = -\frac{\partial}{\partial x_i}\langle \rho_1 v_{1i}\rangle \; . \qquad (2.5.38)$$

From (2.5.34), we obtain

$$\rho_0 \frac{\partial}{\partial x_j}\left\langle v_{1i}v_{1j}\right\rangle + \frac{\partial}{\partial x_i}\langle p_2\rangle$$

$$= \mu\nabla^2\left\langle v_{2i}\right\rangle + \left(\mu' + \frac{\mu}{3}\right)\frac{\partial}{\partial x_i}\left(\frac{\partial}{\partial x_j}\left\langle v_{2j}\right\rangle\right),$$

$$i = 1,\ 2,\ 3\; . \qquad (2.5.39)$$

Since $\frac{\rho_1}{\rho_0} \approx \frac{v_1}{c}$, thus from (2.5.38), we have $\frac{\partial}{\partial x_i}\left\langle v_{2i}\right\rangle \approx \frac{\partial}{\partial x_i}\frac{\left\langle v_{1i}^2\right\rangle}{c}$.
If the viscosity coefficients are not large, then the terms associated with $\frac{\partial}{\partial x_i}\left\langle v_{2i}\right\rangle$ on the right hand side of (2.5.39) can be neglected.
Thus (2.5.39) becomes

$$-\mu\nabla \times \nabla \times \left\langle \mathbf{v}_2\right\rangle = \mathbf{F} + \nabla\langle p_2\rangle \; , \qquad (2.5.40)$$

where the components of **F** are

$$F_i = \rho_0 \frac{\partial}{\partial x_j}\left\langle v_{1i}v_{1j}\right\rangle, \quad i = 1,2,3\; . \qquad (2.5.41)$$

F can be decomposed into the dilatational part $\mathbf{F}_d$ and the rotational part $\mathbf{F}_r$ with

$$\nabla \times \mathbf{F}_d = 0 \quad \text{and} \quad \nabla\cdot\mathbf{F}_r = 0\; . \qquad (2.5.42)$$

Then we have

$$-\mu\nabla \times \nabla \times \langle \mathbf{v}_2 \rangle = \mathbf{F}_r \, , \tag{2.5.43}$$

which yields the solution :

$$\langle v_{2i} \rangle = -\frac{3}{16\pi}\int \frac{\left\{ F_{ri}(\mathbf{x}')\, |\mathbf{x}-\mathbf{x}'|^2 + F_{rj}(\mathbf{x}')\left[x_j - x'_j\right]\left[x_i - x'_j\right]\right\} d^3\mathbf{x}'}{|\mathbf{x}-\mathbf{x}'|^3} \, . \tag{2.5.44}$$

It is worth noting that the acoustic streaming, besides being a nonlinear phenomenon, comes about only when the damping is present in the sound field.

2.6 Simple Wave and Shock Wave

We discuss now nonlinear waves in fluids. We shall begin with nonlinear waves in one spatial dimension. Neglecting the body force, Equations (1.1.1) and (1.1.6) can be written as

$$\frac{\partial \rho}{\partial t} + \rho\frac{\partial v}{\partial x} + v\frac{\partial \rho}{\partial x} = 0 \, , \tag{2.6.1}$$

$$\frac{\partial v}{\partial t} + v\frac{\partial v}{\partial x} + \frac{1}{\rho}\frac{\partial p}{\partial x} = \overline{\upsilon}\frac{\partial^2 v}{\partial x^2} \, , \tag{2.6.2}$$

where

$$\overline{\upsilon} = \frac{4}{3}\upsilon + \upsilon' \, .$$

For ideal fluids, we have $\overline{\upsilon} = 0$, since there is no dissipation. Furthermore, $p = p(\rho)$. Let

$$c^2 = \frac{dp}{d\rho} . \tag{2.6.3}$$

Then (2.6.2) can be rewritten as

$$\frac{\partial v}{\partial t} + v\frac{\partial v}{\partial x} + \frac{c^2}{\rho}\frac{\partial \rho}{\partial x} = 0 . \tag{2.6.4}$$

We shall presently discuss the solution of the system (2.6.1) and (2.6.4). First, we discuss the simpler case such that

$$v = v(\rho) . \tag{2.6.5}$$

Particular solutions that satisfy the condition (2.6.5) are known as simple waves. For simple waves, (2.6.1) and (2.6.4) can be rewritten in the following manner :

$$\frac{\partial \rho}{\partial t} + \left(v + \rho\frac{dv}{d\rho} \right)\frac{\partial \rho}{\partial x} = 0 , \tag{2.6.6}$$

$$\frac{\partial v}{\partial t} + \left(v + \frac{c^2}{\rho\dfrac{dv}{d\rho}} \right)\frac{\partial v}{\partial x} = 0 . \tag{2.6.7}$$

Since $\left(\frac{dx}{dt}\right)_\rho = \left(\frac{dx}{dt}\right)_v$, thus we obtain from (2.6.6) and (2.6.7) that

$$v + \rho\frac{dv}{d\rho} = v + \frac{c^2}{\rho\dfrac{dv}{d\rho}} ,$$

or

$$\rho \frac{dv}{d\rho} = \pm c \,, \tag{2.6.8}$$

or

$$v = \pm \int^{\rho} \frac{c}{\rho} d\rho \,. \tag{2.6.9}$$

Substitute (2.6.8) into (2.6.6) and (2.6.7), we obtain

$$\frac{\partial \rho}{\partial t} + (v \pm c) \frac{\partial \rho}{\partial x} = 0 \,, \tag{2.6.10}$$

and

$$\frac{\partial v}{\partial t} + (v \pm c) \frac{\partial v}{\partial x} = 0 \,, \tag{2.6.11}$$

where $c = c(v)$, since c is function of ρ only.

Denote

$$u = v \pm c \,. \tag{2.6.12}$$

Then (2.6.11) can be rewritten as

$$\frac{\partial u}{\partial t} + u \frac{\partial u}{\partial x} = 0 \,. \tag{2.6.13}$$

Equation (2.6.13), though simple, is a very important equation in the study of nonlinear waves. It exhibits one of the most important features of nonlinear waves, i.e., the development of shock. We shall study the initial value problem of this nonlinear wave equation, i.e., the problem with initial value given by :

$$u(x,0) = u_0(x) \,. \tag{2.6.14}$$

From (2.6.13), it is clear that the characterisitics are given by :

$$\left(\frac{dx}{dt}\right)_u = -\frac{\dfrac{\partial u}{\partial t}}{\dfrac{\partial u}{\partial x}} = u \ . \qquad (2.6.15)$$

Thus, the characteristics are all straight lines, for on each characteristics the slope u does not change. u, of course, is different for different characteristics. For a single hump shaped $u_0(x)$, the characteristics are schematically represented in Fig. 2-1. In Fig. 2-1, x_i is the point of inflection of the initial profile $u_0(x)$. It is illuminating to treat this problem by introducing the Lagrange coordinates (ξ, τ) with

$$x = x(\xi, \tau) \ , \quad t = \tau \ , \qquad (2.6.16)$$

or

$$\xi = \xi(x,t) \ , \quad \tau = t \ , \qquad (2.6.17)$$

and

$$u = \frac{\partial x}{\partial \tau} \ . \qquad (2.6.18)$$

We shall also set the initial correspondence :

$$x(\xi,0) = \xi \ , \quad \xi(x,0) = x \ . \qquad (2.6.19)$$

In terms of the Lagrange coordinates, Equation (2.6.13) is simply

$$\frac{\partial u}{\partial \tau} = 0 \ . \qquad (2.6.20)$$

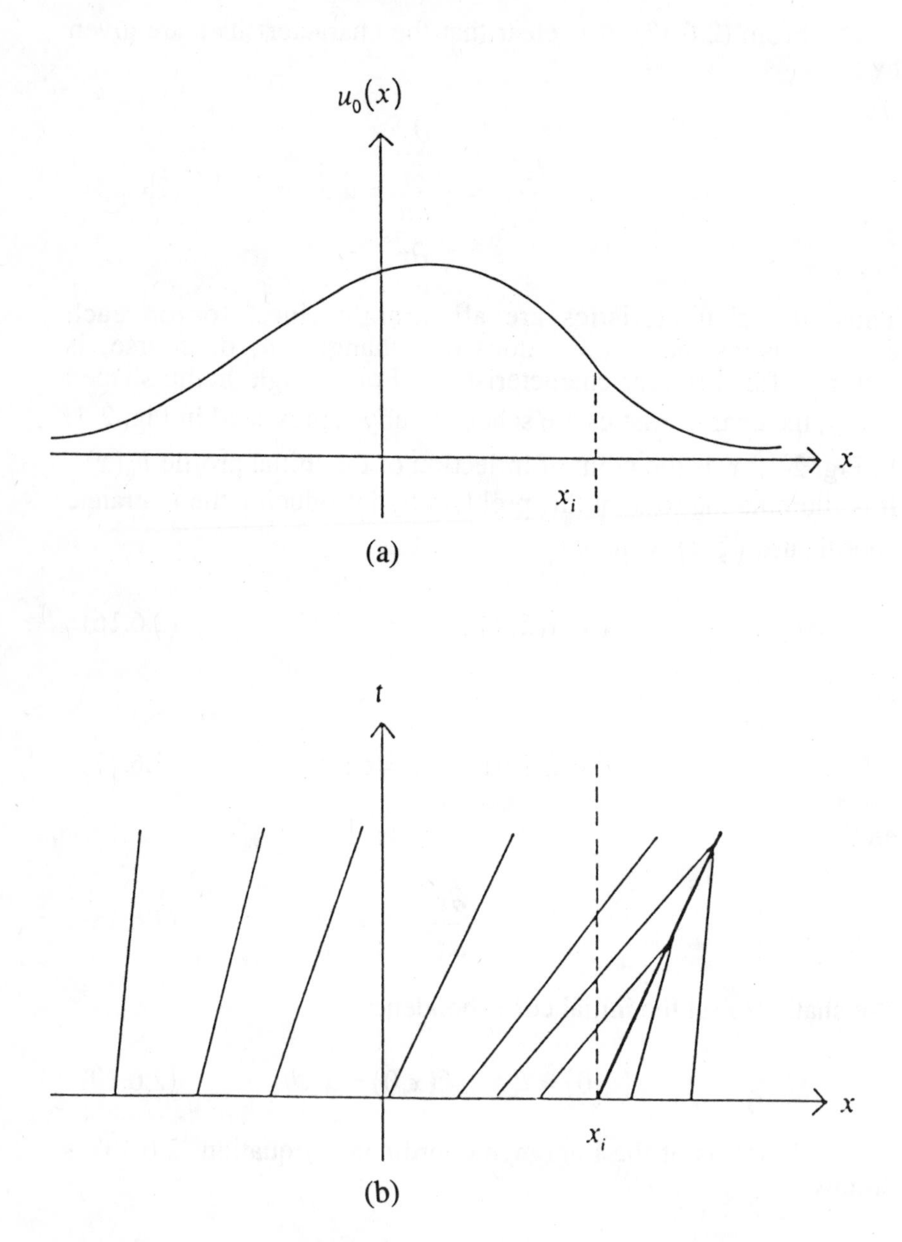

Fig. 2-1 Schematic representation of characteristics.
(a) The initial profile $u_0(x)$. (b) The straight line characteristics.

Hence the solution is

$$u = F(\xi) \ . \tag{2.6.21}$$

From (2.6.14) and (2.6.19) we obtain

$$u_0(x) = F(\xi(x,0)) = F(x) \ . \tag{2.6.22}$$

Thus we have

$$u = u_0(\xi) \ , \tag{2.6.23}$$

and

$$\xi = u_0^{-1}(\mathrm{u}) \ . \tag{2.6.24}$$

(2.6.18) and (2.6.23) lead also to

$$\frac{\partial x}{\partial \tau} = u_0(\xi) \ . \tag{2.6.25}$$

Integrating the last equation, we thus obtain

$$x = \xi + u_0(\xi)\tau \ . \tag{2.6.26}$$

Equation (2.6.26) is the equation for the characteristics. If $u_0(x)$ is the type as represented in Fig. 2-1, then the characteristics will intersect one another sooner or later. Denote t_s the earliest time that the characteristics intersect. Then $u(x)$ will no longer be single-valued for any $t > t_s$. Physically, this means that shock has formed.

Fig. 2-2 depicts the development from the initial profile to shock. Fig. 2-2(a) shows the initial profile $u_0(x)$. ξ_1 , ξ_0 and ξ_2 are the Lagrange coodinates. According to (2.6.23) and (2.6.25), the "particle" associated with a particular ξ has the same u throughout the development and is moving in x-space with velocity u. ξ_0 is the point of inflection of the initial profile. The pair (ξ_1, ξ_2) are chosen so that the area $\xi_0\xi_1\xi_0$ is the same $\xi_0\xi_2\xi_0$. As t progresses, the right hand side of the profile $u(x,t)$ tends to steepen. Eventually, for $t > t_s$, the profile becomes multiple-valued.

The time t_s, i.e., the time that shock starts to form, as shown in Fig. 2-2, can be determined from the following two equations :

$$\left(\frac{\partial x}{\partial u}\right)_t = 0 , \tag{2.6.27}$$

and

$$\left(\frac{\partial^2 x}{\partial u^2}\right)_t = 0 . \tag{2.6.28}$$

From (2.6.24) and (2.6.26), we have

$$x = u_0^{-1}(u) + ut . \tag{2.6.29}$$

Thus (2.6.27) and (2.6.28) become

$$t = -\frac{d}{du}\left[u_0^{-1}(u)\right] \tag{2.6.30}$$

and

$$\frac{d^2}{du^2}\left[u_0^{-1}(u)\right] = 0 . \tag{2.6.31}$$

Solving (2.6.31), we obtain $u = u_s$. Then

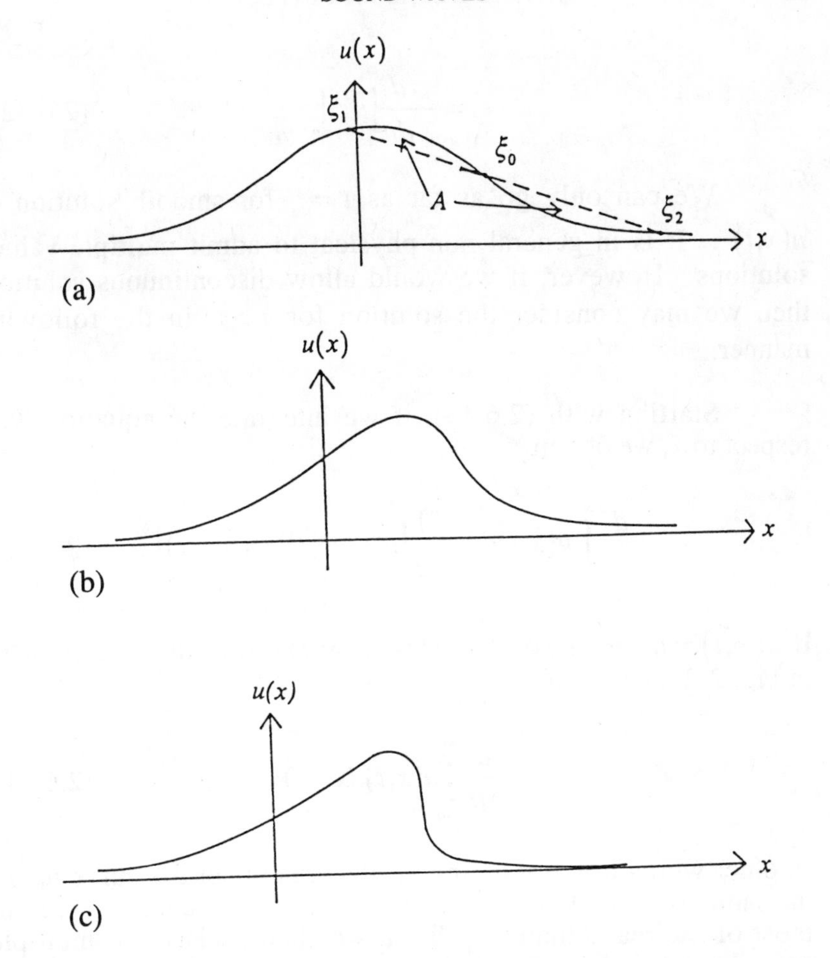

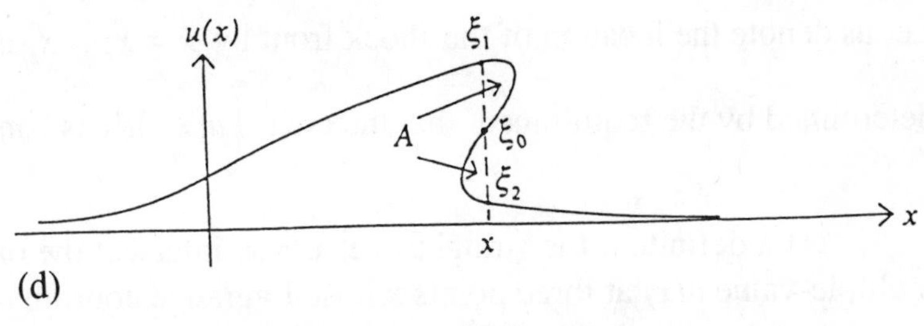

Fig. 2-2 The profile of $u(x,t)$ at various t showing the development of shock formation.

$$t_s = -\frac{d}{du}\left[u_0^{-1}\right]_{u=u_s} . \qquad (2.6.32)$$

We can only go as far as $t = t_s$ for smooth solution of $u(x,t)$. It is in general non-physical to admit multiple-valued solutions. However, if we would allow discontinuous solution, then we may consider the solution for $t > t_s$ in the following manner.

Starting with (2.6.13), if we integrate the equation with respect to x, we obtain

$$\frac{d}{dt}\int_{-\infty}^{\infty} u(x,t)dx = \frac{1}{2}\left[u^2(-\infty,t) - u^2(\infty,t)\right] . \qquad (2.6.33)$$

If $u(\infty,t) = u(-\infty,t)$, or $u(\infty,t) = u(-\infty,t) = 0$ as the case depicted in Fig. 2-2, then we have

$$\frac{d}{dt}\int_{-\infty}^{\infty} u(x,t)dx = 0 . \qquad (2.6.34)$$

In other words, for any definite t, the area under the curve $u(x)$ is the same for all t. Let us consider just this case, since it represents most of the real situations. Then we shall replace the multiple-valued solution by a discontinuous solution as shown in Fig. 2-2 (d) with the discontinuous front representing the shock for $t > t_s$. Let us denote the location of the shock front by $x = x_s$. x_s is to be determined by the requirement that the area $\int_{-\infty}^{\infty} u(x,t)dx$ is constant.

At a definite t, the straight line $x = x_s$ intersect the original multiple-value $u(x)$ at three points whose Lagrange coordinates are designated by ξ_1 , ξ_0 and ξ_2 . These Lagrange coordinates of course change with t. The vertical straight line $\xi_1\xi_0\xi_2$ at $x = x_s$ in

Fig. 2-2(d), which represents the shock front, corresponds to the straight line $\xi_1\xi_0\xi_2$ in Fig. 2-2(a) at t=0. The areas $\xi_0\xi_1\xi_0$ and $\xi_0\xi_2\xi_0$ are both A in Fig. 2-2(a) and Fig. 2-2(d). Referring to Fig. 2-2(a), we thus have

$$\int_{\xi_1}^{\xi_2} u_0(\xi)d\xi = \frac{1}{2}\left[u_0(\xi_1) - u_0(\xi_2)\right](\xi_2 - \xi_1) . \quad (2.6.35)$$

Using (2.6.29), we have

$$x_1 = \xi_1 + u_0(\xi_1)t , \quad (2.6.36)$$

and

$$x_2 = \xi_2 + u_0(\xi_2)t . \quad (2.6.37)$$

But $x_1 = x_2 = x_s$ when $x = x_s$. Thus we obtain from (2.6.36) and (2.6.37) :

$$\frac{1}{t} = -\frac{u_0(\xi_2) - u_0(\xi_1)}{\xi_2 - \xi_1} . \quad (2.6.38)$$

From (2.6.35), we obtain the relation between ξ_1 and ξ_2 , say, $\xi_2 = \xi_2(\xi_1)$. Then we can obtain $\xi_1 = \xi_1(t)$ from (2.6.38). Now $x_s(t)$ may be obtained from

$$x_s = \xi_1 + u_0(\xi_1)t . \quad (2.6.39)$$

Another way to look at the solution is to note that since it is also true that

$$x_s = \xi_2 + u_0(\xi_2)t , \quad (2.6.40)$$

we have

$$2\frac{dx_s}{dt} = u_0(\xi_2) + \left[1 + \left(\frac{du_0}{d\xi}\right)_2 t\right]\frac{d\xi_2}{dt}$$

$$+ u_0(\xi_1) + \left[1 + \left(\frac{du_0}{d\xi}\right)_1 t\right]\frac{d\xi_1}{dt} . \qquad (2.6.41)$$

Now differentiate (2.6.35) with respect to t, we obtain

$$u_0(\xi_2)\frac{d\xi_2}{dt} - u_0(\xi_1)\frac{d\xi_1}{dt} = \frac{1}{2}\left[u_0(\xi_1) + u_0(\xi_2)\right]\left(\frac{d\xi_2}{dt} - \frac{d\xi_1}{dt}\right)$$

$$+ \frac{1}{2}(\xi_2 - \xi_1)\left[\left(\frac{du_0}{d\xi}\right)_1 \frac{d\xi_1}{dt} + \left(\frac{du_0}{d\xi}\right)_2 \frac{d\xi_2}{dt}\right],$$

or

$$\left[u_0(\xi_2) - u_0(\xi_1) - (\xi_2 - \xi_1)\left(\frac{du_0}{d\xi}\right)_2\right]\frac{d\xi_2}{dt}$$

$$= \left[u_0(\xi_1) - u_0(\xi_2) + (\xi_2 - \xi_1)\left(\frac{du_0}{d\xi}\right)_1\right]\frac{d\xi_1}{dt} . \qquad (2.6.42)$$

Using (2.6.38), (2.6.42) can be rewritten as

$$-\left[\frac{1}{t} + \left(\frac{du_0}{d\xi}\right)_2\right]\frac{d\xi_2}{dt} = \left[\frac{1}{t} + \left(\frac{du_0}{d\xi}\right)_1\right]\frac{d\xi_1}{dt} . \qquad (2.6.43)$$

Substitute (2.6.43) into (2.6.41), we obtain

$$\frac{dx_s}{dt} = \frac{1}{2}\left[u_0(\xi_2) + u_0(\xi_1)\right] . \qquad (2.6.44)$$

The speed at which the shock front is moving is thus just the mean of the maximum and the minimum of u that constitute the shock front.

We shall also consider a special case that the initial velocity profile has compact support, as schematically represented by Fig. 2-3. Thus,

$$u_0(\xi) = 0\ , \quad \text{for } \xi < 0 \quad \text{and} \quad \xi > b\ . \tag{2.6.45}$$

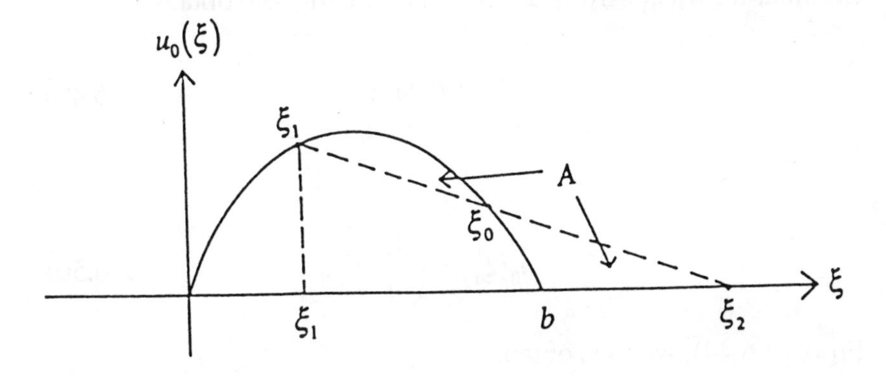

Fig. 2-3 Initial profile with compact support.

The straight line $\xi_1\xi_0\xi_2$ in Fig. 2-3, will become the shock front at some later time t. Now as $t \to \infty$, the straight line $\xi_1\xi_0\xi_2$ in Fig. 2-3 which would correspond to the shock front has the following properties :

$$\xi_2 \to \infty\ , \quad u(\xi_2) = 0\ , \quad \text{and} \quad \xi_1 \to 0\ .$$

Denote the area under the velocity profile by S, i.e.

$$S = \int_0^b u_0(\xi)d\xi\ . \tag{2.6.46}$$

Thus, as $t \to \infty$, we have approximately

$$S \approx \int_{\xi_1}^{\xi_2} u_0(\xi)d\xi \approx \frac{1}{2}\xi_2 u_0(\xi_1) . \tag{2.6.47}$$

Now, corresponding to (2.6.38), we have as $t \to \infty$.

$$\frac{1}{t} = \frac{u_0(\xi_1)}{\xi_2} . \tag{2.6.48}$$

Eliminating $u_0(\xi_1)$ from (2.6.47) and (2.6.48), we obtain

$$\xi_2 \approx (2St)^{\frac{1}{2}} . \tag{2.6.49}$$

Hence

$$u_0(\xi_1) \approx \left(\frac{2S}{t}\right)^{\frac{1}{2}} . \tag{2.6.50}$$

From (2.6.44), we thus obtain

$$\frac{dx_s}{dt} \approx \left(\frac{S}{2t}\right)^{\frac{1}{2}} , \tag{2.6.51}$$

or

$$x_s \approx (2St)^{\frac{1}{2}} . \tag{2.6.52}$$

(2.6.52) gives the location of the shock, while (2.6.50) gives the strength of the shock.

2.7 Structure of Shock - Burgers Equation

The solution of the equation (2.6.13), $u(x,t)$, becomes multiple-valued for $t > t_s$, i.e., after the appearance of shock. The multiple-valuedness is in general physically inadmissible. We have devised a mathematical scheme to admit discontinuous solutions so that we may proceed beyond t_s. However, the deficiency in mathematical soundness often implies the inadequacy of the physical theory. In other words, the mathematical model is too idealized, and it may have neglected some essential mechanisms relevant to the problem.

Let us go back to (2.6.1) and (2.6.2) :

$$\frac{\partial \rho}{\partial t} + \rho \frac{\partial v}{\partial x} + v \frac{\partial \rho}{\partial x} = 0 , \tag{2.7.1}$$

$$\frac{\partial v}{\partial t} + v \frac{\partial v}{\partial x} + \frac{1}{\rho} \frac{\partial p}{\partial x} = \overline{\upsilon} \frac{\partial^2 v}{\partial x^2} . \tag{2.7.2}$$

We neglected the term associated with viscosity in (2.7.2) in the previous development. When $\overline{\upsilon}$ is small, the term can be justifiably neglected indeed under normal circumstances. However, in the neighborhood of shock, where $\frac{\partial v}{\partial x}$ is extremely large, the term $\left(\overline{\upsilon} \frac{\partial^2 v}{\partial x^2} \right)$ will not be small compared with other terms. Therefore, this term cannot be neglected; in particular not in the neighborhood of the shock. The inclusion of the viscous effect would amount to deal with not the equation (2.6.13), but a modified equation of the following form :

$$\frac{\partial u}{\partial t} + u \frac{\partial u}{\partial x} = \frac{\overline{\upsilon}}{2} \frac{\partial^2 u}{\partial x^2} . \tag{2.7.3}$$

Equation (2.7.3) is known as the Burgers Equation. It is the simplest type of wave equations, which incorporates both nonlinearity and dissipation. It is intuitively plausible that (2.7.3) should be the generalization of (2.6.13) with the inclusion of the viscous effect. However, we shall present a systematic derivation of (2.7.3) for the weakly nonlinear wave using the method of multiple-scale expansion.

Derivation of Burgers Equation

Let ε be a small parameter. We make the following expansions for ρ and v :

$$\rho = \rho^{(0)} + \varepsilon\rho^{(1)} + \varepsilon^2\rho^{(2)} , \tag{2.7.4}$$

and

$$v = \varepsilon v^{(1)} + \varepsilon^2 v^{(2)} . \tag{2.7.5}$$

Thus

$$\frac{c^2}{\rho} = \frac{c_0^2}{\rho^{(0)}} + \varepsilon\alpha\rho^{(1)} + 0\left(\varepsilon^2\right) , \tag{2.7.6}$$

where

$$\alpha = -\left(\frac{c_0}{\rho^{(0)}}\right)^2 + \rho^{(0)}\left(\frac{dc^2}{d\rho}\right)_0 . \tag{2.7.7}$$

Denote

$$\xi = x - c_0 t , \tag{2.7.8}$$

$$\eta = \varepsilon t , \tag{2.7.9}$$

and assign

$$\overline{\upsilon} = 0(\varepsilon) . \tag{2.7.10}$$

With these assignments of small parameter, we are essentially following the wave which propagates with the linear wave speed c_0. We now investigate the deviations relative to this frame of reference.

From (2.7.8) and (2.7.9), we obtain

$$\frac{\partial}{\partial t} = -c_0 \frac{\partial}{\partial \xi} + \varepsilon \frac{\partial}{\partial \eta}, \tag{2.7.11}$$

and

$$\frac{\partial}{\partial x} = \frac{\partial}{\partial \xi}. \tag{2.7.12}$$

Substitute (2.7.4) - (2.7.12) into (2.7.1) and (2.7.2), we obtain the following equations for $0(\varepsilon)$ and $0(\varepsilon^2)$:

$0(\varepsilon)$:

$$-c_0 \rho_\xi^{(1)} + \rho^{(0)} v_\xi^{(1)} = 0 , \tag{2.7.13}$$

$$-c_0 v_\xi^{(1)} + \frac{c_0^2}{\rho^{(0)}} \rho_\xi^{(1)} = 0 . \tag{2.7.14}$$

$0(\varepsilon^2)$:

$$-c_0 \rho_\xi^{(2)} + \rho^{(0)} v_\xi^{(2)} = -\left(\rho_\eta^{(1)} + v^{(1)} \rho_\xi^{(1)} + \rho^{(1)} v_\xi^{(1)}\right), \tag{2.7.15}$$

$$-c_0 v_\xi^{(2)} + \frac{c_0^2}{\rho^{(0)}} \rho_\xi^{(2)} = -\left(v_\eta^{(1)} + v^{(1)} v_\xi^{(1)} + \alpha \rho^{(1)} \rho_\xi^{(1)}\right) + \overline{\upsilon} v_{\xi\xi}^{(1)} . \tag{2.7.16}$$

We shall impose the condition that

$$\rho^{(1)} = v^{(1)} = 0 \, , \quad \text{as} \quad \xi \to \infty \, . \tag{2.7.17}$$

Then we obtain from (2.7.13) and (2.7.14) that

$$\rho^{(1)} = \frac{\rho^{(0)}}{c_0} v^{(1)} \, . \tag{2.7.18}$$

Multiply (2.7.15) by c_0 , and (2.7.16) by ρ_0 then add, we obtain

$$c_0\left(\rho_\eta^{(1)} + v^{(1)}\rho_\xi^{(1)} + \rho^{(1)}v_\xi^{(1)}\right)$$

$$+\rho^{(0)}\left(v_\eta^{(1)} + v^{(1)}v_\xi^{(1)} + \alpha\rho^{(1)}\rho_\xi^{(1)}\right) - \rho^{(0)}\overline{\upsilon}v_{\xi\xi}^{(1)} = 0 \, . \tag{2.7.19}$$

Using (2.7.18), then (2.7.19) can be rewritten as

$$2v_\eta^{(1)} + 2v^{(1)}v_\xi^{(1)} + \left[1 + \alpha\left(\frac{\rho^{(0)}}{c_0}\right)^2\right]v^{(1)}v_\xi^{(1)} = \overline{\upsilon}v_{\xi\xi}^{(1)} \, . \tag{2.7.20}$$

Denote

$$\beta = 1 + \frac{1}{2}\left(\frac{\rho^{(0)}}{c_0^2}\right)\left(\frac{dc^2}{d\rho}\right)_0 \, . \tag{2.7.21}$$

Then (2.7.20) can be rewritten as

$$v_\eta^{(1)} + \beta v^{(1)}v_\xi^{(1)} = \frac{\overline{\upsilon}}{2}v_{\xi\xi}^{(1)} \, . \tag{2.7.22}$$

Let

$$u = \beta v^{(1)} \, . \tag{2.7.23}$$

Then (2.7.22) becomes

$$u_\eta + uu_\xi = \frac{\overline{\upsilon}}{2} u_{\xi\xi} \,, \qquad (2.7.24)$$

which is exactly the Burgers equation (2.7.3).

We could also derive the Burgers equation directly as follows. Let $\rho = \rho_0 + \rho_1$, and retain terms only up to the second order. Thus from (2.7.1) and (2.7.2), we obtain

$$\frac{\partial \rho_1}{\partial t} + \rho_0 \frac{\partial v}{\partial x} + \rho_1 \frac{\partial v}{\partial x} + v \frac{\partial \rho_1}{\partial x} = 0 \,, \qquad (2.7.25)$$

and

$$\frac{\partial v}{\partial t} + v \frac{\partial v}{\partial x} + \frac{c_0^2}{\rho_0} \frac{\partial \rho_1}{\partial x} + \alpha \rho_1 \frac{\partial \rho_1}{\partial x} = \overline{\upsilon} \frac{\partial^2 v}{\partial x^2} \,, \qquad (2.7.26)$$

where ρ_1, v and $\overline{\upsilon}$ again are considered to be first order small quantities.

Now, $\rho_1 \approx \frac{\rho_0}{c_0} v$ from the linear theory. Then, multiply (2.7.25) by $\left(\frac{c_0}{\rho_0}\right)$ and add (2.7.26), we obtain :

$$\left(\frac{\partial}{\partial t} + c_0 \frac{\partial}{\partial x}\right) v + \beta v \frac{\partial v}{\partial x} = \frac{\overline{\upsilon}}{2} \frac{\partial^2 v}{\partial x^2} \,, \qquad (2.7.27)$$

which is the same as (2.7.22).

Solution of Burgers Equation

Let us now try to solve the Burgers equation :

$$\frac{\partial u}{\partial t}+u\frac{\partial u}{\partial x}=\frac{\overline{\upsilon}}{2}\frac{\partial^2 u}{\partial x^2}\,, \tag{2.7.3}$$

with the initial conditon :

$$u(x,0)=u_0(x)\;. \tag{2.7.28}$$

Conventional methods of solution are not applicable for this equation with both nonlinearity and dissipation. Methods of separation of variables and transforms do not work because the equation is nonlinear. Method of characteristics is also not useful because of the diffusive term. An ingenious transformation, independently devised by E. Hopf and J. Cole, turns out to be the key to solve the problem.

Let us introduce a new function $\phi(x,t)$ such that

$$u=\frac{\partial \varphi}{\partial x}\,. \tag{2.7.29}$$

Then (2.7.3) becomes

$$\frac{\partial^2 \varphi}{\partial t\partial x}+\frac{\partial \varphi}{\partial x}\frac{\partial^2 \varphi}{\partial x^2}=\frac{\overline{\upsilon}}{2}\frac{\partial^3 \varphi}{\partial x^3}\,. \tag{2.7.30}$$

Integrating (2.7.30) over x, we obtain

$$\frac{\partial \varphi}{\partial t}+\frac{1}{2}\left(\frac{\partial \varphi}{\partial x}\right)^2=\frac{\overline{\upsilon}}{2}\frac{\partial^2 \varphi}{\partial x^2}\,. \tag{2.7.31}$$

An arbitrary function *f(t)* should be included in general in (2.7.31). As the addition of an arbitrary function of t to $\phi(x,t)$ would not affect the uniqueness of $u(x,t)$, there is no loss of generality to choose *f(t)* = 0. To solve (2.7.31), let us introduce $\psi(x,t)$ such that

$$\phi=-\overline{\upsilon}\ln\psi\;. \tag{2.7.32}$$

Then (2.7.31) becomes

$$\frac{\partial \psi}{\partial t} = \frac{\overline{\upsilon}}{2}\frac{\partial^2 \psi}{\partial x^2} , \qquad (2.7.33)$$

which is the familiar diffusion equation and can be readily solved. For initial condition :

$$\psi(x,0) = \psi_0(x) , \qquad (2.7.34)$$

the solution is

$$\psi(x,t) = (2\pi\overline{\upsilon}t)^{-\frac{1}{2}} \int_{-\infty}^{\infty} \psi_0(\eta) e^{-\frac{(x-\eta)^2}{2\overline{\upsilon}t}} d\eta \ . \qquad (2.7.35)$$

As

$$u(x,t) = \frac{\partial \varphi}{\partial x} = -\frac{\overline{\upsilon}}{\psi}\frac{\partial \psi}{\partial x} , \qquad (2.7.36)$$

therefore

$$\psi(x,t) = \exp\left\{\frac{1}{\overline{\upsilon}}\int_x^{\infty} u(\eta,t) d\eta\right\} . \qquad (2.7.37)$$

The initial value $\psi_0(x)$ is thus

$$\psi_0(x) = \exp\left\{\frac{1}{\overline{\upsilon}}\int_x^{\infty} u_0(\eta) d\eta\right\} . \qquad (2.7.38)$$

From (2.7.35), (2.7.36) and (2.7.38), we thus obtain

$$u(x,t) = \frac{\int_{-\infty}^{\infty} \frac{x-\eta}{t} \psi_0(\eta) \exp\left\{-\frac{(x-\eta)^2}{2\overline{\upsilon}t}\right\} d\eta}{\int_{-\infty}^{\infty} \psi_0(\eta) \exp\left\{-\frac{(x-\eta)^2}{2\overline{\upsilon}t}\right\} d\eta} . \quad (2.7.39)$$

An Example to Illustrate the Shock Structure

We shall present a simple example to illustrate the essential features of the solution (2.7.39). Let us take

$$u_0(x) = S\delta(x) . \quad (2.7.40)$$

Thus

$$\psi_0(x) = \begin{cases} 1, & x > 0 , \\ e^R & x < 0 , \end{cases} \quad (2.7.41)$$

where $R = \frac{S}{\overline{\upsilon}}$ is in fact the Reynolds number. Substitute (2.7.41) into (2.7.39), we obtain

$$u(x,t) = \frac{N}{D} , \quad (2.7.42)$$

where

$$N = e^R \int_{-\infty}^{0} \frac{x-\eta}{t} \exp\left\{-\frac{(x-\eta)^2}{2\overline{\upsilon}t}\right\} d\eta$$

$$+ \int_{0}^{\infty} \frac{x-\eta}{t} \exp\left\{-\frac{(x-\eta)^2}{2\overline{\upsilon}t}\right\} d\eta , \quad (2.7.43)$$

$$D = e^R \int_{-\infty}^{0} \exp\left\{-\frac{(x-\eta)^2}{2\overline{\upsilon}t}\right\} d\eta + \int_{0}^{\infty} \exp\left\{-\frac{(x-\eta)^2}{2\overline{\upsilon}t}\right\} d\eta \ . \quad (2.7.44)$$

Denote

$$a = \frac{x}{\sqrt{\overline{\upsilon}t}} \ . \quad (2.7.45)$$

Then, after changing variable η to $y = \dfrac{x-\eta}{\sqrt{\overline{\upsilon}t}}$, we obtain :

$$u(x,t) = \frac{\left(\dfrac{\overline{\upsilon}}{t}\right)^{\frac{1}{2}} \left(e^R - 1\right) e^{-\frac{a^2}{2}}}{\left[\left(e^R - 1\right) \int_a^{\infty} e^{-\frac{y^2}{2}} dy + \sqrt{2\pi}\right]} \ . \quad (2.7.46)$$

Let us now concentrate on two limiting cases.

(i) *R<<1.* The case with small Reynolds number corresponds to that $\overline{\upsilon}$ is large. Therefore the Burgers equation behaves very much like the diffusion equation.

As $e^R - 1 \approx R << 1$, (2.7.46) becomes approximately

$$u(x,t) \approx \left[\frac{\overline{\upsilon}}{2\pi t}\right]^{\frac{1}{2}} \mathrm{Re}^{-\frac{a^2}{2}} = \frac{S}{\sqrt{2\pi\overline{\upsilon}t}} e^{-\frac{x^2}{2\overline{\upsilon}t}} \ , \quad (2.7.47)$$

which is indeed the fundamental solution of the diffusion equation.

(ii) *R>>1.* For this case of large Reynolds number, the nonlinear term will play an important

role, and we should expect the development of the shock. Let

$$Z = \frac{a}{\sqrt{2R}} = \frac{x}{\sqrt{2St}} \ . \qquad (2.7.48)$$

Since $e^R >> 1$, therefore (2.7.46) can be expressed approximately :

$$u(x,t) \approx \frac{\left(\frac{S}{Rt}\right)^{\frac{1}{2}} e^{-RZ^2}}{\left[\int_{Z\sqrt{2R}}^{\infty} e^{-\frac{y^2}{2}} dy + \sqrt{2\pi}e^{-R}\right]} \ . \qquad (2.7.49)$$

The denominator of (2.7.49) needs to be treated differently depending on whether Z is positive or negative.

(1) $Z < 0$. In the denominator of (2.7.49), the term $\int_{Z\sqrt{2R}}^{\infty} e^{-\frac{y^2}{2}} dy$ will dominate, and its value lies between $\left(\frac{\pi}{2}\right)^{\frac{1}{2}}$ and $(2\pi)^{\frac{1}{2}}$. Except for very small $|Z|$, since $R>>1$ its value is close to $(2\pi)^{\frac{1}{2}}$. We shall take this value. Thus,

$$u(x,t) \approx \left(\frac{S}{2\pi Rt}\right)^{\frac{1}{2}} e^{-RZ^2} \ , \qquad \text{for} \qquad Z < 0 \ . \qquad (2.7.50)$$

(2) $Z > 0$. When Z is not extremely small, the term $\int_{Z\sqrt{2R}}^{\infty} e^{-\frac{y^2}{2}} dy$ is also small, and comparable with the

other small term $\sqrt{2\pi}e^{-R}$. Using integration by parts, we obtain approximately that

$$\int_{Z\sqrt{2R}}^{\infty} e^{-\frac{y^2}{2}}\,dy \approx \frac{1}{Z\sqrt{2R}}e^{-RZ^2}, \qquad (2.7.51)$$

since $Z\sqrt{2R} >> 1$. Therefore, we obtain from (2.7.49),

$$u(x,t) \approx \frac{\left(\frac{2S}{t}\right)^{\frac{1}{2}} Z}{\left[1+2\sqrt{\pi R}Ze^{(Z^2-1)R}\right]}. \qquad (2.7.52)$$

As $R >> 1$, we may again obtain from (2.7.52) that

$$u(x,t) \approx \left(\frac{2S}{t}\right)^{\frac{1}{2}} Z = \frac{x}{t}, \quad \text{for} \quad 0 < Z < 1, \qquad (2.7.53)$$

and

$$u(x,t) \approx \left(\frac{S}{2\pi Rt}\right)^{\frac{1}{2}} e^{-(Z^2-1)R}, \quad \text{for} \quad Z > 1. \qquad (2.7.54)$$

For $Z \approx 1$, (2.7.52) needs to be used to connect smoothly the expressions given in (2.7.53) and (2.7.54).

To put the various expressions of $u(x,t)$ together, a schematic representation is shown in Fig. 2-4.

The initial profile at $t = 0^+$ is shown in Fig. 2-4(a). As t increases the initial profile develops into a wave with a shock front. The location of the shock is at $x_s = (2St)^{\frac{1}{2}}$,

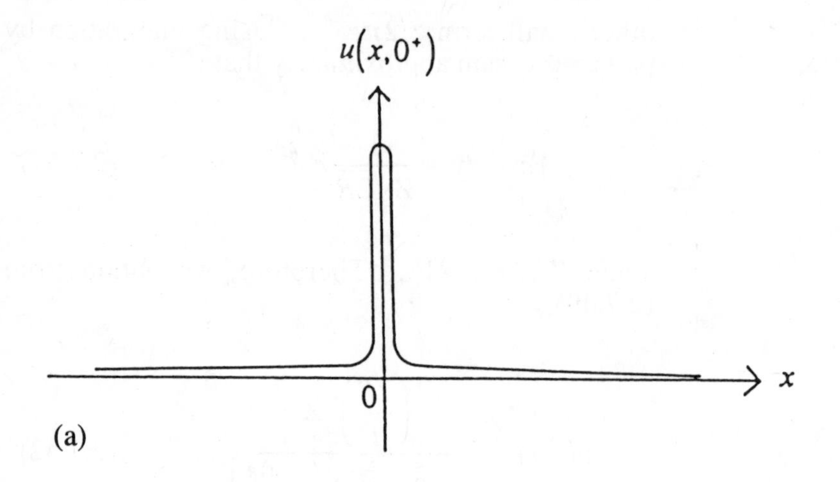

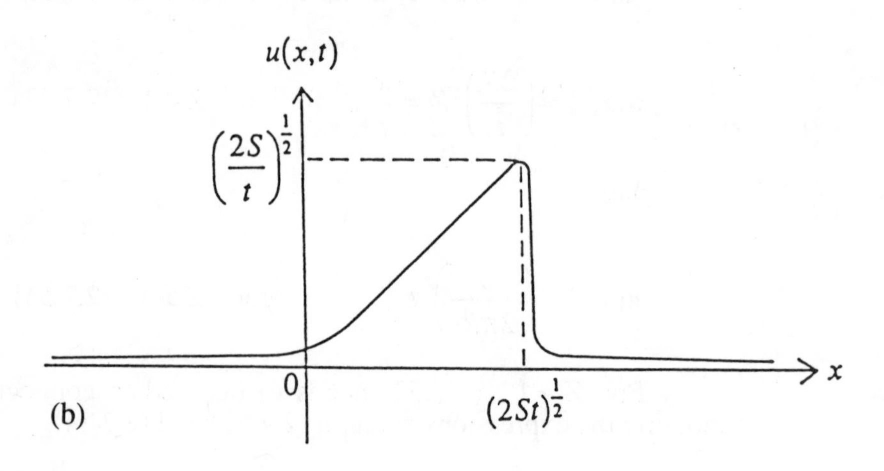

Fig. 2-4 (a) The profile of $u(x,t)$ for $t = 0^+$.
(b) The profile of $u(x,t)$ at a definite later t.

and the strength is about $\left(\frac{2S}{t}\right)^{\frac{1}{2}}$, which agree with (2.6.50) and (2.6.52). u is extremely small for $x > x_s$. u increases linearly with x, for $0 < x < x_s$. There is also a small diffusive tail for $x<0$. There is a sharp drop of u at $x = x_s$ showing the presence of shock. But this steep drop is not discontinuous. It has a continuous structure given by the expression (2.7.52). Thus with the inclusion of dissipation, $u(x,t)$ is no longer multiple-valued, nor discontinuous. Shock will develop because of the nonlinearity. However, the shock has a continuous structure. Still, the treatment in Section 2.6 has captured the essence of shock formation.

2.8 One-Dimensional Nonlinear Wave - Riemann Invariants

When we deal with the one-dimensional nonlinear waves in ideal fluids, the simple wave is a special solution. Now we shall discuss the general case. For ideal fluids, Equation (2.6.1) and (2.6.2) become

$$\frac{\partial \rho}{\partial t} + \rho \frac{\partial v}{\partial x} + v \frac{\partial \rho}{\partial x} = 0 , \tag{2.8.1}$$

$$\frac{\partial v}{\partial t} + v \frac{\partial v}{\partial x} + \frac{c^2}{\rho} \frac{\partial \rho}{\partial x} = 0 . \tag{2.8.2}$$

We now define the characteristics in the (x-t)-space as follows : The curve $\phi(x,t) =$ constant is characteristic, if ρ and v are allowed to be discontinuous, i.e., the increments $\Delta\rho$ and Δv may be finite across the characteristics. It is clear that the wave front which separates a region of disturbance from a calm region is a characteristics.

Let us now change variables (x-t) to $\{\phi(x,t), \psi(x,t)\}$. $\psi(x,t)$ and $\phi(x,t)$ are independent functions. In terms of the new independent variables (ϕ, ψ) , we have

$$\frac{\partial \rho}{\partial t} = \frac{\partial \rho}{\partial \varphi}\frac{\partial \varphi}{\partial t} + \frac{\partial \rho}{\partial \psi}\frac{\partial \psi}{\partial t} \text{ , etc.} \tag{2.8.3}$$

Thus the equations (2.8.1) and (2.8.2) become

$$\left(\frac{\partial \varphi}{\partial t} + v\frac{\partial \varphi}{\partial x}\right)\frac{\partial \rho}{\partial \psi} + \rho\frac{\partial \varphi}{\partial x}\frac{\partial v}{\partial \varphi}$$

$$= -\left[\left(\frac{\partial \psi}{\partial t} + v\frac{\partial \psi}{\partial x}\right)\frac{\partial \rho}{\partial \psi} + \rho\frac{\partial \psi}{\partial x}\frac{\partial v}{\partial \psi}\right], \tag{2.8.4}$$

$$\frac{c^2}{\rho}\frac{\partial \varphi}{\partial x}\frac{\partial \rho}{\partial \varphi} + \left(\frac{\partial \varphi}{\partial t} + v\frac{\partial \varphi}{\partial x}\right)\frac{\partial v}{\partial \varphi}$$

$$= -\left[\frac{c^2}{\rho}\frac{\partial \psi}{\partial x}\frac{\partial \rho}{\partial \psi} + \left(\frac{\partial \psi}{\partial t} + v\frac{\partial \psi}{\partial x}\right)\frac{\partial v}{\partial \psi}\right]. \tag{2.8.5}$$

Let us now perform the integration $\int_{\varphi-\varepsilon}^{\varphi+\varepsilon} d\varphi$ on (2.8.4) and (2.8.5) and let $\varepsilon \to 0$. The integrals on the right hand sides of the equations vanish since the integrands are continuous across the characteristics $\varphi = \text{constant}$. Since $\int_{\varphi-\varepsilon}^{\varphi+\varepsilon} d\varphi \frac{\partial \rho}{\partial \varphi} = \Delta\rho$, and $\int_{\varphi-\varepsilon}^{\varphi+\varepsilon} d\varphi \frac{\partial v}{\partial \varphi} = \Delta v$, we obtain

$$\left(\frac{\partial \varphi}{\partial t}+v\frac{\partial \varphi}{\partial x}\right)\Delta\rho+\rho\frac{\partial \varphi}{\partial x}\Delta v=0\ , \qquad (2.8.6)$$

$$\frac{c^2}{\rho}\frac{\partial \varphi}{\partial x}\Delta\rho+\left(\frac{\partial \varphi}{\partial t}+v\frac{\partial \varphi}{\partial x}\right)\Delta v=0\ . \qquad (2.8.7)$$

$(\Delta\rho)$ and (Δv) have non-trivial solutions if $\varphi = \text{constant}$ is characteristic. Thus we obtain the equation for characteristics :

$$\left(\frac{\partial \varphi}{\partial t}+v\frac{\partial \varphi}{\partial x}\right)^2-c^2\left(\frac{\partial \varphi}{\partial x}\right)^2=0\ , \qquad (2.8.8)$$

or

$$\frac{\partial \varphi}{\partial t}+(v\pm c)\frac{\partial \varphi}{\partial x}=0\ . \qquad (2.8.9)$$

Let us denote the two families of solutions of (2.8.9) by $\alpha(x,t)$ and $\beta(x,t)$. Specifically :

$$\frac{\partial \alpha}{\partial t}+(v-c)\frac{\partial \alpha}{\partial x}=0\ , \qquad (2.8.10)$$

$$\frac{\partial \beta}{\partial t}+(v+c)\frac{\partial \beta}{\partial x}=0\ . \qquad (2.8.11)$$

The family of characteristics $\alpha(x,t) = \text{constant}$ will be denoted as C_- , while the family $\beta(x,t) = \text{constant}$, C_+ . In other words :

$$\beta=\text{constant}\ , \quad \left(\frac{dx}{dt}\right)_+=v+c\ , \quad \text{on } C_+\ , \qquad (2.8.12)$$

$$\alpha=\text{constant}\ , \quad \left(\frac{dx}{dt}\right)_-=v-c\ , \quad \text{on } C_-\ . \qquad (2.8.13)$$

Since $\alpha(x,t)$ and $\beta(x,t)$ are independent, they can also serve as independent coordinates. In terms of these characteristic coordinates, Equations (2.8.1) and (2.8.2) can be simplified extensively. We now proceed to do so.

As

$$d\alpha = \alpha_x dx + \alpha_t dt \ ,$$

$$d\beta = \beta_x dx + \beta_t dt \ ,$$

thus we have

$$dx = \frac{1}{j}(\beta_t d\alpha - \alpha_t d\beta) \ ,$$

$$dt = \frac{1}{j}(-\beta_x d\alpha + \alpha_x d\beta) \ ,$$

so long as

$$j = \alpha_x \beta_t - \alpha_t \beta_x \neq 0 \ . \tag{2.8.14}$$

Thus, we have

$$\alpha_x = j t_\beta \ , \qquad \alpha_t = -j x_\beta \ , \tag{2.8.15}$$

$$\beta_x = -j t_\alpha \ , \qquad \beta_t = j x_\alpha \ . \tag{2.8.16}$$

Hence from (2.8.12), we have

$$\frac{dx}{dt} = -\frac{\beta_t}{\beta_x} = \frac{x_\alpha}{t_\alpha} = v + c \ , \quad \text{on } C_+ \ ,$$

or

$$x_\alpha - (v+c)t_\alpha = 0 \ , \quad \text{on } C_+ \ . \tag{2.8.17}$$

Similarly, we have

$$x_\beta - (v - c)t_\beta = 0 \,, \quad \text{on } C_- \,. \qquad (2.8.18)$$

To express (2.8.1) and (2.8.2) in terms of characteristic coordinates (α, β), we need to perform such calculations as :

$$\rho_t = \rho_\alpha \alpha_t + \rho_\beta \beta_t = j\left(-\rho_\alpha x_\beta + \rho_\beta x_\alpha\right),$$

$$\rho_x = \rho_\alpha \alpha_x + \rho_\beta \beta_x = j\left(\rho_\alpha t_\beta - \rho_\beta t_\alpha\right),$$

and similar expressions for v_t and v_x. Substitute these relations into (2.8.1) and (2.8.2), we can obtain :

$$-\rho_\alpha x_\beta + \rho_\beta x_\alpha + (\rho v_\alpha + v\rho_\alpha)t_\beta - \left(\rho v_\beta + v\rho_\beta\right)t_\alpha = 0 \,, \quad (2.8.19)$$

$$-v_\alpha x_\beta + v_\beta x_\alpha + \left(vv_\alpha + \frac{c^2}{\rho}\rho_\alpha\right)t_\beta - \left(vv_\beta + \frac{c^2}{\rho}\rho_\beta\right)t_\alpha = 0 \,. \quad (2.8.20)$$

Now let us investigate the situation on C_+. On C_+, $\beta = \text{constant}$. Hence everything is a function of α only. But on C_+, we have (2.8.17), i.e. $x_\alpha = (v + c)t_\alpha$. We can thus eliminate x_α and t_α from (2.8.17), (2.8.19) and (2.8.20), and obtain the following relation :

$$\left(v_\alpha + \frac{c}{\rho}\rho_\alpha\right)x_\beta + \left(cv_\alpha + \frac{c}{\rho}v\rho_\alpha + vv_\alpha + \frac{c^2}{\rho}\rho_\alpha\right)t_\beta = 0 \,,$$

which can be put in the form :

$$\left(v_\alpha + \frac{c}{\rho}\rho_\alpha\right)\left[x_\beta - (v + c)t_\beta\right] = 0 \,. \qquad (2.8.21)$$

Now, the relation $x_\beta = (v-c)t_\beta$ defines C_- . Thus in general $x_\beta \neq (v+c)t_\beta$ on C_+ . Therefore, we obtain

$$v_\alpha + \frac{c}{\rho}\rho_\alpha = 0 , \quad \text{on } C_+ . \tag{2.8.22}$$

Define

$$r = v + \int \frac{c}{\rho} d\rho . \tag{2.8.23}$$

We thus conclude r is constant on C_+ , i.e. r does not change on any C_+ - characteristics. The family of C_+ characteristics are designated by the parameter β. Thus we conclude :

$$r = r(\beta) . \tag{2.8.24}$$

A similar development would lead to the definition of the function s :

$$s = v - \int \frac{c}{\rho} d\rho , \tag{2.8.25}$$

and we have

$$s = s(\alpha) , \tag{2.8.26}$$

i.e. s is constant on any C_- - characteristics.

The functions r and s are known as Riemann invariants, because r is invariant on C_+ and s is invariant on C_- .

Conceptually, with characteristics in place, the solution can be readily obtained by making use of the Riemann invariants. Fig. 2-5 shows schematically the procedure of the solution.

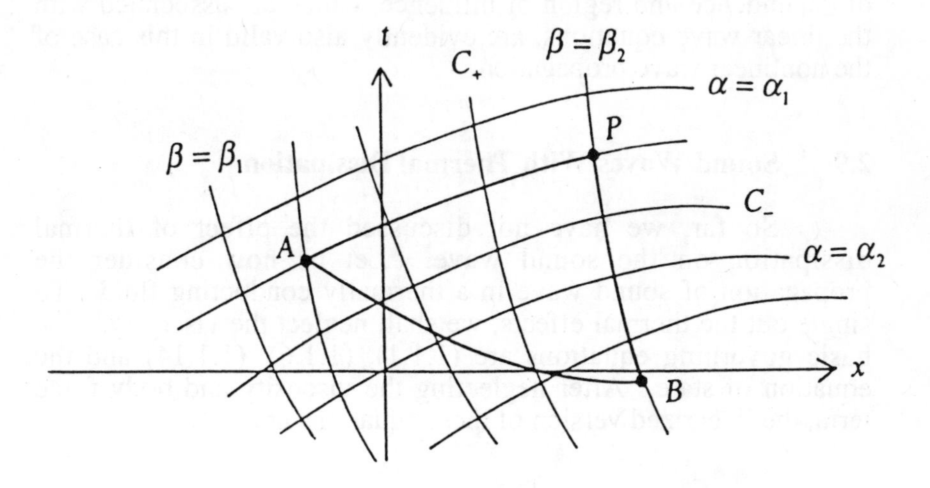

Fig. 2-5 The characteristics in x-t plane.

If ρ and v are given on the initial curve AB. Then the values of ρ and v inside the region PAB can be completely determined if the characteristics have been mapped out. To be specific, as r and s are Riemann invariants, we have

$$r(P) = r(B) \ , \tag{2.8.27}$$

since r is invariant on C_+ . Similarly,

$$s(P) = s(A) \ . \tag{2.8.28}$$

Now $r(B)$ can be obtained from $\rho(B)$ and $v(B)$ which are prescribed as initial conditions, and $s(A)$ can be obtained from $\rho(A)$ and $v(A)$. Thus $r(P)$ and $s(P)$ can be obtained, and $\rho(P)$ and $v(P)$ in turn can be found. The only trouble is that because the wave equations are nonlinear, C_+ and C_- cannot be predetermined. The characteristics need to be computed step by step from the initial values along the entire curve AB. For the characteristics to reach P, we need in fact to have the solution throughout the region PAB. Still, with this conceptual construction, the ideas of domain of dependence and region of influence, which are associated with the linear wave equations, are evidently also valid in this case of the nonlinear wave propagation.

2.9 Sound Waves With Thermal Dissipation

So far, we have not discussed the effect of thermal dissipation on the sound wave. Let us now consider the propagation of sound wave in a thermally conducting fluid. To single out the thermal effects, we shall neglect the viscosity. The basic governing equations are (1.1.1), (1.1.6), (1.1.14) and the equation of state. After neglecting the viscosity and body force term, the linearized version of these equations are

$$\frac{\partial \rho}{\partial t} + \rho_0 (\nabla \cdot \mathbf{v}) = 0 , \qquad (2.9.1)$$

$$\frac{\partial \mathbf{v}}{\partial t} = -\frac{1}{\rho_0} \nabla p , \qquad (2.9.2)$$

$$\rho_0 c_v \frac{\partial T}{\partial t} = \kappa \nabla^2 T - p_0 (\nabla \cdot \mathbf{v}) , \qquad (2.9.3)$$

and

$$\frac{p}{p_0} = \frac{\rho}{\rho_0} + \frac{T}{T_0} , \qquad (2.9.4)$$

where ρ_0 , p_0 and T_0 are equilibrium quantities and ρ , p , T and $\mathbf{v}$ are respective small perturbed quantities. Equation (2.9.4) is the linearized version of the equation of state for the perfect gas.

Differentiate (2.9.1) with respect to t, and make use of (2.9.2) and (2.9.4), we obtain the following relation :

$$\frac{\partial^2 \rho}{\partial t^2} = \frac{p_0}{\rho_0} \nabla^2 \rho + \frac{p_0}{T_0} \nabla^2 T . \qquad (2.9.5)$$

Thus if there is no temperature variation, i.e. for the isothermal case, (2.9.5) becomes

$$\frac{\partial^2 \rho}{\partial t^2} = c_i^2 \nabla^2 \rho \,, \tag{2.9.6}$$

where

$$c_i^2 = \frac{p_0}{\rho_0} \,, \tag{2.9.7}$$

is the isothermal sound speed.

For the general case, let us differentiate (2.9.5) with respect t again, and obtain

$$\frac{\partial}{\partial t}\left[\frac{\partial^2 \rho}{\partial t^2} - \frac{p_0}{\rho_0}\nabla^2 \rho\right] = \frac{p_0}{T_0}\nabla^2 \frac{\partial T}{\partial t} \,. \tag{2.9.8}$$

The term $\nabla^2 \dfrac{\partial T}{\partial t}$ on the right-hand side of (2.9.8) can be expressed in terms of $\nabla^2 T$ and $(\nabla \cdot \mathbf{v})$ by taking the Laplacian of (2.9.3). These terms can again be expressed in terms of ρ by using (2.9.1) and (2.9.5). The end result is the following equation :

$$\frac{\partial}{\partial t}\left[\frac{\partial^2 \rho}{\partial t^2} - \frac{p_0}{\rho_0}\left(1 + \frac{p_0}{\rho_0 T_0 c_v}\right)\nabla^2 \rho\right]$$

$$= \frac{\kappa}{\rho_0 c_v}\nabla^2\left[\frac{\partial^2 \rho}{\partial t^2} - \frac{p_0}{\rho_0}\nabla^2 \rho\right]. \tag{2.9.9}$$

Denote the thermal diffusion coefficient D by

$$D = \frac{\kappa}{\rho_0 c_v} . \tag{2.9.10}$$

Now the equilibrium equation of state for perfect gas is $p_0 = R\rho_0 T_0$, and $R = c_p - c_v$. Denote the ratio of the specific heats, following the common convention, by γ :

$$\gamma = \frac{c_p}{c_v} . \tag{2.9.11}$$

Then

$$\gamma = 1 + \frac{p_0}{\rho_0 T_0 c_v} . \tag{2.9.12}$$

Thus, (2.9.9) can be rewritten as

$$\frac{\partial}{\partial t}\left[\frac{\partial^2 \rho}{\partial t^2} - c_a^2 \nabla^2 \rho\right] = D\nabla^2\left[\frac{\partial^2 \rho}{\partial t^2} - c_i^2 \nabla^2 \rho\right] , \tag{2.9.13}$$

where

$$c_a^2 = \frac{\gamma p_0}{\rho_0} , \tag{2.9.14}$$

c_a is known as the adiabatic sound speed, because it can be shown that

$$c_a^2 = \left(\frac{dp}{d\rho}\right)_s . \tag{2.9.15}$$

To show that (2.9.15) is valid, we start with the equation of state of the perfect gas :

$$dp = R(\rho dT + T d\rho) . \tag{2.9.16}$$

Now, for adiabatic change of thermodynamic states, equation (1.1.10) becomes

$$c_p dT = \frac{1}{\rho} dp \ . \tag{2.9.17}$$

Then, it can be readily established that c_a^2 is indeed the adiabatic sound speed.

It is interesting to note that if $c_a = c_i$, then the structure of the equation (2.9.13) is the successive operations of a diffusive operator and a wave operator.

To get a more direct feeling about the significance of the equation (2.9.13), let us consider the one-dimensional sinusoidal waves.

Let $\rho = ae^{i(kx-\omega t)}$ and substitute into (2.9.13), we obtain the following dispersion equation :

$$i\omega\left(\omega^2 - c_a^2 k^2\right) = Dk^2\left(\omega^2 - c_i^2 k^2\right) , \tag{2.9.18}$$

or

$$k^4 - k^2\left(\frac{\omega^2}{c_i^2} + \frac{i\omega c_a^2}{Dc_i^2}\right) + \frac{i\omega^3}{Dc_i^2} = 0 \ , \tag{2.9.19}$$

or

$$k^2 = \frac{1}{2}\left[\frac{\omega^2}{c_i^2} + \frac{i\omega c_a^2}{Dc_i^2} \pm \left\{\left(\frac{\omega^2}{c_i^2} + \frac{i\omega c_a^2}{Dc_i^2}\right)^2 - \frac{4i\omega^3}{Dc_i^2}\right\}^{\frac{1}{2}}\right] . \tag{2.9.20}$$

Let us consider two limiting cases :

(i) $\omega << \dfrac{c_a^2}{D}$

It is clear that one branch of the solutions of (2.9.20) yields

$$k^2 \approx \frac{i\omega c_a^2}{Dc_i^2}, \tag{2.9.21}$$

which leads to heavily damped wave, and is not the one of interest to wave propagation. We shall pay attention to the other branch. Now, for $\omega << \dfrac{c_a^2}{D}$, we have approximately :

$$\left\{\left(\frac{\omega^2}{c_i^2}+\frac{i\omega c_a^2}{c_i^2 D}\right)^2-\frac{4i\omega^3}{Dc_i^2}\right\}^{\frac{1}{2}} \approx \frac{i\omega c_a^2}{Dc_i^2}+\frac{\omega^2}{c_i^2}-\frac{2\omega^2}{c_a^2}. \tag{2.9.22}$$

Substitute (2.9.22) into (2.9.20), we obtain

$$k^2 \approx \frac{\omega^2}{c_a^2}. \tag{2.9.23}$$

Thus the wave is propagating with the adiabatic sound speed. To find the damping coefficient, we write

$$k = \left(\frac{\omega}{c_a}\right)(1+\eta_a), \tag{2.9.24}$$

and substitute into (2.9.19), and retaining terms only up to $0(\eta_a)$. Then it can be readily found that

$$\eta_a = \frac{iD\omega}{2c_a^4}\left(c_a^2-c_i^2\right). \tag{2.9.25}$$

(ii) $\omega >> \frac{c_a^2}{D}$

Similar computation shows again that one branch of the solutions of (2.9.20) is heavily damped, while for the other branch, we obtain

$$k \approx \left(\frac{\omega}{c_i}\right)(1+\eta_i) \, , \tag{2.9.26}$$

and

$$\eta_i = \frac{i}{2D\omega}\left(c_a^2 - c_i^2\right) . \tag{2.9.27}$$

Thus the wave is propagating with the isothermal sound speed.

To gain a clearer understanding of the physical meaning of the previous results, we note that the diffusion length

$$\lambda_D = \left(\frac{D}{\omega}\right)^{\frac{1}{2}} \tag{2.9.28}$$

is a measure of range of thermal diffusion, while the wavelength

$$\lambda = \frac{c_a}{\omega} \, , \tag{2.9.29}$$

is a measure of region of compression and rarefaction in the time interval of a period. For $\omega << \frac{c_a^2}{D}$, we have $\lambda_D << \lambda$. Thus the heat generated is unable to diffuse out of the region of compression, and the process is essentially adiabatic. Therefore the propagation speed is the adiabatic sound speed. On the other hand, for $\omega >> \frac{c_a^2}{D}$, because $\lambda_D >> \lambda$, the heat generated is diffused and temperature equalized over many wavelengths. Therefore the process is isothermal and the wave speed is the isothermal sound speed. Laplace was the first one to point out that sound speed in air under the normal circumstance is the adiabatic sound speed. But his reasoning is erroneous. "Laplace considered that the condensations and rarefactions concerned in the propagation of sound take place with such rapidity that the heat and cold produced have not time to pass away, and that therefore the relation between volume and pressure is sensibly the same as if the air were confined in an absolutely non-conducting vessel".(1) If ω is taken to be a measure of rapidity of the process, then we see that speed is adiabatic not because the process is rapid, but because the process is sufficiently slow. For air under normal circumstances, $c_a^2 \approx 10^9\ (\mathrm{cm/sec})^2$, $D \approx 0.29\ \mathrm{cm}^2/\mathrm{sec}$. Thus $\frac{c_a^2}{D} \approx 3 \times 10^9\ \mathrm{sec}^{-1}$. For ordinary sound propagation, ω is smaller in comparison with $3 \times 10^9\ \mathrm{sec}^{-1}$. Therefore the propagation speed is the adiabatic sound speed c_a.

(1) Lord Rayleigh : "Theory of Sound", Vol. II, p. 19, Dover, New York (1945).

CHAPTER 3

Water Waves

In contrast with the sound waves, Water waves relate to waves in incompressible fluids. The sound wave speed in a perfectly incompressible fluid is infinite. The basic equation governing fluid flow in incompressible fluids is the Laplace equation rather than the wave equation. Water waves are possible only because of the presence of the free surface. Thus, both physically and mathematically, water waves are quite different from the sound waves.

3.1 Governing Equations of Water Waves

The water waves to be discussed are reasonably idealized. Water, of course, is compressible. But we shall treat it as an incompressible fluid. Water is also viscous. But we shall treat it as an inviscid fluid. Over the surface of the water, there is in general the air, which is another fluid. However, we shall treat it as empty. So the water surface is a free surface. Water is normally situated in a gravity field which we shall take into consideration. We shall also sometimes take into consideration the effect of surface tension. We make these simplification in order to bring forth clearly the essential characteristics of the water waves. The effects of compressibility, viscousity and the superposed air can be incorporated as corrections if we so wish.

The general configuration of the water wave problem is schematically presented in Fig. 3-1 The free surface is defined by

$$S(\mathbf{x},t) = y - \eta(x,z,t) = 0 \ , \qquad (3.1.1)$$

and the fixed bottom is given by

$$y = h(x,z) \ . \qquad (3.1.2)$$

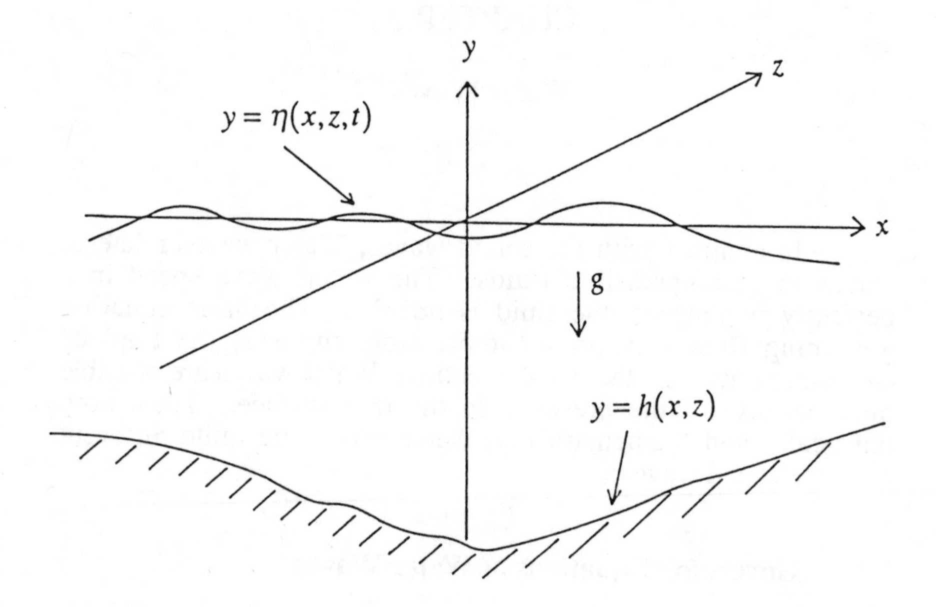

Fig. 3-1 Schematic representation of the water wave problem.

The water occupies the region : $h(x,z) < y < \eta(x,z,t)$. Since the fluid is incompressible and inviscid, we shall consider the case that the flow is irrotational. Thus the flow velocity field $\mathbf{v}(\mathbf{x},t)$ can be expressed in terms of the velocity potential, i.e.

$$\mathbf{v} = \nabla \varphi \ . \tag{3.1.3}$$

The continuity equation (1.1.1) for incompressible fluids is

$$\nabla \cdot \mathbf{v} = 0 \ . \tag{3.1.4}$$

From (3.1.3) and (3.1.4), we thus obtain

$$\nabla^2 \varphi = 0 \ , \quad h(x,z) < y < \eta(x,z,t) \ . \tag{3.1.5}$$

The momentum equation (1.1.6) for this incompressible and inviscid fluid in the gravitation field has the following form :

$$\frac{\partial \mathbf{v}}{\partial t}+(\mathbf{v}\cdot\nabla)\mathbf{v}=-\frac{1}{\rho}\nabla p-g\mathbf{e}_y\,, \qquad (3.1.6)$$

where g is the gravitation constant. Mathematically, (3.1.5) plays the active role as the primary equation to be solved, while (3.1.6) plays the relatively passive role to obtain p, once φ and hence $\mathbf{v}$ are obtained by the solution of (3.1.5). For potential flows, (3.1.6) can be integrated to yield the Bernoulli's equation :

$$\frac{p}{\rho}+\frac{1}{2}(\nabla\rho)^2+\frac{\partial\varphi}{\partial t}+gy=C(t)\,, \quad h(x,z)<y<\eta(x,z,t)\,, \quad (3.1.7)$$

where $C(t)$ is an arbitrary function of t.

The boundary of the fluid region has two parts : the bottom and the free surface. At the fixed bottom, the condition is that the fluid velocity normal to the surface has to vanish. Since the normal to the bottom surface is proportional to the vector $\left(\frac{\partial h}{\partial x},\ -1,\ \frac{\partial h}{\partial z}\right)$, thus we have

$$\frac{\partial\varphi}{\partial x}\frac{\partial h}{\partial x}+\frac{\partial\varphi}{\partial z}\frac{\partial h}{\partial z}-\frac{\partial\varphi}{\partial y}=0\,, \qquad y=h(x,z)\,. \quad (3.1.8)$$

On the free surface we have both the dynamic and kinematic conditions to be satisfied. The dynamic condition is such that the pressure over the surface is zero since it is a free surface. Therefore the fluid pressure at the free surface needs only to balance the surface tension and any externally imposed pressure. Using the Bernoulli's equation (3.1.7), we obtain

$$-\frac{\partial\varphi}{\partial t}-\frac{1}{2}(\nabla\varphi)^2-g\eta+C(t)$$

$$=\frac{p^{(0)}}{\rho}+\frac{\sigma}{\rho}\left(\frac{1}{R_1}+\frac{1}{R_2}\right), \qquad y=\eta(x,z,t)\,, \qquad (3.1.9)$$

where σ is the surface tension coefficient, R_1 and R_2 are the principal radii of curvature taken to be positive if the center of curvature is in the water, and $p^{(0)}$ is the pressure at the free surface.

The kinematic free surface condition can be stated simply that the fluid particle once on the free surface will remain on the free surface. As the free surface is given by (3.1.1), the condition is just

$$\frac{\partial S}{\partial t}+(\mathbf{v}\cdot\nabla)S=0\ , \quad \text{on } S=0, \tag{3.1.10}$$

or

$$\frac{\partial\eta}{\partial t}+\frac{\partial\varphi}{\partial x}\frac{\partial\eta}{\partial x}+\frac{\partial\varphi}{\partial z}\frac{\partial\eta}{\partial z}-\frac{\partial\varphi}{\partial y}=0\ , \quad \text{on} \quad y=\eta(x,z,t)\ . \tag{3.1.11}$$

In Section 5.1, we shall deal with the interfacial conditions in much more details. The free surface conditions are special cases of the more general conditions at the interface between two fluids.

Equations (3.1.5), (3.1.7), (3.1.8), (3.1.9) and (3.1.11) form the governing system of the equations for water waves.

Often we are interested in the small amplitude waves over a calm surface. The calm surface is defined by

$$\varphi=0\ , \quad \eta=0\ , \quad p^{(0)}=gC(t)\ , \quad \text{and} \quad R_1,\, R_2\to\infty\ . \tag{3.1.12}$$

It can be easily verified that the state (3.1.12) satisfies the governing system of the water wave equations. Let us denote ε as a small parameter. Then, for small amplitude waves, we have

$$\varphi=O(\varepsilon)\ , \quad \eta=O(\varepsilon)\ , \tag{3.1.13}$$

and

$$p_e = p^{(0)} - gC(t) = O(\varepsilon) \, , \qquad (3.1.14)$$

where $p_e(x,z,t)$ is the externally imposed pressure on the surface.

With the small order requirements (3.1.13) and (3.1.14), neglecting the terms of $O(\varepsilon^2)$ or higher, equations (3.1.5), (3.1.8), (3.1.9) and (3.1.11) become respectively :

$$\nabla^2\varphi = 0 \, , \quad h(x,z) < y < 0 \, , \qquad (3.1.15)$$

$$\frac{\partial\varphi}{\partial x}\frac{\partial h}{\partial x} + \frac{\partial\varphi}{\partial z}\frac{\partial h}{\partial z} - \frac{\partial\varphi}{\partial y} = 0 \, , \quad y = h(x,z) \, , \qquad (3.1.16)$$

$$\frac{\partial\eta}{\partial t} - \frac{\partial\varphi}{\partial y} = 0 \, , \quad y = 0 \, , \qquad (3.1.17)$$

$$\frac{\partial\varphi}{\partial t} + g\eta = -\frac{1}{\rho}p_e(x,z,t) + \frac{\sigma}{\rho}\left(\frac{\partial^2\eta}{\partial x^2} + \frac{\partial^2\eta}{\partial z^2}\right) , \quad y = 0 \, . \qquad (3.1.18)$$

We should remark that the difficulty associated with original governing system of equations is not only the nonlinearity of the problem, but also that the free surface, which is the boundary of the fluid region, is unknown and variable and is part of the solution to be found. For small amplitude waves, the problem is now linear, and moreover the boundary becomes known and fixed due to the linearization. The problem is thus greatly simplified.

We shall proceed to consider the small amplitude waves first.

3.2 Small Amplitude Waves

We shall confine most of our discussions on waves in two spatial dimensions with a flat horizontal bottom. Thus, all the quantities are independent of z, and $h(x,z) = -h$. These simplifications will bring out the essential features of the water waves more clearly. Thus, equations (3.1.15) - (3.1.18) become now :

$$\frac{\partial^2 \varphi}{\partial x^2} + \frac{\partial^2 \varphi}{\partial y^2} = 0 , \quad -h < y < 0 , \tag{3.2.1}$$

$$\frac{\partial \varphi}{\partial y} = 0 , \quad y = -h, \tag{3.2.2}$$

$$\frac{\partial \eta}{\partial t} - \frac{\partial \varphi}{\partial y} = 0 , \quad y = 0 , \tag{3.2.3}$$

and

$$\frac{\partial \varphi}{\partial t} + g\eta = -\frac{1}{\rho} p_e(x,t) + \frac{\sigma}{\rho}\frac{\partial^2 \eta}{\partial x^2} , \quad y = 0. \tag{3.2.4}$$

We may proceed to solve the problem by Fourier transform. Thus we could write

$$\eta(x,t) = \int_{-\infty}^{\infty} d\omega \int_{-\infty}^{\infty} dk A(k,\omega) e^{-i(kx-\omega t)} , \tag{3.2.5}$$

and a corresponding expression for φ , and substitute into (3.2.1) - (3.2.4) to find the solutions. Since these equations are linear, it is equivalent to consider just the monochromatic waves. Hence we shall take

$$\eta(x,t) = A \sin(kx - \omega t) . \tag{3.2.6}$$

For this monochromatic wave, A is the amplitude, ω the frequency, and k the wave number. Hence $\lambda = \dfrac{2\pi}{k}$ is the wavelength, and $\tau = \dfrac{2\pi}{\omega}$ is the period; while the phase velocity is

$$c = \frac{\omega}{k} . \tag{3.2.7}$$

With η specified by (3.2.6), it is clear from (3.2.3) and (3.2.4), $\varphi(x, y, t)$ is to be expressed in the following form :

$$\varphi = \Phi(y)\cos(kx - \omega t) . \tag{3.2.8}$$

Substituting (3.2.8) into (3.2.1), and making use of (3.2.2), we obtain

$$\varphi = B\cosh k(y + h)\cos(kx - \omega t) . \tag{3.2.9}$$

We consider now the case that there is no externally applied pressure, i.e. $p_e = 0$. Thus, (3.2.3) and (3.2.4) become

$$-\omega A - kB\sinh kh = 0 , \tag{3.2.10}$$

and

$$\omega B\cosh kh + gA = -\frac{\sigma}{\rho}k^2 A . \tag{3.2.11}$$

The existence of nontrivial solutions from (3.2.10) and (3.2.11) leads to the following dispersion relation :

$$\omega^2 = gk\left(1 + \frac{\sigma k^2}{g\rho}\right)\tanh kh . \tag{3.2.12}$$

Using (3.2.7), the dispersion relation can also be written as :

$$c^2 = \frac{g}{k}\left(1 + \frac{\sigma k^2}{g\rho}\right)\tanh kh . \tag{3.2.13}$$

Waves whose phase velocity changes with wave number are called dispersive waves. It is clear that water waves are in general dispersive waves. However, if the wavelength is long, i.e. when $kh << 1$, (3.2.13) becomes approximately :

$$c^2 = gh \ , \tag{3.2.14}$$

and the wave is nondispersive.

Let us introduce the non-dimensional quantities :

$$c_*^2 = \frac{c^2}{gh} \ , \qquad k_* = kh \ , \qquad \sigma_* = \frac{\sigma}{\rho g h^2} \ . \tag{3.2.15}$$

Then (3.2.13) can be rewritten as

$$c_*^2 = \left(\frac{1}{k_*} + \sigma_* k_* \right) \tanh k_* \ . \tag{3.2.16}$$

The relation between c_* and k_* is schematically shown in Fig. 3-2.

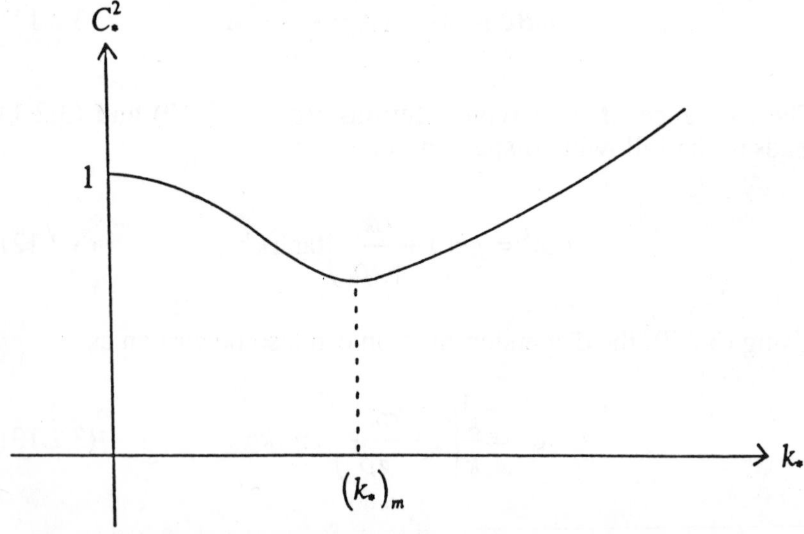

Fig. 3-2 Schematic diagram of the dispersion relation (3.2.16).

As shown in Fig. 3-2, $c_* = 1$ at $k_* = 0$, and c_*^2 has a minimum at $k = (k_*)_m$. Approximately, we have $(k_*)_m \approx \left(\frac{1}{\sigma^*}\right)^{\frac{1}{2}}$. In normal daily life circumstance, we have in cgs units, $\sigma = 73$, g = 980, thus $\sigma^* \approx \frac{1}{13h^2}$. Hence $(k_*)_m \approx 3.5h$, or $(k)_m \approx 3.5\text{cm}^{-1}$, or $(\lambda)_m \approx 1.8$ cm .

We commonly denote those water waves with $k_* << (k_*)_m$ the gravity waves, since the gravity effect is dominant, while those waves with $k_* >> (k_*)_m$ the capillary waves, since the effect of surface tension is dominant. We now discuss these two cases in more details.

(i) *Gravity Wave.* The gravity waves refer to those cases that $\sigma_* k_*^2 << 1$. Under ordinary circumstances, these are the waves whose wavelength $\lambda >> 2$ cm. For this case, (3.2.16) becomes approximately.

$$c_*^2 = \frac{1}{k^*} \tanh k_* , \tag{3.2.17}$$

or

$$\omega^2 = gk \tanh kh . \tag{3.2.18}$$

If, furthermore we have $k_* = kh << 1$, then we obtain

$$c_*^2 = 1 , \quad \text{or} \quad c^2 = gh . \tag{3.2.19}$$

This is the case for shallow water wave, since the wavelength is long compared with the depth of the water. Shallow water waves are non-dispersive.

If, on the other hand, we have $k_* >> 1$, then it is the case for deep water wave, and we have

$$c_*^2 = \frac{1}{k_*}, \quad \text{or} \quad c = \left(\frac{g}{k}\right)^{\frac{1}{2}}. \tag{3.2.20}$$

$k_* >> 1$ implies that $\lambda << h$. However λ still has to be long enough to be in the regime of gravity wave.

The deep water wave is dispersive. An important concept in discussing the dispersive waves is the concept of group velocity. We define the group velocity c_g by

$$c_g = \frac{d\omega}{dk}. \tag{3.2.21}$$

Since for the deep water wave, we have from (3.2.20) that

$$\omega = (gk)^{\frac{1}{2}}, \tag{3.2.22}$$

the corresponding group velocity is

$$c_g = \frac{1}{2}\left(\frac{g}{k}\right)^{\frac{1}{2}} = \frac{c}{2}. \tag{3.2.23}$$

In contrast, for the shallow water waves, we have $c = c_g$, since it is non-dispersive. We shall discuss group velocity and dispersive waves in more details in the next section.

(ii) *Capillary Wave.* The capillary waves refer to those cases that $\sigma_* k_*^2 >> 1$. These are waves with wavelength $\lambda << 2$ cm ordinarily. For this case, we have approximately

$$c_*^2 = \sigma_* k_* \tanh k_* \ , \tag{3.2.24}$$

or

$$\omega^2 = \frac{\sigma}{\rho} k^3 \tanh kh \ . \tag{3.2.25}$$

Let us consider the case that $k_* >> 1$, i.e. $h >> \lambda$. Then we have, from (3.2.24), approximately :

$$c_*^2 = \sigma_* k_* \ , \quad \text{or} \quad c = \left(\frac{\sigma k}{\rho}\right)^{\frac{1}{2}} . \tag{3.2.26}$$

This is the usual case of capillary wave. It is dispersive, and the corresponding group velocity is

$$c_g = \frac{3}{2}\left(\frac{\sigma k}{\rho}\right)^{\frac{1}{2}} = \frac{3}{2} c \ . \tag{3.2.27}$$

From (3.2.23) and (3.2.27), we see that for water with considerable depth, ordinarily for waves with $\lambda > 2$ cm, it is the deep water gravity wave; while for $\lambda < 2$ cm, it is the capillary wave. If the wavelength becomes so large as to be long compared with the depth of the water layer, then the wave is the shallow water wave. It is to be noted that the group velocity for the deep water wave is smaller than the phase velocity, while the group velocity of the capillary wave is larger than the phase velocity.

The other relatively unusual case of the capillary wave is that for which $k_* << 1$. Since we also have $\sigma_* k_*^2 >> 1$, it is the case that σ_* is very large, or equivalently, very shallow water layer. In other words, it is the capillary waves on a film. For this case, (3.2.24) becomes approximately :

$$c_*^2 = \sigma_* k_*^2 \,, \quad \text{or} \quad c = \left(\frac{\sigma h}{\rho}\right)^{\frac{1}{2}} k \,. \tag{3.2.28}$$

It is again dispersive, and the corresponding group velocity is

$$c_g = 2\left(\frac{\sigma h}{\rho}\right)^{\frac{1}{2}} k = 2c \,. \tag{3.2.29}$$

Thus the group velocity is again larger than the phase velocity. This type of capillary wave, because its depth h is so small, the effect of viscosity should be important. Thus the wave is more difficult to sustain.

3.3 Dispersion of Waves - Group Velocity

The analysis of monochromatic waves led to the dispersion relation (3.2.12). In general, the dispersion relation for wave propagation is of the form :

$$\omega = \omega(k) \,. \tag{3.3.1}$$

Therefore, the equation (3.2.5) can be written as

$$\eta(x,t) = \int_{-\infty}^{\infty} A(k) e^{-i[kx - \omega(k)t]} dk \,. \tag{3.3.2}$$

In other words, in (3.2.5) we have $A(k,\omega) = A(k)\delta[\omega - \omega(k)]$.

To understand the meaning of dispersion of waves, let us first consider two trains of monochromatic waves of the same amplitude :

$$\eta_1 = A\sin(kx - \omega t) \,, \tag{3.3.3}$$

$$\eta_2 = A\sin[(k+\Delta k)x - (\omega+\Delta\omega)t] . \qquad (3.3.4)$$

We require that $\frac{\Delta k}{k} << 1$, and $\frac{\Delta\omega}{\omega} << 1$. Thus these two wavetrains have only slightly different wave numbers and frequencies. Now let us consider the combination of these two wavetrains :

$$\eta = \eta_1 + \eta_2 . \qquad (3.3.5)$$

Substitute (3.3.3) and (3.3.4) into (3.3.5) we obtain

$$\eta = 2A\sin\left\{(kx-\omega t)\left[1+0\left(\frac{\Delta k}{k}\right)\right]\right\}\cos\left[\frac{\Delta k}{2}\left(x-\frac{\Delta\omega}{\Delta k}\right)t\right] . \qquad (3.3.6)$$

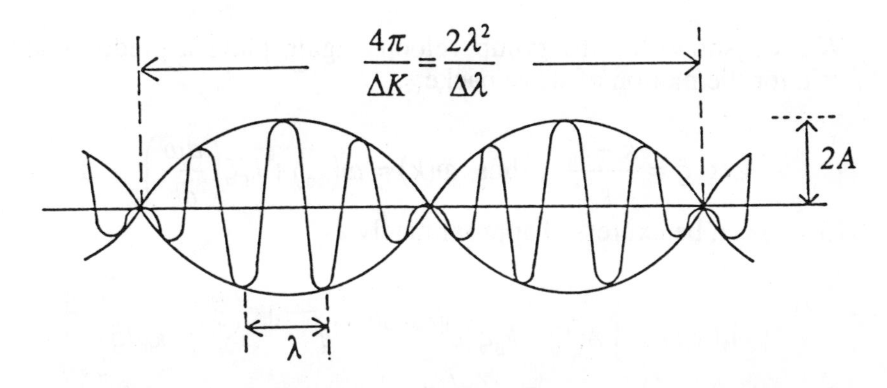

Fig. 3.3 Schematic representation of the envelope wave.

At a definite t , η is schematically represented in Fig. 3.3 It shows many waves with short wavelengths λ enveloped in a long wave with wavelength $\frac{4\pi}{\Delta k}$, or $\frac{2\lambda^2}{\Delta\lambda}$. Here we have used the relation that $|\Delta k| = \frac{2\pi}{\lambda^2}|\Delta\lambda|$, since $k = \frac{2\pi}{\lambda}$. The short waves are progressing with speed $c = \dfrac{\omega}{k}$. But the envelope wave is moving forward with the group velocity $c_g = \dfrac{d\omega}{dk}$. Looking at the whole phenomenon, the envelope wave with group velocity is more representative for the general picture. For nondispersive waves, $c_g = c$. Thus both the constituent short wave and the envelope wave move with the same velocity.

Pure monochromatic waves seldom exist in reality. What we often encounter are wave packets composed of waves with wave number clustered around a definite k_0. In other words, we have

$$\eta(x,t) = \int_{k_0(1-\varepsilon)}^{k_0(1+\varepsilon)} A(k) e^{-i[kx-\omega(k)t]} dk \,, \qquad \varepsilon << 1 \,. \tag{3.3.7}$$

We can show that the group velocity again plays a predominant role for the motion of wave packets.

Let $\xi = \dfrac{k-k_0}{k_0}$, thus $\omega(k) \approx \omega(k_0) + k_0 \xi \left(\dfrac{d\omega}{dk}\right)_0$. Hence (3.3.7) can be expressed approximately as

$$\eta(x,t) = \int_{-\varepsilon}^{\varepsilon} A(k_0 + k_0\xi) e^{i[k_0 x - \omega(k_0)t]} e^{-ik_0\xi\left[x - \left(\frac{d\omega}{dk}\right)_0 t\right]} k_0 d\xi$$

$$= F\left[x - \left(\frac{d\omega}{dk}\right)_0 t\right] e^{i[k_0 x - \omega(k_0)t]} \,, \tag{3.3.8}$$

where

$$F(z) = \int_{-k_0\varepsilon}^{k_0\varepsilon} A(k_0 + u) e^{-iuz} du \,. \tag{3.3.9}$$

Thus the envelope wave $F\left[x-\left(\frac{d\omega}{dk}\right)_0 t\right]$ is again progressing with the group velocity $\left(\frac{d\omega}{dk}\right)_0$. It is to be noted that the width of $F(z)$ is proportional to $\left(\frac{1}{\varepsilon}\right)$.

For general wave profiles, we should consider (3.3.2). Let us rewrite (3.3.2) as

$$\eta(x,t) = \int_{-\infty}^{\infty} A(k) e^{it\psi\left(k;\frac{x}{t}\right)} dk \,, \tag{3.3.10}$$

where

$$\psi\left(k;\frac{x}{t}\right) = \omega(k) - \left(\frac{x}{t}\right)k \,. \tag{3.3.11}$$

We shall now find the asymptotic expression of $\eta(x,t)$ as $t \to \infty$. We have $x \to \infty$ also as $t \to \infty$. But $\left(\frac{x}{t}\right)$ remains finite. We use the method of stationary phase to find the asymptotic expression. First we locate the stationary point. From $\frac{d\psi}{dk} = 0$, we obtain

$$\frac{d\omega}{dk} = \frac{x}{t} \,, \tag{3.3.12}$$

which gives k as a function of $\left(\frac{x}{t}\right)$, i.e.

$$k = k_0\left(\frac{x}{t}\right) . \tag{3.3.13}$$

Now, we expand the function $\psi\left(k;\frac{x}{t}\right)$ in the neighborhood of $k = k_0$. Thus

$$\psi\left(k;\frac{x}{t}\right) = \left[\omega(k_0) - \frac{k_0 x}{t}\right] + \frac{1}{2}(k-k_0)^2 \omega''(k_0)$$

$$+ 0(k-k_0)^3 \,. \tag{3.3.14}$$

Since the major contribution to the integral (3.3.10) as $t \to \infty$ comes from the part of the integrand that is in the neighborhood of $k = k_0$, we have approximately :

$$\eta \approx A(k_0) e^{i[\omega(k_0)t - k_0 x]} \int_{-\infty}^{\infty} e^{\frac{1}{2}\omega''(k_0)t(k-k_0)^2} dk \,. \tag{3.3.15}$$

Or

$$\eta(x,t) \approx A(k_0) \left[\frac{2\pi}{t|\omega''(k_0)|}\right]^{\frac{1}{2}} e^{i\left[k_0 x - \omega(k_0)t - \frac{\pi}{4} sgn\omega''(k_0)\right]} \,. \tag{3.3.16}$$

In the expression (3.3.16), $k_0\left(\frac{x}{t}\right)$ is determined from (3.3.12). In other words, if the desired wave number k_0 is specified first, then the location of the dominant profile (x,t) is to be determined from the relation $\frac{x}{t} = \left(\frac{d\omega}{dk}\right)_{k=k_0}$. On the other hand, at a given (x,t), the same relation will determine the prevalent wave number of the wave there. In any case, if we concentrate on a specific portion of the spectrum, the location is then determined by the group velocity, and the amplitude is given by (3.3.16).

Velocity of Energy Transport

The group velocity is usually also the velocity of energy transport in waves. Let us demonstrate this property for the case of water waves.

From (3.2.9), we see that the velocity potential can be expressed as follows :

$$\varphi(x,y,t)=\int_0^\infty B(k)\cosh k(y+h)\cos(kx-\omega t)dk\ , \quad (3.3.17)$$

while the surface profile is given by

$$\eta(x,t)=\int_0^\infty A(k)\sin(kx-\omega t)dk\ . \quad (3.3.18)$$

Let us just consider the case of gravity waves, then we have the dispersion relation :

$$\omega=\omega(k)=(gk\tanh kh)^{\frac{1}{2}}\ . \quad (3.3.19)$$

Also, from (3.2.10), we have

$$A=-\frac{Bk}{\omega}\sinh kh\ . \quad (3.3.20)$$

Let us denote e as the total energy per unit horizontal area, then

$$e=e_k+e_p\ , \quad (3.3.21)$$

where

$$e_k=\int_{-h}^{0}\frac{1}{2}\rho v^2 dy\ , \quad (3.3.22)$$

and

$$e_p = \int_0^\eta \rho g y dy \; . \tag{3.3.23}$$

e_k is the kinetic energy per unit horizontal area, and e_p the unit potential energy, computed up to terms of $0(\eta^2)$.

Now, let us discuss the case of monochromatic waves. Denote the symbol < > as the time average or space average. Because of the factors of $\sin(kx - \omega t)$ and $\cos(kx - \omega t)$, we obtain readily that

$$\langle e_k \rangle = \left\langle \frac{\rho}{2} \int_{-h}^{0} \left[\left(\frac{\partial \varphi}{\partial x} \right)^2 + \left(\frac{\partial \varphi}{\partial y} \right)^2 \right] dy \right\rangle$$

$$= \frac{\rho}{4} B^2 k^2 \int_{-h}^{0} \left[\sinh^2 k(y+h) + \cosh^2 k(y+h) \right] dy$$

$$= \frac{\rho}{8} k B^2 \sinh 2kh \; , \tag{3.3.24}$$

and

$$\langle e_p \rangle = \frac{\rho g}{4} A^2 = \frac{\rho g}{4} \left(\frac{k}{\omega} \right)^2 B^2 \sinh^2 kh$$

$$= \frac{\rho}{8} k B^2 \sinh 2kh \; . \tag{3.3.25}$$

Thus we obtain

$$\langle e_k \rangle = \langle e_p \rangle \; , \tag{3.3.26}$$

which satisfies the principle of equipartition of kinetic and potential energies. The total energy is thus

$$\langle e \rangle = \frac{\rho}{4} k B^2 \sinh 2kh \ . \tag{3.3.27}$$

Let us now introduce the energy flux f, i.e., the work done by the fluid per unit time and unit width in the z-direction. Thus up to $0(\eta^2)$, we have

$$f = \int_{-h}^{0} p \left(\frac{\partial \varphi}{\partial x} \right) dy \ . \tag{3.3.28}$$

From (3.1.7), for small amplitude waves, the pressure p is given by

$$p = -\rho g y - \rho \frac{\partial \varphi}{\partial t} = -\rho g y + \rho \frac{\omega}{k} \frac{\partial \varphi}{\partial x} \ . \tag{3.3.29}$$

Thus, we obtain

$$\langle f \rangle = \left\langle \rho \frac{\omega}{k} \int_{-h}^{0} \left(\frac{\partial \varphi}{\partial x} \right)^2 dy \right\rangle$$

$$= \frac{\rho}{2} \omega k B^2 \int_{-h}^{0} \cosh^2 k(y+h) dy$$

$$= \frac{\rho}{4} \omega B^2 \left(kh + \frac{1}{2} \sinh 2kh \right) . \tag{3.3.30}$$

We now define the velocity of energy transport, c_E as the ratio of energy flux and the unit total energy, i.e.

$$c_E = \frac{\langle f \rangle}{\langle e \rangle} \ . \tag{3.3.31}$$

Then from (3.3.27) and (3.3.30), we obtain

$$c_E = \frac{c}{2}\left(1 + \frac{2kh}{\sinh 2kh}\right). \tag{3.3.32}$$

If we calculate $c_g = \dfrac{d\omega}{dk}$ from (3.3.19), we find that

$$c_E = c_g \ .$$

Thus the velocity of energy transport is the same as the group velocity.

The group velocity can also be viewed from another perspective. Suppose the wave profile can be represented approximately by the following form :

$$\eta(x,t) = \eta_1(x,t)e^{i\alpha(x,t)} \ , \tag{3.3.33}$$

where $\eta_1(x,t)$ is slowly varying function of x and t, and $\alpha(x,t)$ is the phase function. The expression is a generalization of monochromatic waves. We define the phase velocity c as the velocity moving with the constant phase, i.e.,

$$c = \left(\frac{dx}{dt}\right)_\alpha = -\frac{\dfrac{\partial\alpha}{\partial t}}{\dfrac{\partial\alpha}{\partial x}} \ . \tag{3.3.34}$$

For the case that $\alpha(x,t) = \omega t - kx$, then $c = \dfrac{\omega}{k}$, and

$$\omega = \frac{\partial\alpha}{\partial t} \ , \tag{3.3.35}$$

and

$$k = -\frac{\partial \alpha}{\partial x} . \tag{3.3.36}$$

We shall define frequency ω and wavenumber k by (3.3.35) and (3.3.36) even for non-monochromatic waves. Then ω and k will no longer be constants. For most cases of interest, we require that ω and k be slowly varying function of x and t.

From (3.3.35) and (3.3.36), we have the compatibility relation :

$$\frac{\partial k}{\partial t} + \frac{\partial \omega}{\partial x} = 0 . \tag{3.3.37}$$

If there is a dispersion relation $\omega = \omega(k)$, then (3.3.37) can be rewritten as

$$\frac{\partial k}{\partial t} + c_g \frac{\partial k}{\partial x} = 0 . \tag{3.3.38}$$

Thus

$$c_g = -\frac{\frac{\partial k}{\partial t}}{\frac{\partial k}{\partial x}} = \left(\frac{dx}{dt}\right)_k . \tag{3.3.39}$$

We have thus shown again that the group velocity is the velocity following the group of waves with the same wavenumber (or wavelength). When there is no exchange of energy between waves with different wavelengths, then the velocity of energy transport should be the group energy.

3.4 Wave Produced by a Moving Source

To model the waves produced by the motion of ships, we can consider the problem of the waves due to a moving source. On ocean or lake, the water depth is usually large, thus we may take

$h \to \infty$. The characteristic length of the problem is also large. Hence, the effect of surface tension can be neglected. Therefore, the governing equations (3.1.15) - (3.1.18) are simplifed to :

$$\nabla^2 \varphi = 0 , \qquad y < 0 , \qquad (3.4.1)$$

$$\frac{\partial \varphi}{\partial y} = 0 , \qquad y \to -\infty , \qquad (3.4.2)$$

$$\frac{\partial \eta}{\partial t} = \frac{\partial \varphi}{\partial y} , \qquad y = 0 , \qquad (3.4.3)$$

and

$$\frac{\partial \varphi}{\partial t} + g\eta = -\frac{1}{\rho} p_e(x,z,t) , \qquad y = 0 . \qquad (3.4.4)$$

The effect of the moving source is contained in the term of $p_e(x,z,t)$. For the problem of motion of real ships, the equations (3.4.1) - (3.4.4) can only be valid outside the ship's hull. We also need to impose the boundary condition that the relative normal velocity vanishes on the contact surface of the hull with water. Here we have replaced the interaction between the ship's hull and water by the term $p_e(x,z,t)$. The model of ship motion by $p_e(x,z,t)$ is adequate if we are only interested in the wave motion far away from the ship. A further simplification of the model is to consider the moving source a point source.

Let the trajectory of the moving point source be given by

$$x = \alpha(t) , \qquad z = \beta(t) . \qquad (3.4.5)$$

Then we may write :

$$\begin{aligned} p_e(x,z,t) &= p(t)\delta[x - \alpha(t)]\delta[z - \beta(t)] \\ &= \int d\tau \delta(t-\tau)\delta[x - \alpha(\tau)]\delta[z - \beta(\tau)]p(\tau) . \qquad (3.4.6) \end{aligned}$$

Since the problem is linear, the waves produced by the general motion of the source can be considered as the superposition of instantaneous sources over the times. Thus we may seek first the fundamental solution, i.e., the solution due to the following $p_e(x,z,t)$:

$$p_e(x,z,t) = \delta(x)\delta(z)\delta(t)$$

$$= \frac{\delta(r)}{2\pi r}\delta(t) \,, \tag{3.4.7}$$

where $r^2 = x^2 + z^2$.

We shall actually discuss the more general case in the following, and let

$$p_e = I(r)\delta(t) \,. \tag{3.4.8}$$

Thus, we shall deal in fact with a special kind of distributed sources and not just a point source.

The equation (3.4.4) for this case now becomes

$$\frac{\partial \varphi}{\partial t} + g\eta = -\frac{1}{\rho} I(r)\delta(t) \,, \quad y = 0 \,. \tag{3.4.9}$$

The last equation, which is valid for all *t*, because of the $\delta(t)$ term, can be transformed into an initial condition at $t = 0$. Integrating (3.4.9) with respect to *t* from $t = 0^-$ to $t = 0^+$, we obtain

$$\varphi = -\frac{1}{\rho} I(r) \,, \quad \text{on } y = 0, \text{ at } t = 0^+ \,. \tag{3.4.10}$$

Also, we obtain from (3.4.9)

$$\frac{\partial\varphi}{\partial t}+g\eta=0\ , \quad \text{on } y=0\ , \text{for } t>0\ . \qquad (3.4.11)$$

Combine (3.4.3) and (3.4.11), we thus obtain

$$\frac{\partial^2\varphi}{\partial t^2}+g\frac{\partial\varphi}{\partial y}=0\ , \quad \text{on } y=0\ , \text{for } t>0\ . \qquad (3.4.12)$$

(3.4.9) also implies that $\dfrac{\partial\varphi}{\partial t}=-g\eta$ for $t=0^+$ on $y=0$. We shall set

$$\frac{\partial\varphi}{\partial t}=\eta=0\ , \quad \text{for } t=0^+ \text{ on } y=0\ . \qquad (3.4.13)$$

In this way, the wave is produced solely by $I(r)$ from (3.4.10). If $I(r)=0$, then the surface is calm. Free waves are permissible within the framework of our formulation. The imposition of (3.4.13) removes the free waves from our consideration. To summarize, the problem we are facing is the following :

$$\nabla^2\varphi=0\ , \qquad y<0\ , \qquad (3.4.14)$$

$$\frac{\partial\varphi}{\partial y}=0\ , \quad \text{as } y\to-\infty\ , \qquad (3.4.15)$$

$$\frac{\partial^2\varphi}{\partial t^2}+g\frac{\partial\varphi}{\partial y}=0\ , \quad \text{for } t>0, \quad \text{on } y=0, \qquad (3.4.16)$$

$$\varphi=-\frac{1}{\rho}I(r)\ , \quad \text{at } t=0, \quad \text{on } y=0, \qquad (3.4.17)$$

$$\frac{\partial\varphi}{\partial t}=0\ , \quad \text{at } t=0, \quad \text{on } y=0. \qquad (3.4.18)$$

Since the problem possesses the axial symmetry, we shall employ the cylindrical coordinates. Thus (3.4.14) is

$$\frac{\partial^2 \varphi}{\partial r^2} + \frac{1}{r}\frac{\partial \varphi}{\partial r} + \frac{\partial^2 \varphi}{\partial y^2} = 0 , \quad y < 0, r > 0. \quad (3.4.19)$$

Now, from the Fourier-Bessel theorem, we know that a function *F(r)* can be expressed in the following way :

$$F(r) = \int_0^\infty \mathcal{F}(k) J_0(kr) k dk , \quad (3.4.20)$$

where

$$\mathcal{F}(k) = \int_0^\infty F(\alpha) J_0(k\alpha) \alpha d\alpha . \quad (3.4.21)$$

Making use of this kind of representation, and using (3.4.19), we obtain the following representation of φ :

$$\varphi(r, y, t) = \int_0^\infty dk A(k,t) e^{ky} J_0(kr) . \quad (3.4.22)$$

We have made use of (3.4.15) and the requirement that φ is bounded at $r = 0$ to arrive at (3.4.22). Substitute (3.4.22) into (3.4.16), we obtain

$$\frac{\partial^2 A}{\partial t^2} + gkA = 0. \quad (3.4.23)$$

Thus, using (3.4.18), we obtain

$$A(k,t) = B(k) \cos \omega t , \quad (3.4.24)$$

where

$$\omega = (gk)^{\frac{1}{2}} . \tag{3.4.25}$$

We have taken $\omega = +(gk)^{\frac{1}{2}}$ for definiteness. We could have taken $\omega = -(gk)^{\frac{1}{2}}$, so long as it is consistent throughout. Thus, (3.4.22) becomes

$$\varphi(r,y,t) = \int_0^\infty dk B(k) \cos\left[(gk)^{\frac{1}{2}} t\right] e^{ky} J_0(kr) . \tag{3.4.26}$$

Substitute (3.4.26) into (3.4.17), we then obtain

$$\int_0^\infty dk B(k) J_0(kr) = -\frac{1}{\rho} I(r) . \tag{3.4.27}$$

From the Fourier-Bessel theorem (3.4.20) and (3.4.21), we thus have :

$$B(k) = -\frac{k}{\rho} \int_0^\infty d\alpha I(\alpha) \alpha J_0(k\alpha) . \tag{3.4.28}$$

Then, (3.4.26) becomes :

$$\varphi(r,y,t)$$

$$= -\frac{1}{\rho} \int_0^\infty dk e^{ky} \cos\left[(gk)^{\frac{1}{2}} t\right] J_0(kr) k \int_0^\infty d\alpha I(\alpha) \alpha J_0(k\alpha). \tag{3.4.29}$$

From (3.4.3), we thus also obtain the expression for η :

$$\eta(r,t) = \int_0^\infty dk k^{\frac{3}{2}} \sin\left[(gk)^{\frac{1}{2}} t\right] J_0(kr) G(k) , \tag{3.4.30}$$

where

$$G(k) = -\frac{1}{\rho g^{\frac{1}{2}}} \int_0^\infty dr I(r) r J_0(kr) . \qquad (3.4.31)$$

For the special case of point source, i.e., for the case that $I(r) = \frac{A\delta(r)}{2\pi r}$, then we have

$$G(k) = -\frac{A}{2\pi \rho g^{\frac{1}{2}}} . \qquad (3.4.32)$$

We shall now investigate the wave pattern represented by (3.4.30).

Wave Pattern Due to a Moving Source

The disturbance represented by (3.4.30) and (3.4.32) is the disturbance at (x,z,t) with the source point at origin and the pressure applied at the instant $t = 0$. If the pressure is applied at $t = \tau$ and at the location (ξ, ζ) , then at a later time $t > \tau$, the disturbance at *(x, z)* will be given by :

$$\eta(x,z,t;\xi,\zeta,\tau) = \int_0^\infty dk k^{\frac{3}{2}} \sin\left[(gk)^{\frac{1}{2}}(t-\tau)\right] J_0(kr) G(k) , \qquad (3.4.33)$$

where now

$$r^2 = (x-\xi)^2 + (z-\zeta)^2 . \qquad (3.4.34)$$

Suppose now we have a moving source, like a ship, which starts to move at time $\tau = 0$, following a trajectory $(\xi(\tau), \zeta(\tau))$

with a variable pressure intensity given by $A(\tau)$, then with $G(k)$ now changed to $G(k,\tau) = -\frac{A(\tau)}{2\pi\rho g^{\frac{1}{2}}}$, we can get from (3.4.33) :

$$\eta(x,z,t) = -\frac{1}{2\pi\rho g^{\frac{1}{2}}}\int_0^t d\tau A(\tau)\int_0^\infty dk k^{\frac{3}{2}} \sin\left[(gk)^{\frac{1}{2}}(t-\tau)\right]J_0(kr), \quad (3.4.35)$$

where

$$r^2 = [x-\xi(\tau)]^2 + [z-\zeta(\tau)]^2 \; . \qquad (3.4.36)$$

The expression (3.4.35) may be evaluated approximately using the asymptotic methods. Usually, we are interested in the wave profiles at a distance from the source. Thus we have $kr >> 1$, and hence

$$J_0(kr) \approx \left(\frac{2}{\pi kr}\right)^{\frac{1}{2}} \cos\left(kr - \frac{\pi}{4}\right) = \left(\frac{2}{\pi kr}\right)^{\frac{1}{2}} \sin\left(kr + \frac{\pi}{4}\right) . \quad (3.4.37)$$

We shall first find the asymptotic expression for the second integral in (3.4.35), i.e. the integral K :

$$K = \int_0^\infty dk k^{\frac{3}{2}} \sin\left[(gk)^{\frac{1}{2}} s\right] J_0(kr) \; . \qquad (3.4.38)$$

Using (3.4.37), we have approximately :

$$\sin\left[(gk)^{\frac{1}{2}} s\right] J_0(kr) = \left(\frac{2}{\pi kr}\right)^{\frac{1}{2}} \sin\left[(gk)^{\frac{1}{2}} s\right] \sin\left(kr + \frac{\pi}{4}\right)$$

$$= \left(\frac{1}{2\pi kr}\right)^{\frac{1}{2}} \left\{\cos\left[(gk)^{\frac{1}{2}} s - kr - \frac{\pi}{4}\right] - \cos\left[(gk)^{\frac{1}{2}} s + kr + \frac{\pi}{4}\right]\right\}$$

$$= \left(\frac{1}{2\pi kr}\right)^{\frac{1}{2}} Rl\left\{e^{i\left[(gk)^{\frac{1}{2}}s-kr-\frac{\pi}{4}\right]} - e^{i\left[(gk)^{\frac{1}{2}}s+kr+\frac{\pi}{4}\right]}\right\}.$$

Thus, the integral K can be expressed approximately :

$$K = \left(\frac{1}{2\pi r}\right)^{\frac{1}{2}} Rl\int_0^\infty dkk\left\{e^{i\left[(gk)^{\frac{1}{2}}s-kr-\frac{\pi}{4}\right]} - e^{i\left[(gk)^{\frac{1}{2}}s+kr+\frac{\pi}{4}\right]}\right\}. \quad (3.4.39)$$

The exponents of the terms in the integrand in (3.4.39) are large. We may thus use the method of stationary phase to find the asymptotic expression of K. Let us denote the phases of the two exponential terms by m and n, thus

$$m(k) = (gk)^{\frac{1}{2}}s - kr - \frac{\pi}{4}, \quad (3.4.40)$$

and

$$n(k) = (gk)^{\frac{1}{2}}s + kr + \frac{\pi}{4}. \quad (3.4.41)$$

The stationary points are given by $m'(k) = 0$ and $n'(k) = 0$. From $m'(k) = 0$, we find that the stationary point k_0 is given by

$$k_0 = \frac{gs^2}{4r^2}. \quad (3.4.42)$$

The stationary point from $n'(k) = 0$ is $(-k_0)$ which lies outside the range of integration in (3.4.39). Hence this term does not contribute to the asymptotic expression. With k_0 given by (3.4.42), we readily obtain

$$m(k_0) = \frac{gs^2}{4r} - \frac{\pi}{4}, \quad (3.4.43)$$

and

$$m''(k_0) = -\frac{2r^3}{gs^2} . \qquad (3.4.44)$$

Therefore, we obtain the asymptotic expression of K :

$$K = \left(\frac{1}{2\pi r}\right)^{\frac{1}{2}} Rl\left\{\left(\frac{2\pi gs^2}{2r^3}\right)^{\frac{1}{2}}\left(\frac{gs^2}{4r^2}\right)e^{i\left(\frac{gs^2}{4r}-\frac{\pi}{2}\right)}\right\}$$

$$= \left(\frac{g^3}{2^5}\right)^{\frac{1}{2}} \frac{s^3}{r^4}\sin\left(\frac{gs^2}{4r}\right) . \qquad (3.4.45)$$

Putting K back into (3.4.35), and change the integration varaible from τ to $s = t - \tau$, we now obtain :

$$\eta(x,z,t) = -\frac{g^2}{2^{\frac{7}{2}}\pi\rho}\int_0^t dsA(t-s)\frac{s^3}{r^4}\sin\left(\frac{gs^2}{4r}\right), \qquad (3.4.46)$$

where

$$r^2 = [x - \xi(t-s)]^2 + [y - \zeta(t-s)]^2 . \qquad (3.4.47)$$

Now, from (3.4.42), we see that $\frac{gs^2}{4r} = k_0 r$, which is large. Thus we may again use the method of stationary phase to find the asymptotic expression of the integral (3.4.46).

In the integral (3.4.46), the phase function is

$$\psi = \frac{gs^2}{4r} . \qquad (3.4.48)$$

Prominant contributions to the wave profile η come from the stationary points of ψ . As

$$\frac{d\psi}{ds}=\frac{g}{4}\left(\frac{2s}{r}-\frac{s^2}{r^2}\frac{dr}{ds}\right),$$

thus the condition that $\dfrac{d\psi}{ds}=0$ yields :

$$\frac{dr}{ds}=\frac{2r}{s}. \tag{3.4.49}$$

The relation (3.4.49) together with (3.4.47) establish a correspondence between the field point *P (x, z, t)* with the source point $Q(\xi_0,\zeta_0,\tau)$. Because with given *(x, z t)*, (3.4.47) and (3.4.49) can be solved to yield *r* and *s*. Then we obtain $\tau=t-s$, $\xi_0=\xi(\tau)$, $\zeta_0=\zeta(\tau)$. This correspondence can be made clearer with the aid of the schematic diagram shown in Fig. 3-4.

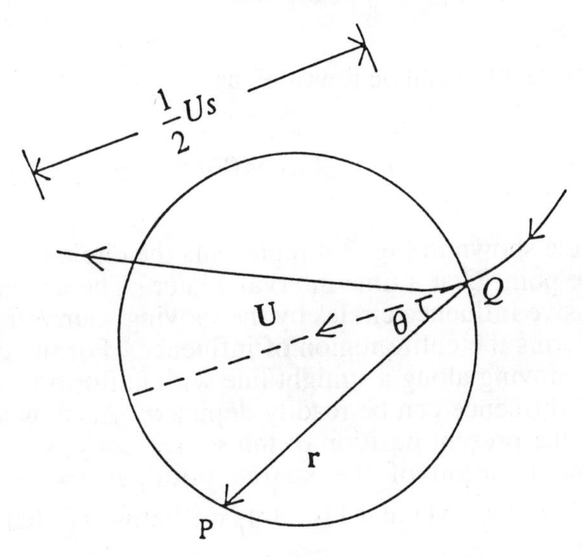

Fig. 3-4 Relationship between the field point *P* and the source point *Q*.

In Fig. 3-4, the field point is located at (x, z) at time t; while the source point Q is located at $(\xi(\tau), \zeta(\tau))$. Thus $\overline{PQ} = r$. We shall designate the vector $\overline{QP} = \mathbf{r}$, i.e. $\mathbf{r} = (x - \xi(t-s), z - \zeta(t-s))$. Let the velocity of the source at Q be $\mathbf{U} = \left(\frac{d\xi}{d\tau}, \frac{d\zeta}{d\tau}\right)$. Hence $\mathbf{U} = \frac{d\mathbf{r}}{ds}$. Now, let θ be the angle between $\mathbf{r}$ and $\mathbf{U}$, then we obtain

$$r\frac{dr}{ds} = \frac{1}{2}\frac{d}{ds}\left(r^2\right) = \frac{1}{2}\frac{d}{ds}(\mathbf{r}\cdot\mathbf{r})$$

$$= \mathbf{r}\cdot\mathbf{U} = rU\cos\theta .$$

Hence

$$\frac{dr}{ds} = U\cos\theta . \tag{3.4.50}$$

Therefore, the relation can be rewritten as

$$r = \frac{1}{2}Us\cos\theta . \tag{3.4.51}$$

Thus, the circle shown in Fig. 3-4 represents the circle of influence by the source point Q at a time interval s later. The accumulation of the successive influence circles by the moving source from $s = t$ up to $s = 0$ forms the entire region of influence. For the case that the source is moving along a straight line with uniform velocity U, the region of influence can be readily depicted. As shown in Fig. 3.5, let O be the present position of the source point, say at $t = 0$. Let Q be the location of the source point at $t = -s$. Thus $\overline{OQ} = Us$. From (3.4.51) and Fig. 3-4, we know the diameter of the influence circle by Q is $\frac{1}{2}Us$. The center of the influence circle is located at O_1, and $\overline{OO_1} = \frac{3}{4}Us$. Thus $\frac{\overline{O_1P}}{\overline{OO_1}} = \frac{1}{3}$, and this relation is independent of s. Therefore, we conclude that the region of influence is a wedge with the wedge angle β, and

$$\beta = \sin^{-1}\frac{1}{3} = 19°28' \ . \tag{3.4.52}$$

The disturbances outside wedge are not significant.

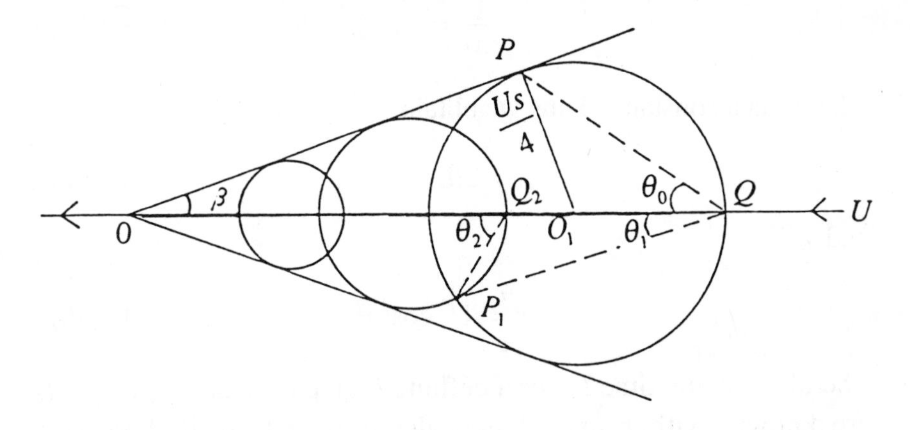

Fig. 3-5 Region of influence by a source moving with uniform velocity.

Within the wedge of influence, every point is the intersection of two circles of influence. Taking the point P_1 in Fig. 3-5. It is the intersection of the influence circles associated with Q and Q_2. The wavefront of the disturbance due to Q is perpendicular to QP_1, while that due to Q_2 perpendicular to Q_2P_1. In other words, on calculating η using (3.4.46), there are two contributions from two stationary points arising from (3.4.49). At the edge of the wedge, e.g. at point P, these two stationary points merge into one, and we shall find that $\frac{d^2\psi}{ds^2} = 0$ there. Then we need to find $\dfrac{d^3\psi}{ds^3}$ in order to obtain the asymptotic expression. The amplitudes at the edge of the wedge is thus larger than that at the interior points.

To get the wave patterns, we note that wave crests or troughs are defined by lines of the same phase. Thus wave patterns are determined by (3.4.51) together with equation of constant phase, i.e.

$$\frac{s^2}{4r} = l \ , \tag{3.4.53}$$

where l is a constant. Thus we obtain

$$s = 2lU\cos\theta \ , \tag{3.4.54}$$

and

$$r = lU^2\cos^2\theta \ . \tag{3.4.55}$$

Therefore at any time t, for a definite l, given s, since ξ, ζ and U are known, with r and θ now determined from (3.4.54) and (3.4.55), we can trace the pattern $\left(x(s),\, z(s)\right)$ as s varies.

Consider the special example that the source is moving uniformly with velocity U along x axis as depicted in Fig. 3-5. We have $\xi = Us$, $\zeta = 0$. Now

$$x = \xi - r\cos\theta = Us - r\cos\theta \ ,$$

$$z = \zeta + r\sin\theta = r\sin\theta \ .$$

Using (3.4.54) and (3.4.55), we thus obtain

$$x = lU^2\left(2\cos\theta - \cos^3\theta\right) . \tag{3.4.56}$$

$$z = lU^2 \cos^2 \theta \sin \theta \ . \tag{3.4.57}$$

The pattern of wave crests are schematically shown in Fig. 3-6. At each point, two types of waves are present. The ones corresponding to QP_1 in Fig. 3-5 are known as transverse waves; while the ones corresponding to Q_2P_1, the diverging waves. As we may see from Fig. 3-5, $\theta_0 = \frac{\pi}{4} - \frac{\beta}{2} = 35°16'$. The transverse waves are those associated with $0 < |\theta| < \theta_0$; while the divergent waves with $\theta_0 < |\theta| < \frac{\pi}{2}$.

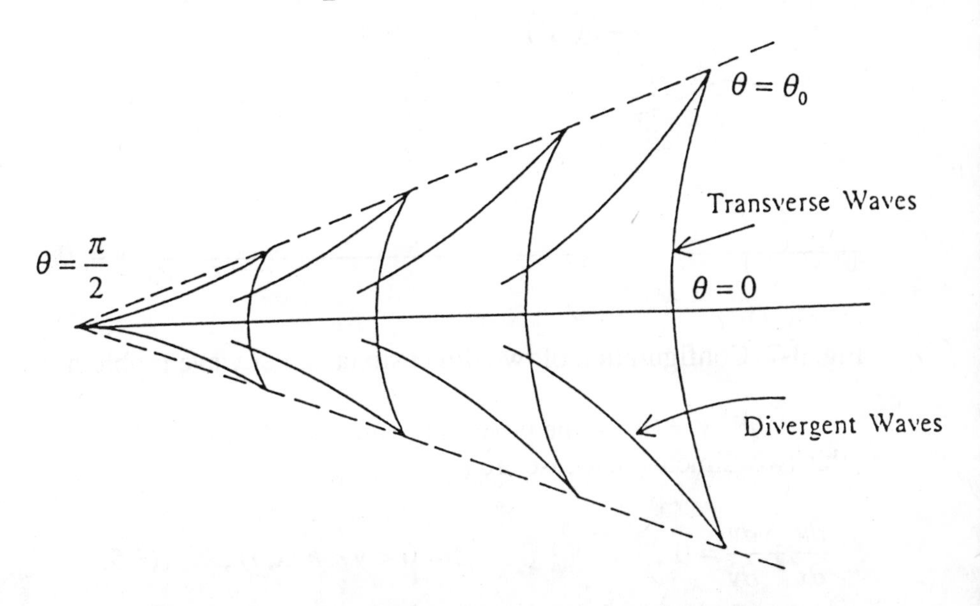

Fig. 3-6 Patterns of wave crests for a straight course.

3.5 Weakly Nonlinear Waves and Korteweg-DeVries Equation

We concentrate our discussion in this section on water waves in two spatial dimension with flat bottom. Just for convenience in writing, the bottom will be designated by $y = 0$, while the free surface be given by

$$y = h(x, t). \tag{3.5.1}$$

Let h_0 be the equilibrium water depth. We designate the deviation of the surface elevation from the equilibrium height by η :

$$\eta(x,t) = h(x,t) - h_0 . \tag{3.5.2}$$

The configuration is schematically represented in Fig. 3-7.

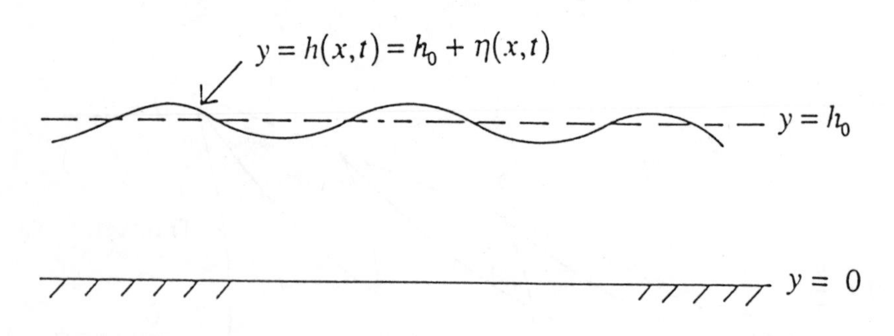

Fig. 3-7 Configuration of two dimensional water wave problem.

With $\mathbf{v} = (u, v)$, the governing equations (3.1.4) and (3.1.6) in this two-dimensional case are :

$$\frac{\partial u}{\partial x} + \frac{\partial v}{\partial y} = 0 , \qquad \text{for } 0 < y < h(x, t) , \tag{3.5.3}$$

$$\frac{\partial u}{\partial t} + u\frac{\partial u}{\partial x} + v\frac{\partial u}{\partial y} = -\frac{1}{\rho}\frac{\partial p}{\partial x} , \qquad \text{for } 0 < y < h(x, t) , \tag{3.5.4}$$

$$\frac{\partial v}{\partial t} + u\frac{\partial v}{\partial x} + v\frac{\partial v}{\partial y} = -\frac{1}{\rho}\frac{\partial p}{\partial y} - g , \qquad \text{for } 0 < y < h(x, t) . \tag{3.5.5}$$

We shall neglect the surface tension, and set the pressure on free surface zero. Then the boundary and free surface conditions (3.1.8), (3.1.9) and (3.1.11) can be rewritten respectively as :

$$v(x, 0, t) = 0, \tag{3.5.6}$$

$$p(x, h, t) = 0, \tag{3.5.7}$$

$$\frac{\partial h}{\partial t} + u(x,h,t)\frac{\partial h}{\partial x} - v(x,h,t) = 0 \ . \tag{3.5.8}$$

In the following, we shall consider the case that h is small in comparison with the characterisitic wavelength λ. Thus the wave amplitude η is also small. Hence, we expect $v(x, h, t)$ to be also small. Since $v(x, 0, t) = 0$ from (3.5.6), we expect that $v(x, y, t)$ to be small throughout. Then (3.5.5) can be approximated by :

$$-\frac{1}{\rho}\frac{\partial p}{\partial y} - g = 0 \ . \tag{3.5.9}$$

Integrating with respect to y , and using (3.5.7), we obtain from (3.5.9).

$$p = \rho g h(x,t) - \rho g y \ . \tag{3.5.10}$$

Thus (3.5.4) can be rewritten as :

$$\frac{\partial u}{\partial t} + u\frac{\partial u}{\partial x} + v\frac{\partial u}{\partial y} + g\frac{\partial h}{\partial x} = 0 \ . \tag{3.5.11}$$

Integrating (3.5.3) with respect to y from $y = 0$ to $y = h$, we obtain

$$\frac{\partial}{\partial x}\int_0^{h(x,t)} u\,dy - u(x,h,t)\frac{\partial h}{\partial x} + v(x,h,t) - v(x,0,t) = 0 \ ,$$

which, after using (3.5.6) and (3.5.8), becomes

$$\frac{\partial h}{\partial t}+\frac{\partial}{\partial x}\int_0^h u dy = 0 . \tag{3.5.12}$$

If we denote $\frac{D}{Dt}$ as the derivative following the particle, then (3.5.11) can be written as

$$\frac{Du}{Dt} = -g\frac{\partial h}{\partial x} . \tag{3.5.13}$$

The right hand side of (3.5.13) is independent of y. Therefore if u is independent of y initially, u will remain so. Thus $u = u(x, t)$. We shall limit our consideration to this case. Then (3.5.11) and (3.5.12) become

$$\frac{\partial u}{\partial t}+u\frac{\partial u}{\partial x}+g\frac{\partial h}{\partial x} = 0 , \tag{3.5.14}$$

and

$$\frac{\partial h}{\partial t}+u\frac{\partial h}{\partial x}+h\frac{\partial u}{\partial x} = 0 . \tag{3.5.15}$$

Equations (3.5.14) and (3.5.15) are also approximately valid if $u(x, y, t)$ is a slowly varying function of y. They form the basis of study of nonlinear shallow water waves. We shall come back later for more detailed discussions in Section 3.9. The structure of the system (3.5.14) and (3.5.15) is the same as that of the nonlinear one-dimensional sound waves (2.6.1) and (2.6.4). The correspondence reveals that for this case

$$c^2 = gh , \tag{3.5.16}$$

and for simple waves we have

$$h\frac{du}{dh} = \pm c = \pm(gh)^{\frac{1}{2}} , \tag{3.5.17}$$

which leads to

$$c = c_0 \pm \frac{u}{2} , \tag{3.5.18}$$

or

$$(gh)^{\frac{1}{2}} = (gh_0)^{\frac{1}{2}} \pm \frac{u}{2} . \tag{3.5.19}$$

Thus, corresponding to (2.6.10) we have the simple wave equation:

$$\frac{\partial h}{\partial t} \pm \left[3(gh)^{\frac{1}{2}} - 2(gh_0)^{\frac{1}{2}} \right] \frac{\partial h}{\partial x} = 0 . \tag{3.5.20}$$

Now we have $h = h_0 + \eta$. Thus

$$h^{\frac{1}{2}} = h_0^{\frac{1}{2}} \left[1 + \frac{1}{2} \left(\frac{\eta}{h_0} \right) + \cdots \right] .$$

Hence if $\left(\frac{\eta}{h_0} \right)$ is small, (3.5.20) can be written approximately :

$$\frac{\partial \eta}{\partial t} + (gh_0)^{\frac{1}{2}} \frac{\partial \eta}{\partial x} = 0 , \tag{3.5.21}$$

if we just follow one branch of the simple waves.

When the next order of $\left(\frac{\eta}{h_0} \right)$ is taken into account, we then have

$$\frac{\partial \eta}{\partial t} + (gh_0)^{\frac{1}{2}} \left(1 + \frac{3}{2} \frac{\eta}{h_0} \right) \frac{\partial \eta}{\partial x} = 0 . \tag{3.5.22}$$

(3.5.21) is the linear wave equation in one direction. For monochromatic waves, the dispersion relation is

$$\omega = (gh_0)^{\frac{1}{2}} k \; . \tag{3.5.23}$$

However, we know that for linear gravity waves, the general dispersion relation is given by (3.2.18) :

$$\omega^2 = gk \tanh kh \; , \tag{3.5.24}$$

or

$$\omega = (gh_0)^{\frac{1}{2}} k \left[1 - \frac{1}{6} (kh_0)^2 + \cdots \right] . \tag{3.5.25}$$

A generalization of (3.5.21) to accommodate for shorter waves would be an equation of the following form

$$\frac{\partial \eta}{\partial t} + (gh_0)^{\frac{1}{2}} \frac{\partial \eta}{\partial x} + (gh_0)^{\frac{1}{2}} \frac{h_0^2}{6} \frac{\partial^3 \eta}{\partial x^3} = 0 \; , \tag{3.5.26}$$

which will lead approximately to the dispersion relation (5.3.25). Combining both the corrections of the dispersion (3.5.26), and the nonlinearity (3.5.22), we obtain the following equation :

$$\frac{\partial \eta}{\partial t} + (gh_0)^{\frac{1}{2}} \left(1 + \frac{3}{2} \frac{\eta}{h} \right) \frac{\partial \eta}{\partial x} + (gh_0)^{\frac{1}{2}} \frac{h_0^2}{6} \frac{\partial^3 \eta}{\partial x^3} = 0 \; . \tag{3.5.27}$$

Equation (3.5.27) is known as the Korteweg-DeVries equation or KdV equation. It is the simplest equation which combines both the effects of nonlinearity and dispersion.

3.6 Solitary and Cnoidal Waves

Take the equation (3.5.27), and write

$$u=-6^{-\frac{2}{3}}\left(1+\frac{3}{2}\frac{\eta}{h_0}\right), \quad \bar{x}=6^{\frac{1}{3}}\frac{x}{h_0}, \quad \bar{t}=\left(\frac{g}{h_0}\right)^{\frac{1}{2}}t,$$

then (3.5.27) can be rewritten as

$$\frac{\partial u}{\partial \bar{t}}-6u\frac{\partial u}{\partial \bar{x}}+\frac{\partial^3 u}{\partial \bar{x}^3}=0 .$$

For later convenience in writing, we shall write x and t instead of $\bar{x}$ and $\bar{t}$. Thus the equation we shall study is

$$\frac{\partial u}{\partial t}-6u\frac{\partial u}{\partial x}+\frac{\partial^3 u}{\partial x^3}=0 . \tag{3.6.1}$$

(3.6.1) is the conventional form of the Korteweg-DeVries equation.

In order to understand various aspects of the Korteweg-DeVries equation, let us discuss first its linearized version :

$$\frac{\partial u}{\partial t}+\frac{\partial^3 u}{\partial x^3}=0 . \tag{3.6.2}$$

Let us find the solution of (3.6.2) with the initial condition :

$$u(x,0)=u_0(x) . \tag{3.6.3}$$

We may make use of the method of Fourier transform. Let

$$u_0(x)=\int_{-\infty}^{\infty}F(k)e^{ikx}dk , \quad F(k)=\frac{1}{2\pi}\int_{-\infty}^{\infty}u_0(x)e^{-ikx}dx . \tag{3.6.4}$$

Then the solution is given by

$$u(x,t)=\int_{-\infty}^{\infty}F(k)e^{i\left(kx+k^3t\right)}dk . \tag{3.6.5}$$

It is illuminating to consider the special case that

$$u_0(x) = \delta(x) . \tag{3.6.6}$$

Then, for this case,

$$F(k) = \frac{1}{2\pi} . \tag{3.6.7}$$

Hence

$$u(x,t) = \frac{1}{2\pi}\int_{-\infty}^{\infty} e^{i\left(kx+k^3t\right)} dk$$

$$= \frac{1}{\pi}\int_{0}^{\infty} \cos\left(kx + k^3 t\right) dk$$

$$= \frac{1}{\pi(3t)^{\frac{1}{3}}}\int_{0}^{\infty} \cos\left\{\left[\frac{x}{(3t)^{\frac{1}{3}}}\right] s + \frac{s^3}{3}\right\} ds .$$

Or

$$u(x,t) = \frac{1}{(3t)^{\frac{1}{3}}} Ai\left[\frac{x}{(3t)^{\frac{1}{3}}}\right] , \tag{3.6.8}$$

where $Ai(z)$ is the Airy function.

Figure 3-8 shows schematically the Airy function. We may also take note of some of its asymptotic expressions :

$$Ai(z) \approx \frac{1}{2}\pi^{-\frac{1}{2}} z^{-\frac{1}{4}} \exp\left(-\frac{2}{3} z^{\frac{3}{2}}\right) , \quad \text{as } z \to +\infty , \tag{3.6.9}$$

$$Ai(z) \approx \pi^{-\frac{1}{2}}|z|^{-\frac{1}{4}} \sin\left(\frac{2}{3}|z|^{\frac{3}{2}} + \frac{\pi}{4}\right), \text{ as } z \to -\infty . \quad (3.6.10)$$

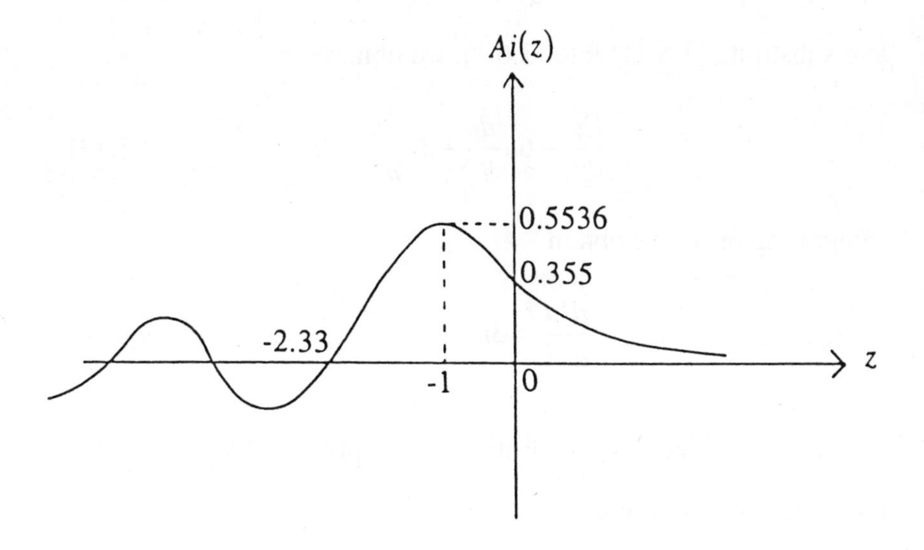

Fig. 3-8 Schematic representation of $Ai(z)$.

The solution (3.6.8) has thus the following properties : At time t, the maximum height is located approximately at $x = -(3t)^{\frac{1}{3}}$, with amplitude $u_m \approx \dfrac{0.5536}{(3t)^{\frac{1}{3}}}$; the width of the segment around the maximum amplitude is $\Delta x \approx 2(3t)^{\frac{1}{3}}$.

It is difficult to find solutions of nonlinear equations in general. Let us first consider some particular solutions. One class of the particular solutions is the travelling wave. The travelling wave solution has the following form :

$$u(x,t) = u(x - Ut) , \quad (3.6.11)$$

where U is a constant. This particular form of the solution implies that it is a wave moving with U in the $+x$ direction. Denote

$$\xi = x - Ut , \qquad (3.6.12)$$

and substitute (3.6.11) into (3.6.1), we obtain

$$\frac{d^3u}{d\xi^3} - 6u\frac{du}{d\xi} - U\frac{du}{d\xi} = 0 . \qquad (3.6.13)$$

Integrating once, we obtain

$$\frac{d^2u}{d\xi^2} = 3u^2 + Uu + A , \qquad (3.6.14)$$

where A is an integration constant. Mulitply (3.6.14) by $\frac{du}{d\xi}$, and then integrate, we obtain

$$\frac{1}{2}\left(\frac{du}{d\xi}\right)^2 = u^3 + \frac{U}{2}u^2 + Au + B , \qquad (3.6.15)$$

where B is another integration constant. In (3.6.15), there are three constants, i.e. U, A and B, to be specified. Let us write the right hand side of (3.6.15) as

$$F(u) = u^3 + \frac{U}{2}u^2 + Au + B = (u-a)(u-b)(u-c) . \qquad (3.6.16)$$

Given U, A and B, we can determine the constants a, b and c, and vice versa. However, while U, A and B are all real, a, b and c may all be real or only one of them is real. If a, b and c are all real, we shall order them such that $a \geq b \geq c$. The schematic representation of $F(u)$ is shown in Fig. 3-9.

Integrating (3.6.15) once more, we obtain

$$\xi = \int_{\alpha}^{u} \frac{du}{[2F(u)]^{\frac{1}{2}}} . \qquad (3.6.17)$$

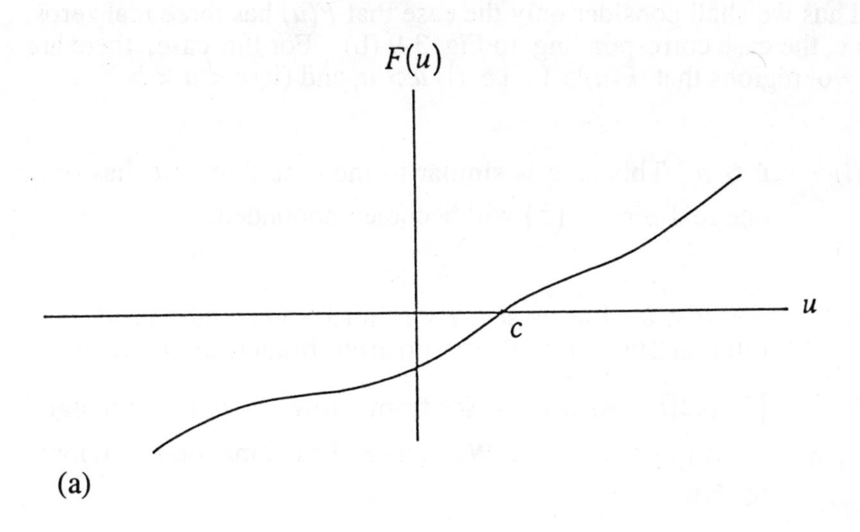

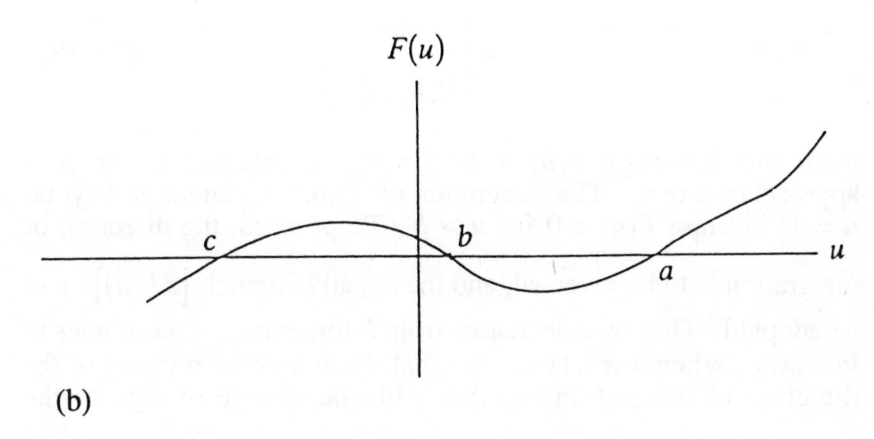

Fig. 3-9 Schematic representation of $F(u)$:
(a) a, b complex conjugates, c real; (b) a, b and c all real.

Since ξ is real, the solution is admissible only if $F(u) > 0$. For the case that $F(u)$ has only one real zero, i.e. the case that corresponds to Fig. 3-9 (a), the admissible range of u is $u > c$, and $u(\xi)$ will ultimately be unbounded. This is not the case of interest to us. Thus we shall consider only the case that $F(u)$ has three real zeros, i.e, the case corresponding to Fig. 3-9 (b). For this case, there are two regions that $F(u) > 0$, i.e. (i) $u > a$, and (ii) $c < u < b$.

(i) $u > a$. This case is similar to the case that $F(u)$ has only one real zero. $u(\xi)$ will become unbounded.

(ii) $c < u < b$. Let us set $\alpha = c$ in (3.6.17), and consider the integral starting with the positive branch of the quantity $[2F(u)]^{\frac{1}{2}}$. As u increases from c towards b, in the integral ξ will increase also. When u reaches b, the total increment of ξ is

$$L = \int_c^b \frac{du}{[2F(u)]^{\frac{1}{2}}} . \tag{3.6.18}$$

Note that although $F(b) = F(c) = 0$, the integral exists as u approaches b or c. The integration of (3.6.17) cannot go beyond $u = b$, because $F(u) < 0$ for $u > b$. To proceed, the direction of integration is to be reversed, and the negative branch $[2F(u)]^{\frac{1}{2}}$ will be adopted. Thus as u decreases from b towards c, ξ continues to increase. When u reaches c, we shall have another reversal of the direction of integration together with the change of sign of the quantity $[2F(u)]^{\frac{1}{2}}$. Therefore $u(\xi)$ will oscillate between c and b, and it is a periodic function with period $2L$.

Many nonlinear travelling wave solutions can be put in the form of (3.6.17). They all have the same qualitative features as described above. For the specific $F(u)$ as given by (3.6.16), we may even obtain the explicit solutions.

Let

$$u = b - (b-c)v^2 , \quad k^2 = \frac{b-c}{a-c} .$$

Then we found that

$$1 - v^2 = \frac{u-c}{b-c} ,$$

$$1 - k^2 + k^2 v^2 = \frac{a-u}{a-c} ,$$

$$du = -2\left[(b-c)(b-u)\right]dv ,$$

and

$$\frac{dv}{\left[(1-v^2)(1-k^2+k^2v^2)\right]^{\frac{1}{2}}} = -\frac{(a-c)^{\frac{1}{2}}du}{2\left[(a-u)(u-c)(b-u)\right]^{\frac{1}{2}}} .$$

Thus, we obtain

$$\left(\frac{a-c}{2}\right)^{\frac{1}{2}} \xi = \int_v^1 \frac{dv}{\left[(1-v^2)(1-k^2+k^2v^2)\right]^{\frac{1}{2}}} . \qquad (3.6.19)$$

In terms of the Jacobian elliptic functions, we can express (3.6.19) as

$$v = cn\left[\left(\frac{a-c}{2}\right)^{\frac{1}{2}} \xi\right], \tag{3.6.20}$$

or

$$u(x,t) = b - (b-c)cn^2\left[\left(\frac{a-c}{2}\right)^{\frac{1}{2}}(x - Ut)\right]. \tag{3.6.21}$$

We recall that the cnoidal function has the following properties :

$$cn(0) = 1 , \quad cn(u + 4K) = cnu ,$$

$$K = \int_0^1 \frac{dx}{\left[(1-x^2)(1-k^2x^2)\right]^{\frac{1}{2}}} .$$

Therefore with given a, b and c, the amplitude $(b-c)$ the propagation velocity U, and the period can be determined. The period is directly related to K which depends on the value of $k = \left(\frac{b-c}{a-c}\right)^{\frac{1}{2}}$. Therefore the period is related to the amplitude.

The periodic waves represented by (3.6.21) are expressed in terms of cnoidal functions. Therefore, they are conventionally called cnoidal waves. The amplitude, propagation speed, and period are all interrelated, in contrast with the situation for linear waves whose amplitude, speed and period are all independent.

A degenerate case of the cnoidal wave is the one for which $a = b$. Thus the curve $F(u)$ is tangent to the u-axis as shown in Fig. 3-10.

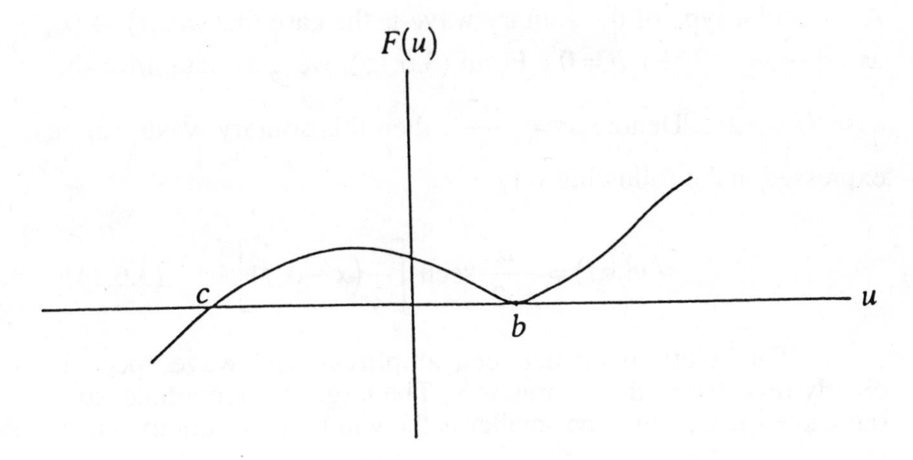

Fig. 3-10 Schematic representation of *F(u)* for the degenerate case.

Now in the neighborhood of $u = b$, $F(u) \propto (u-b)^2$. Thus the integral (3.6.17) is divergent as $u \to b$. In other words, $\xi \to \pm\infty$, as $u \to b$. Therefore when we perform the integration of (3.6.17), as u progresses from b to c and back to b, ξ progresses from $-\infty$ to ∞ , covering the whole real axis. Therefore, the period of this wave is infinite, and it is a solitary wave. For our specific problem, $b = a$ implies that $k = 1$. Thus (3.6.19) becomes

$$\left(\frac{b-c}{2}\right)^{\frac{1}{2}} \xi = \int_{v}^{1} \frac{dv}{v\left(1-v^2\right)^{\frac{1}{2}}}$$

$$= \operatorname{sech}^{-1} v \ . \qquad (3.6.22)$$

Therefore we have explicitly the expression of the solitary wave :

$$u(x,t) = b-(b-c)\operatorname{sech}^2\left[\left(\frac{b-c}{2}\right)^{\frac{1}{2}}(x-Ut)\right] . \qquad (3.6.23)$$

A particular type of the solitary wave is the case that $u(x,t) \to 0$, as $|x| \to \infty$. Then $b=0$. From (3.6.16), we thus obtain for this case $U=-2c$. Denote $c=-\dfrac{\alpha^2}{2}$, then this solitary wave can be expressed in the following way :

$$u(x,t) = -\frac{\alpha^2}{2}\mathrm{sech}^2\left[\frac{\alpha}{2}\left(x-\alpha^2 t\right)\right]. \qquad (3.6.24)$$

The relationship between amplitude and wave speed is clearly revealed in this expression. The larger the amplitude, so is the wave speed, while the smaller is the width of the solitary wave.

3.7 Solitons - Inverse Scattering

Let us, for convenience, set $\alpha=2$ in (3.6.24). Then we have the solution of the KdV equation :

$$u(x,t) = -2\mathrm{sech}^2(x-4t), \qquad (3.7.1)$$

which can be considered as the exact solution of the equation with initial value :

$$u(x,0) = -2\mathrm{sech}^2 x. \qquad (3.7.2)$$

Conceivably, if an initial value (3.7.2) is given first, clever people may guess and try, and arrive at the exact solution (3.7.1). However it is not likely that given the initial condition:

$$u(x,0) = -6\mathrm{sech}^2 x, \qquad (3.7.3)$$

one may obtain the following exact solution by guessing :

$$u(x,t) = -\frac{12[3+4\cosh(2x-8t)+\cosh(4x-64t)]}{[3\cosh(x-28t)+\cosh(3x-36t)]^2}. \qquad (3.7.4)$$

For this particular solution, it may be verified that as $t \to \infty$, $|u(x,t)|$ is very small except in the neighborhood of $x = 4t - \frac{1}{2}\ln 3$, and $x = 16t + \frac{1}{4}\ln 3$, where $|u(x,t)|$ attains maximal values of 2 and 8 respectively. Thus the solution (3.7.4) represents two solitary waves, one with amplitude (–2) moving with velocity 4 in x-direction, while the other with amplitude (–8) moving with velocity 16 in x-direction. For $t < 0$, the solitary wave with larger amplitude is behind the solitary wave with smaller amplitude. The wave with larger amplitude catches up the smaller one at $t = 0$. There is strong interaction between these two waves resulting in the combination solution (3.7.3). However, these two waves separate again after the interaction, and except for phase changes, resume the original forms of solitary waves for $t > 0$.

Now the larger wave with higher velocity is in front of the smaller wave. It is rather surprising that the solitary waves would retain their original form after the nonlinear interaction, since KdV equation is indeed nonlinear. Because of this preservation of the identity of local wave pulses through nonlinear interaction, these solitary waves are now termed solitons.

To investigate the behaviors of solitons, we shall address the general initial value problem :

$$u_t - 6uu_x + u_{xxx} = 0 , \quad -\infty < x < \infty , \qquad (3.7.5)$$

with

$$u(x,0) = u_0(x) . \qquad (3.7.6)$$

We shall look for solutions that $u(x, t)$ vanishes rapidly as $|x| \to \infty$.

An approach borrowed from the idea of Hopf-Cole transformation in solving the Burgers equation is to try $u = \frac{\psi_{xx}}{\psi}$. It turns out that the scheme does not work. However a modification of the previous idea indeed leads to the solution of the problem in quite remarkable manner. Let us introduce $\psi(x,t)$ by :

$$u(x,t) = \frac{\psi_{xx}}{\psi} + \lambda(t) , \tag{3.7.7}$$

which can be rewritten as

$$\psi_{xx} - \left[u(x,t) - \lambda(t)\right]\psi = 0 . \tag{3.7.8}$$

If we treat t as a parameter, and $u(x,t)$ as a given function, then (14.7) is a Sturm-Liouville problem with eigenvalue λ. From the perspective of quantum mechanics, (3.7.8) is the one-dimensional stationary Schrödinger equation, with ψ as the wave function, u the potential, and λ the energy level. It is a problem of scattering of particles with energy λ by the potential u. Since $u(x,t)$ is the unknown function to be determined for our problem, our interest is connected with the inverse scattering problem. To proceed, we shall briefly review the scattering problem.

In (3.7.8), t is a parameter, and we shall consider the equation with definite t. Assume $u(x)$ is known, then we may find the eigenvalues and eigenfunctions of the problem. We shall keep in mind that $u(x) \to 0$ as $|x| \to \infty$. There are two separate cases :

(i) $\lambda < 0$.

The problem is equivalent to the potential well problem in quantum mechanics. The particle is in the bound state, and the eigenvalues are discrete. Denote the eigenvalues λ_m , thus

$$\lambda_m = -K_m^2 , \quad m = 1, \cdots, N, \tag{3.7.9}$$

where we may take $K_m > 0$. The corresponding eigenfunctions will be denoted by $\psi_m(x)$, which is bounded as $|x| \to \infty$. Thus $\psi_m(x)$ decays exponentially as $|x| \to \infty$. In other words, we have

$$\psi_m(x) \approx c_m e^{-K_m x} , \quad \text{as } x \to \infty . \tag{3.7.10}$$

With $u(x)$ known, K_m and c_m can be determined. The form of $\psi_m(x)$ is schematically shown in Fig. 3-11(a). To make c_m unique, we usually impose the following normalization condition :

$$\int_{-\infty}^{\infty} \psi_m^2(x) dx = 1 . \tag{3.7.11}$$

(ii) $\lambda > 0$.

This is the case of continuous spectrum, and eigenfunctions are travelling waves, as shown in Fig. 3-11(b). Let

$$\lambda = k^2 . \tag{3.7.12}$$

Then the eigenfunction is a linear combination of e^{ikx} and e^{-ikx} as $|x| \to \infty$.

Since (3.7.8) is a second order equation, its solution may contain two arbitrary constants. They shall be chosen in such a way that :

$$\psi(x) \approx e^{-ikx} + b(k) e^{ikx} , \qquad x \to \infty , \tag{3.7.13}$$

$$\psi(x) \approx a(k) e^{-ikx} , \qquad x \to -\infty . \tag{3.7.14}$$

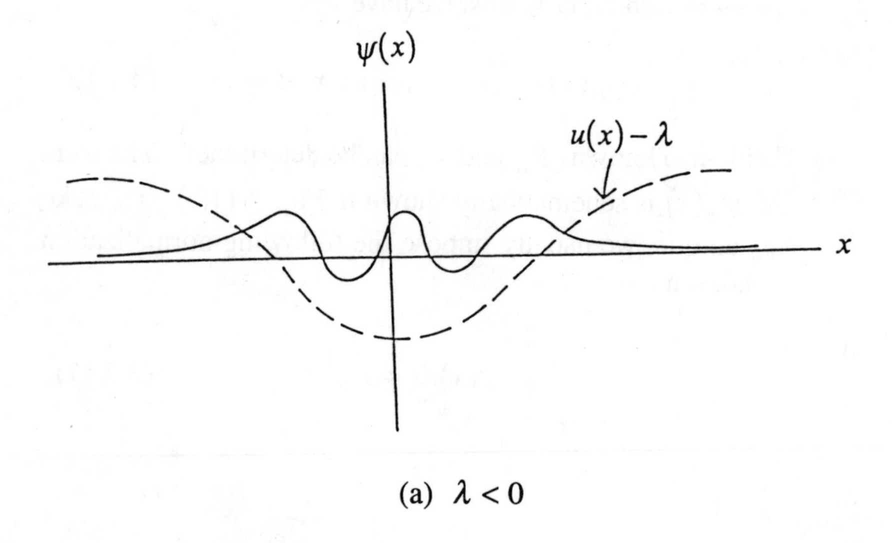

(a) $\lambda < 0$

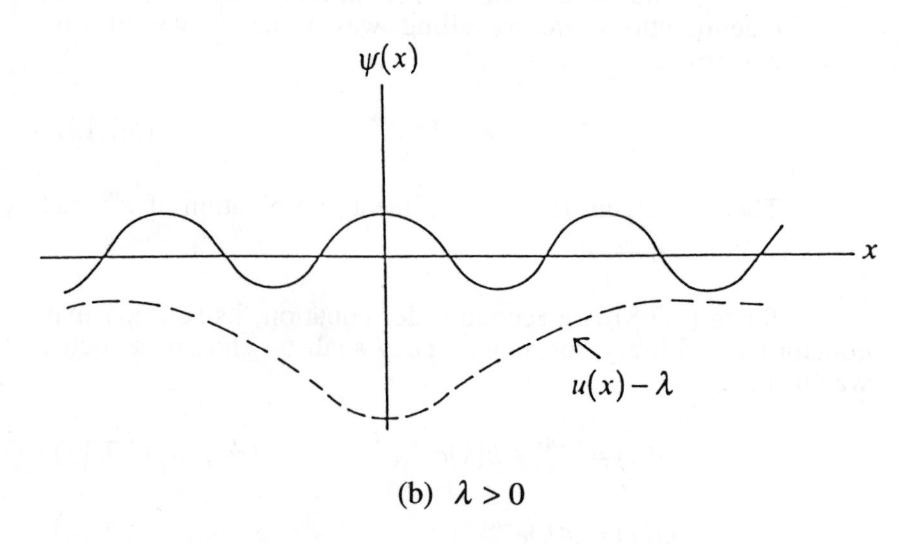

(b) $\lambda > 0$

Fig. 3-11 Schematic representation of $\psi(x)$:
(a) $\lambda < 0$, (b) $\lambda > 0$.

Then $a(k)$ and $b(k)$ can be uniquely determined. The physical interpretation of the above representation is as follows : An incident wave of unit amplitude from $x = \infty$ moves in the $(-x)$ direction, and encounters the scattering potential $u(x)$. The results of the encounter are a reflected wave with amplitude $b(k)$ and a transmitted wave with amplitude $a(k)$. Thus $b(k)$ is the reflection coefficient, and $a(k)$ the transmission coefficient.

If $u(x)$ is known, the eigenvalues and eigenfunction can be determined. In particular, we may determine $\{K_m\}$, $\{c_m\}$, $b(k)$ and $a(k)$. This is the scattering problem. To find $u(x)$ with the complete knowledge of $\{K_m\}$, $\{c_m\}$, $b(k)$ and $a(k)$ is the inverse scattering problem.

The inverse scattering problem was solved by Gel'fand and Levitan (1955). The result may be summarized as follows :

$$u(x) = -2\frac{d}{dx}K(x,x) , \qquad (3.7.15)$$

where $K(x, y)$ satisfies the following integral equation :

$$K(x,y) + B(x+y) + \int_x^\infty B(y+z)K(x,z)dz = 0 , \qquad (3.7.16)$$

where

$$B(\xi) = \sum_{m=1}^{N} c_m^2 e^{-K_m\xi} + \frac{1}{2\pi}\int_{-\infty}^{\infty} b(k)e^{ik\xi}dk . \qquad (3.7.17)$$

The equation (3.7.16) is known as the Gel'fand-Levitan equation, or G-L equation. It is a linear integral equation. For a definite x , it is actually a Fredholm integral equation of the second kind, which can be solved by well-known methods.

We should emphasis that in the previous discussions of scattering and inverse scattering problems, we have omitted the writing of the parameter t. All these discussions are valid for a definite t. As far as the solving the KdV equation is concerned, the given knowledge is $u(x,0)$, and the solution to be found is $u(x,t)$. Given $u(x,0)$, the scattering problem enables us to find $\{K_m(0)\}$, $\{c_m(0)\}$, $\{b(k,0)\}$ and $\{a(k,0)\}$; while $u(x,t)$ may be determined from the knowledge of $\{K_m(t)\}$, $\{c_m(t)\}$ and $\{b(k,t)\}$. If a link can be established between the scattering data at $t = 0$ and $t = t$, then the initial value problem of the KdV equation can be solved. Indeed, this is what Gardner, Greene, Kruskal and Miura (1974) has achieved. The result can be summarized in the following theorem :

If $u(x,t)$ satisfies the KdV equation (3.7.5), and $u(x,t)$ vanishes rapidly as $|x| \to \infty$, then

$$K_m(t) = K_m(0) , \tag{3.7.18}$$

$$c_m(t) = c_m(0)\exp\left(4K_m^3 t\right) , \tag{3.7.19}$$

$$b(k,t) = b(k,0)\exp\left(i8k^3 t\right) , \tag{3.7.20}$$

$$a(k,t) = a(k,0) . \tag{3.7.21}$$

We shall refer the proof, which is fairly straightforward, to the original paper by Gardner et al.

With the general procedure established, let us illustrate the method to some specific examples :

(i) *One sole soliton.*

Consider the initial value

$$u(x,0) = -2\text{sech}^2 x . \tag{3.7.22}$$

We first solve the following eigenvalue problem :

$$\psi_{xx} + \left(2\mathrm{sech}^2 x + \lambda\right)\psi = 0 \ . \qquad (3.7.23)$$

The equation can be transformed into a hypergeometric equation, and it can be found that there is only one discrete eigenvalue, while for all the continuous spectrum, the reflection coefficient is zero. Specifically, we have

$$K_1 = 1 \ , \quad c_1(0) = \sqrt{2} \ , \quad b(k,0) = 0 \ . \quad (3.7.24)$$

Using (3.7.18) - (3.7.20), we obtain from (3.7.17) that

$$B(\xi;t) = 2e^{8t-\xi} \ . \qquad (3.7.25)$$

Thus the G-L equation (3.7.16) becomes

$$K(x,y;t) + 2e^{8t-x-y} + 2e^{8t-y}\int_x^\infty K(x,z;t)e^{-z}dz = 0 \ . \quad (3.7.26)$$

The form of equation (3.7.26) suggests that

$$K(x,y;t) = L(x;t)e^{-y} \ , \qquad (3.7.27)$$

and we find that $L(x;t)$ satisfies the following equation :

$$L(x;t) + 2e^{8t-x} + 2e^{8t}L(x;t)\int_x^\infty e^{-2Z}dz = 0 \ ,$$

for which the solution is

$$L(x;t) = -\frac{2e^x}{1+e^{2x-8t}} \ .$$

Hence

$$K(x,y;t) = -\frac{2e^{x-y}}{1+e^{2x-8t}} \ . \qquad (3.7.28)$$

Substitute (3.7.28) into (3.7.15), we found that

$$u(x,t) = -2\text{sech}^2(x-4t) \ , \qquad (3.7.29)$$

which agrees with (3.7.1).

(ii) *Two solitons only.*

Consider the initial value

$$u(x,0) = -6\text{sech}^2 x \ . \qquad (3.7.30)$$

The eigenvalue problem

$$\psi_{xx} + \left(6\text{sech}^2 x + \lambda\right)\psi = 0 \ , \qquad (3.7.31)$$

leads to the following scattering data with two discrete eigenvalues :

$$K_1 = 1 \ , \quad K_2 = 2 \ , \quad b(k,0) = 0 \ . \qquad (3.7.32)$$

A similar procedure eventually leads to the solution (3.7.4).

(iii) *N solitons only.*

From the previous examples, we may expect that solitons are related to discrete eigenvalues. Indeed, it can be established generally, if $b(k,0) = 0$, i.e. it is a reflectionless potential, and K_1, K_2, $\cdots$, K_N are the discrete eigenvalues, with each $K_m > 0$ and distinct, the solution can be found and it represents N solitons. To be more explicit, it can be shown that

$$u(x,t) \approx -\sum_{p=1}^{N} u_p(x,t)\ , \quad \text{as } t \to \pm\infty\ , \quad (3.7.33)$$

where

$$u_p(x,t) = 2K_p^2 \operatorname{sech}^2\left[K_p\left(x - x_p^{\pm} - 4K_p^2 t\right)\right]\ , \quad (3.7.34)$$

such that $x_p^{\pm}$ is the phase which is different for $t \to \infty$ and $t \to -\infty$.

If $b(k,0) \neq 0$, then $u(x,\ t)$ will contain an oscillatory part corresponding to the decaying wave as represented by the Airy function in the linearized equation. And it can also be shown that this oscillatory part will decay algebraically as $t \to \infty$. Therefore, for large t , only the solitons corresponding to discrete eigenvalues will remain.

3.8 Nonlinear Modulation of Gravity Waves

In deriving the KdV equation, the assumption is that the water depth is small in comparison with the wavelength. We have included only two leading terms in the expansion of the dispersion relation (3.5.24). In order to deal with the more general case as far as the dispersive effect is concerned, we need to sacrifice the generality in other aspects. We shall concentrate on a particular sinusoidal wave and consider its nonlinear modulation.

We shall start with the governing equations (3.1.5), (3.1.8), (3.1.11), (3.1.9) with the flat bottom $y = -h_0$, and neglect the surface tension and external pressure. Thus the governing equations are :

$$\frac{\partial^2 \varphi}{\partial x^2} + \frac{\partial^2 \varphi}{\partial y^2} = 0\ , \quad \text{for } -\infty < x < \infty\ , \ -h_0 < y < \eta(x,t)\ , \quad (3.8.1)$$

$$\frac{\partial \varphi}{\partial y} = 0 \ , \quad \text{on } y = -h_0 \ , \tag{3.8.2}$$

$$\frac{\partial \eta}{\partial t} + \frac{\partial \varphi}{\partial x}\frac{\partial \eta}{\partial x} - \frac{\partial \varphi}{\partial y} = 0 \ , \quad \text{on } y = \eta(x,t) \ , \tag{3.8.3}$$

and

$$\frac{\partial \varphi}{\partial t} + \frac{1}{2}\left[\left(\frac{\partial \varphi}{\partial x}\right)^2 + \left(\frac{\partial \varphi}{\partial y}\right)^2\right] + g\eta = 0 \ , \quad \text{on } y = \eta(x,t) \ . \tag{3.8.4}$$

One approach to deal with the problem of nonlinear modulation is to use the method of multiple scale expansion, and let

$$t_0 = t \ , \qquad t_1 = \varepsilon t \ , \qquad t_2 = \varepsilon^2 t \ , \tag{3.8.5}$$

$$x_0 = x \ , \qquad x_1 = \varepsilon x \ , \tag{3.8.6}$$

$$\eta = \sum_{n=1} \varepsilon^n \eta_n(x_0,\, x_1;\, t_0,\, t_1,\, t_2) \ , \tag{3.8.7}$$

and

$$\varphi = \sum_{n=1} \varepsilon^n \phi_n(x_0,\, x_1;\, y;\, t_0,\, t_1,\, t_2) \ . \tag{3.8.8}$$

where ε is small parameter for book keeping purposes. It is clear that by adopting such an expansion scheme, the basic framework is the linear wave solution, and the modulation is slowly varying in x and t. We shall restrict furthermore the linear solution to be the sinusoidal waves, i.e.

$$\eta_1 = i\frac{\omega_0}{g}\left[\psi(x_1,\, t_1,\, t_2)e^{i(k_0x_0-\omega_0t_0)} - c.c.\right] \tag{3.8.9}$$

where k_0 is the chosen wave number, and

$$\omega_0^2 = gk_0\sigma \ , \qquad \sigma = \tanh k_0 h_0 \ , \tag{3.8.10}$$

and *c.c.* denote the complex conjugate. It can be verified that η_1 is indeed a solution of the first order equations. The detailed analyses of various order of the expansion is fairly straightforward [See Hasimoto and Ono (1972)]. We shall merely state the results up to the third order of expansion :

$$\psi(x_1, t_1, t_2) = \psi(\xi, t_2) \ , \tag{3.8.11}$$

where

$$\xi = x_1 - c_g t_1 \ , \qquad c_g = \frac{d\omega_0}{dk_0} = \frac{c_0}{2}\left[1 + \left(\frac{1-\sigma^2}{\sigma}\right)k_0 h_0\right] , \tag{3.8.12}$$

$$c_0 = \left(\frac{\omega_0}{k_0}\right)^{\frac{1}{2}} = \left(\frac{g\sigma}{k_0}\right)^{\frac{1}{2}} , \tag{3.8.13}$$

and $\psi(\xi, t_2)$ satisfies the equation :

$$\frac{1}{i}\frac{\partial \psi}{\partial t_2} = \mu \frac{\partial^2 \psi}{\partial \xi^2} + \upsilon |\psi|^2 \psi \ , \tag{3.8.14}$$

where

$$\mu = \frac{1}{2}\frac{d^2\omega_0}{dk_0^2}$$

$$= -\frac{g}{8\sigma\omega_0 k_0}\left\{\left[\sigma - (1-\sigma^2)k_0 h_0\right]^2 + 4(1-\sigma^2)(\sigma k_0 h_0)^2\right\} , \tag{3.8.15}$$

$$\upsilon = -\frac{k_0^4}{4\omega_0}\left\{\frac{\left[4c_0^2 + 4(1-\sigma^2)c_0c_g + gh_0(1-\sigma^2)^2\right]}{c_g^2 - gh_0} + \frac{(9-10\sigma^2+9\sigma^4)}{2\sigma^2}\right\}$$

(3.8.16)

(3.8.14) is a nonlinear Schrödinger equation, or cubic Schrödinger equation. The coefficeint μ is always negative, since $\sigma^2 < 1$, but the coefficient υ may be positive or negative.

The correction to η_1 can also be obtained :

$$\eta_2 = \frac{1}{g}\left\{\left[c_g\frac{\partial\psi}{\partial\xi}e^{i(k_0x_0-\omega_0t_0)} + \gamma_2\psi^2e^{2i(k_0x_0-\omega_0t)}\right] + c.c.\right\}$$

$$+\frac{\gamma}{g}|\psi|^2 , \qquad (3.8.17)$$

where

$$\gamma_2 = \frac{(\sigma^2-3)k_0^2}{2\sigma^2} ,$$

$$\gamma = \frac{1}{c_g^2 - gh_0}\left[2\omega_0k_0c_g - (1-\sigma^2)k_0^2gh_0\right] . \qquad (3.8.18)$$

Stokes Waves

We shall not discuss the general analysis of the nonlinear Schrödinger equation, which can also be solved by the method of inverse scattering. We just single out a particular solution which is independent of ξ :

$$\psi^{(0)} = \psi_0 e^{i\alpha_0 t_2} , \qquad (3.8.19)$$

where

$$\alpha_0 = \upsilon |\psi_0|^2 , \tag{3.8.20}$$

and ψ_0 is constant. Take

$$\psi_0 = \frac{ga}{2i\omega_0} , \tag{3.8.21}$$

and substitute (3.8.20) into (3.8.9) and (3.8.17), we obtain

$$\eta = a\cos[k_0x - (\omega_0 - \alpha_0)t]$$

$$+ \frac{a^2}{4\sigma k_0}\{\gamma - \gamma_2 \cos 2[k_0x - (\omega_0 - \alpha_0)t]\} + 0(a^3) . \tag{3.8.22}$$

The wave represented by (3.8.22) is known as Stokes wave. As we may see from (3.8.22), the phase of the fundamental mode is no longer $(k_0x - \omega_0 t)$. The frequency is shifted to ω with

$$\omega = \omega_0 - \alpha_0 = \omega_0 - \frac{\upsilon g^2 a^2}{4\omega_0^2} . \tag{3.8.23}$$

The correction to frequency is proportional to the square of the amplitude. The correction has to be small, thus the small parameter of the problem can be estimated from :

$$\frac{\upsilon g^2 a^2}{4\omega_0^2} << \omega_0 . \tag{3.8.24}$$

Since $\upsilon = 0\left(\frac{k_0^4}{\omega_0}\right)$ from (3.8.16), this criterion for the validity of Stokes waves amounts to

$$k_0 \eta << \tanh k_0 h_0 . \tag{3.8.25}$$

Thus for shallow water waves, since $k_0h_0 << 1$, (3.8.25) is equivalent to

$$\eta << h_0 \,, \tag{3.8.26}$$

while for deep water waves, since $\tanh k_0h_0 \approx 1$, we have

$$k_0\eta << 1 \,. \tag{3.8.27}$$

Stability of Stokes Waves

We may also investigate the stability of the Stokes waves. Let us write

$$\psi = (\psi_0 + \hat{\psi})e^{i(\alpha_0 t_2 + \hat{\theta})} \,, \tag{3.8.28}$$

where $\hat{\psi}(\xi, t_2)$ and $\hat{\theta}(\xi, t_2)$ are perturbations of $\psi^{(0)}$. Both $\hat{\psi}$ and $\hat{\theta}$ are real. Substitute (3.8.28) into (3.8.14), and retain only the first order terms of $\hat{\psi}$ and $\hat{\theta}$, we obtain :

$$\frac{\partial \hat{\psi}}{\partial t_2} + \mu|\psi_0|\frac{\partial^2 \hat{\theta}}{\partial \xi^2} = 0 \,, \tag{3.8.29}$$

$$|\psi_0|\frac{\partial \hat{\theta}}{\partial t_2} - 2\upsilon|\psi_0|^2 \hat{\psi} - \mu\frac{\partial^2 \hat{\psi}}{\partial \xi^2} = 0 \,. \tag{3.8.30}$$

We may solve (3.8.29) and (3.8.30) by Fourier analysis, and let

$$\begin{pmatrix} \hat{\psi} \\ \hat{\theta} \end{pmatrix} = \begin{pmatrix} \hat{\psi}_0 \\ \hat{\theta}_0 \end{pmatrix} e^{i(\hat{k}\xi - \hat{\omega}t_2)} + c.c. \tag{3.8.31}$$

Substitute (3.8.31) into (3.8.29) and (3.8.30) we obtain the following dispersion relation :

$$\hat{\omega}^2 = \mu^2 \hat{k}^2 \left(\hat{k}^2 - \frac{2\upsilon}{\mu} |\psi_0|^2 \right) . \tag{3.8.32}$$

The Stokes wave is stable if $\hat{\omega}$ is real, and unstable if $\hat{\omega}$ is complex. Now $\hat{\omega}$ is real if $\mu\upsilon < 0$. On the other hand, for $\mu\upsilon > 0$, $\hat{\omega}$ is imaginary for $\hat{k} < 2\left(\frac{\upsilon}{\mu}\right)|\psi_0|$, i.e. there are always some $\hat{k}$ for which to make $\hat{\omega}$ imaginary. Now for our problem, $\mu < 0$ always. Therefore we conclude that Stokes waves are stable if and only if

$$\upsilon > 0 . \tag{3.8.33}$$

From (3.8.16), the condition (2.8.33) is equivalent to

$$kh_0 < 1.363 . \tag{3.8.34}$$

The condition (3.8.34) can also be obtained by the variational method and other approaches. It is remarkable that although this is a criterion concerning nonlinear waves, it is independent of wave amplitude.

Gravity Waves in Lagrangian Coordinates

In Section 1.1, we mentioned that fluid motion can be described equally well in terms of Lagrangian coordinates. Let us use (x,y,z,t) as the Eulerian coordinates, and (X,Y,Z,T) the Lagrangian coordinates. Then corresponding to the continuity equation (3.1.4) and the momentum equation (3.1.6), we have respectively :

$$\frac{\partial(x,y,z)}{\partial(X,Y,Z)} = 1 , \tag{3.8.35}$$

$$\frac{\partial^2 x}{\partial T^2}\frac{\partial x}{\partial X}+\frac{\partial^2 y}{\partial T^2}\frac{\partial y}{\partial X}+\frac{\partial^2 z}{\partial T^2}\frac{\partial z}{\partial X}=-\frac{1}{\rho}\frac{\partial p}{\partial X}-g\frac{\partial y}{\partial X}, \quad (3.8.36)$$

$$\frac{\partial^2 x}{\partial T^2}\frac{\partial x}{\partial Y}+\frac{\partial^2 y}{\partial T^2}\frac{\partial y}{\partial Y}+\frac{\partial^2 z}{\partial T^2}\frac{\partial z}{\partial Y}=-\frac{1}{\rho}\frac{\partial p}{\partial Y}-g\frac{\partial y}{\partial Y}, \quad (3.8.37)$$

$$\frac{\partial^2 x}{\partial T^2}\frac{\partial x}{\partial Z}+\frac{\partial^2 y}{\partial T^2}\frac{\partial y}{\partial Z}+\frac{\partial^2 z}{\partial T^2}\frac{\partial z}{\partial Z}=-\frac{1}{\rho}\frac{\partial p}{\partial Z}-g\frac{\partial y}{\partial Z}. \quad (3.8.38)$$

For the two dimensional case we have been considering, the equations are simplified to :

$$\frac{\partial x}{\partial X}\frac{\partial y}{\partial Y}-\frac{\partial x}{\partial Y}\frac{\partial y}{\partial X}=1, \quad (3.8.39)$$

$$\frac{\partial^2 x}{\partial T^2}\frac{\partial x}{\partial X}+\frac{\partial^2 y}{\partial T^2}\frac{\partial y}{\partial X}=-\frac{1}{\rho}\frac{\partial p}{\partial X}-g\frac{\partial y}{\partial X}, \quad (3.8.40)$$

$$\frac{\partial^2 x}{\partial T^2}\frac{\partial x}{\partial Y}+\frac{\partial^2 y}{\partial T^2}\frac{\partial y}{\partial Y}=-\frac{1}{\rho}\frac{\partial p}{\partial Y}-g\frac{\partial y}{\partial Y}. \quad (3.8.41)$$

Since we have been dealing with potential flows, we shall also impose the condition that the vorticity is zero, i.e.,

$$\frac{\partial x}{\partial X}\frac{\partial^2 x}{\partial T\partial Y}-\frac{\partial x}{\partial Y}\frac{\partial^2 x}{\partial T\partial X}=\frac{\partial y}{\partial Y}\frac{\partial^2 y}{\partial T\partial X}-\frac{\partial y}{\partial X}\frac{\partial^2 y}{\partial T\partial Y}. \quad (3.8.42)$$

The solutions we seek are $x(X,Y,T)$ and $y(X,Y,T)$ in the region $-h_0 \le Y \le 0$. $Y=0$ is the free surface and there is no need of any more kinematic condition. At the bottom surface, the condition is

$$y(X,-h_0,T)=-h_0, \quad (3.8.43)$$

this is equivalent to the condition that the normal velocity at bottom is zero, since the fluid particles will remain on $y = -h_0$, if initially they are on $y = -h_0$. The dynamic free surface condition is simply that pressure is zero on free surface, i.e.,

$$p(X,0,T) = 0 \, . \tag{3.8.44}$$

Equations (3.8.39) - (3.8.44) are the Lagrangian equivalent to the Eulerian formulation (3.8.1) - (3.8.4). While the governing equations in Lagrangian coordinates is more complex and nonlinear, the boundary conditions are simpler. Also it is a fixed boundary value problem in contrast with the free boundary value problem in Eulerian formulation.

The equilibrium solution is given by

$$x = X , \tag{3.8.45}$$

$$y = Y , \tag{3.8.46}$$

and

$$p = -\rho g Y \, . \tag{3.8.47}$$

Let the deviation from the equilibrium solution be given by u, v and q, i.e.

$$x = X + u , \tag{3.8.48}$$

$$y = Y + v , \tag{3.8.49}$$

and

$$p = -\rho g Y + q \, . \tag{3.8.50}$$

Then equations (3.8.39) - (3.8.44) become

$$\frac{\partial u}{\partial X} + \frac{\partial v}{\partial Y} = \frac{\partial u}{\partial Y}\frac{\partial v}{\partial X} - \frac{\partial u}{\partial X}\frac{\partial v}{\partial Y} , \tag{3.8.51}$$

$$\frac{\partial^2 u}{\partial T^2}+\frac{1}{\rho}\frac{\partial q}{\partial X}+g\frac{\partial v}{\partial X}=-\frac{\partial u}{\partial X}\frac{\partial^2 v}{\partial T^2}-\frac{\partial v}{\partial X}\frac{\partial^2 v}{\partial T^2}\,, \tag{3.8.52}$$

$$\frac{\partial^2 v}{\partial T^2}+\frac{1}{\rho}\frac{\partial q}{\partial Y}+g\frac{\partial v}{\partial Y}=-\frac{\partial u}{\partial Y}\frac{\partial^2 u}{\partial T^2}-\frac{\partial v}{\partial Y}\frac{\partial^2 v}{\partial T^2}\,, \tag{3.8.53}$$

$$\frac{\partial}{\partial T}\left(\frac{\partial u}{\partial Y}-\frac{\partial v}{\partial X}\right)=\frac{\partial u}{\partial Y}\frac{\partial^2 u}{\partial T\partial X}-\frac{\partial u}{\partial X}\frac{\partial^2 u}{\partial T\partial Y}+\frac{\partial v}{\partial Y}\frac{\partial^2 v}{\partial T\partial X}-\frac{\partial v}{\partial X}\frac{\partial^2 v}{\partial T\partial Y}\,. \tag{3.8.54}$$

$$v(X,-h,T)=0\,, \tag{3.8.55}$$

and

$$q(X,0,T)=0\,. \tag{3.8.56}$$

For this set of equations (3.8.51) - (3.8.56), the linear solution agrees exactly with what we found from the Eulerian formulation. We can also proceed to discuss the nonlinear modulation of a sinusoidal wave. Let us again use the method of multiple scale expansion. Let

$$X_n=\varepsilon^n X\,, \qquad n=0,\,1,\,2,\,\cdots\,, \tag{3.8.57}$$

$$T=\varepsilon^n T\,, \qquad n=0,\,1,\,2,\,\cdots\,, \tag{3.8.58}$$

where ε is a small parameter. Let us also expand u, v, and q in series of the form :

$$u=\sum_{n=1}\varepsilon^n u_n\left(X_0,\,X_1,\,X_2,\,Y,\,T_0,\,T_1,\,T_2\right)\,, \tag{3.8.59}$$

$$v=\sum_{n=1}\varepsilon^n v_n\left(X_0,\,X_1,\,X_2,\,Y,\,T_0,\,T_1,\,T_2\right)\,, \tag{3.8.60}$$

and

$$q = \sum_{n=1} \varepsilon^n q_n(X_0, X_1, X_2, Y, T_0, T_1, T_2) . \qquad (3.8.61)$$

Substitute (3.8.57) - (3.8.61) into (3.8.51) - (3.8.56), we can obtain the equation for various orders of ε. We shall not go into the details of the development (See Hsieh (1990)). We shall just state the main results.

The first order solutions are :

$$u_1 = u_1^{(0)}(\xi_1, X_2, T_2)$$

$$+\left[iA(\xi_1, X_2, T_2)\cosh k(Y+h)e^{i(kX_0-\omega T_0)} + c.c.\right], \qquad (3.8.62)$$

$$v_1 = A(\xi_1, X_2, T_2)\sinh k(Y+h)e^{i(kX_0-\omega T_0)} + c.c. , \qquad (3.8.63)$$

$$q_1 = \rho g[\tanh kh \cosh k(Y+h) - \sinh k(Y+h)]A(\xi_1, X_2, T_2)e^{i(kX_0-\omega T_0)}$$

$$+c.c., \qquad (3.8.64)$$

where

$$\omega^2 = gk \tanh kh , \qquad (3.8.65)$$

$$\xi_1 = X_1 - c_g T_1 , \qquad (3.8.66)$$

$$c_g = \frac{d\omega}{dk} , \qquad (3.8.67)$$

and A satisfies the following equation :

$$\frac{1}{i}\frac{\partial A}{\partial \zeta_2} = \mu \frac{\partial^2 A}{\partial \xi_1^2} + \upsilon |A|^2 A , \qquad (3.8.68)$$

where

$$\frac{\partial}{\partial \zeta_2} = \frac{\partial}{\partial T_2} + c_g \frac{\partial}{\partial X_2}, \tag{3.8.69}$$

$$\mu = \frac{1}{2}\frac{d^2\omega}{dk^2}, \tag{3.8.70}$$

and

$$\upsilon = \omega k^2 \left[\frac{\omega^2\left(\frac{1}{2} + \frac{kh}{\cosh kh \sinh kh}\right)}{k^2\left(gh - c_g^2\right)} - \left(2\sinh^2 kh + \frac{9}{4\sinh^2 kh}\right) \right]. \tag{3.8.71}$$

The comparison of the results from Lagrangian formulation with those from Eulerian formulation shows that they agree in many aspects, but there is also some difference. They agree entirely in the linear regime. The dispersion relations (3.8.10) and (3.8.65) are the same. ξ in (3.8.12) and ξ_1 in (3.8.66) are practically the same. (3.8.14) and (3.8.68) are both cubic Schrödinger equations and with the same coefficient μ . But the coefficient υ in (3.8.68) is different from that of (3.8.16). This difference has nonlinear origin, although υ does not depend on the small parameter. An implication is that the stability criterion for the Stokes waves will be different, and this is disturbing.

The Eulerian and Lagrangian descriptions for the fluid motion are just two descriptions of the same phenomena. If we are able to find the complete solutions, they should be equivalent to each other. But we are solving the problems using perturbation expansion schemes. Although the perturbation schemes look similar, the small parameters are not the same. They agree only to the first order. It is reasonable to expect discrepancies between solutions of higher orders. Somehow, from second order terms, a stability criterion can be established which is independent of the small parameters, and we are facing the difficult question to determine the validity of these assertions. For the physical problem, the question could be settled by experimental observations. Mathematically, the message is that we should be cautions about interpretation of results, derived from perturbation expansions, which has the appearance of general validity.

3.9 Hydraulic Jumps

In Section 3.5, we have derived a set of equations (3.5.14) and (3.5.15) governing the propagation of nonlinear shallow water waves. These equations are :

$$\frac{\partial u}{\partial t}+u\frac{\partial u}{\partial x}+g\frac{\partial h}{\partial x}=0\ , \tag{3.9.1}$$

$$\frac{\partial h}{\partial t}+u\frac{\partial h}{\partial x}+h\frac{\partial u}{\partial x}=0\ . \tag{3.9.2}$$

The structure of this system is the same as that of the nonlinear one-dimensional sound waves (2.6.1) and (2.6.4). Therefore we may follow the similar procedure to study this problem.

First, we shall discuss the case of simple waves. Consider the case that

$$u=u\,(h)\ . \tag{3.9.3}$$

Then (3.9.1) and (3.9.2) can be rewritten as :

$$\frac{\partial u}{\partial t}+\left(u+\frac{g}{\dfrac{\partial u}{\partial h}}\right)\frac{\partial u}{\partial x}=0\ , \tag{3.9.4}$$

and

$$\frac{\partial h}{\partial t}+\left(u+h\frac{\partial u}{\partial h}\right)\frac{\partial h}{\partial x}=0\ . \tag{3.9.5}$$

Thus, since $\left(\dfrac{dx}{dt}\right)_u = \left(\dfrac{dx}{dt}\right)_h$, we obtain from (3.9.4) and (3.9.5) that

$$\frac{\partial u}{\partial h} = \pm\left(\frac{g}{h}\right)^{\frac{1}{2}} . \tag{3.9.6}$$

Thus

$$u = \pm 2\left[(gh)^{\frac{1}{2}} - (gh_0)^{\frac{1}{2}}\right] , \tag{3.9.7}$$

where h_0 is the equilibrium value of h , as $u = 0$ and $h = h_0$ are the equilibrium solutions of (3.9.1) and (3.9.2). Substitute (3.9.6) and (3.9.7) into (3.9.5), we obtain

$$\frac{\partial h}{\partial t} \pm \left[3(gh)^{\frac{1}{2}} - 2(gh_0)^{\frac{1}{2}}\right]\frac{\partial h}{\partial x} = 0 . \tag{3.9.8}$$

Write

$$v = 3(gh)^{\frac{1}{2}} - 2(gh_0)^{\frac{1}{2}} , \tag{3.9.9}$$

then (3.9.8) can be put in the form :

$$\frac{\partial v}{\partial t} + v\frac{\partial v}{\partial x} = 0 , \tag{3.9.10}$$

which is again the same as (2.6.13) and which we have studied in details before. The only difference is that in this water wave problem, $v(x)$ or $h(x)$ can physically admit multiple-value solutions at a definite t . What corresponds to shock is the bore or hydraulic jump.

We may also follow the same procedure as given in Section 2.8 to find the Riemann invariants of the system of (3.9.1) and (3.9.2). The characteristics $\varphi(x, t) = \text{constant}$ can be readily shown to satisfy the following equation :

$$\left(\frac{\partial \varphi}{\partial t} + u\frac{\partial \varphi}{\partial x}\right)^2 - gh = 0 \ . \qquad (3.9.11)$$

Thus, there are two families of characteristics : $\alpha(x, t) = \text{constant}$ and $\beta(x,t) = \text{constant}$. α and β satisfy the following equations :

$$\frac{\partial \alpha}{\partial t} + \left[u - (gh)^{\frac{1}{2}}\right]\frac{\partial \alpha}{\partial x} = 0 \ , \qquad (3.9.12)$$

$$\frac{\partial \beta}{\partial t} + \left[u + (gh)^{\frac{1}{2}}\right]\frac{\partial \beta}{\partial x} = 0 \ . \qquad (3.9.13)$$

Let us now take $\alpha(x, t)$ and $\beta(x, t)$ as independent variables. Then equations (3.9.1) and (3.9.2) can be put in the form :

$$-u_\alpha x_\beta + u_\beta x_\alpha + (uu_\alpha + gh_\alpha)t_\beta - (uu_\beta + gh_\beta)t_\alpha = 0 \ , \quad (3.9.14)$$

$$-h_\alpha x_\beta + h_\beta x_\alpha + (uh_\alpha + hu_\alpha)t_\beta - (uh_\beta + hu_\beta)t_\alpha = 0 \ . \quad (3.9.15)$$

Now equations (3.9.12) and (3.9.13) can also be written equivalently :

$$x_\alpha - \left[u + (gh)^{\frac{1}{2}}\right]t_\alpha = 0 \ , \qquad (3.9.16)$$

and

$$x_\beta - \left[u - (gh)^{\frac{1}{2}}\right]t_\beta = 0 \ . \qquad (3.9.17)$$

Making use of (3.9.16) and (3.9.17), (3.9.14) and (3.9.15) can be rewritten as

$$\left[(gh)^{\frac{1}{2}}u_\beta - gh_\beta\right]t_\alpha + \left[(gh)^{\frac{1}{2}}u_\alpha + gh_\alpha\right]t_\beta = 0\,, \quad (3.9.18)$$

and

$$\left[(gh)^{\frac{1}{2}}h_\beta - hu_\beta\right]t_\alpha + \left[(gh)^{\frac{1}{2}}h_\alpha + hu_\alpha\right]t_\beta = 0\,. \quad (3.9.19)$$

Thus we obtain

$$\left[(gh)^{\frac{1}{2}}h_\alpha + hu_\alpha\right]\left[(gh)^{\frac{1}{2}}u_\beta - gh_\beta\right]$$
$$-\left[(gh)^{\frac{1}{2}}u_\alpha + gh_\alpha\right]\left[(gh)^{\frac{1}{2}}h_\beta - hu_\beta\right] = 0\,,$$

or

$$\left[u_\alpha + \left(\frac{g}{h}\right)^{\frac{1}{2}}h_\alpha\right]\left[u_\beta - \left(\frac{g}{h}\right)^{\frac{1}{2}}h_\beta\right] = 0\,. \quad (3.9.20)$$

Or

$$\frac{\partial r}{\partial \alpha} = 0\,, \quad (3.9.21)$$

and

$$\frac{\partial s}{\partial \beta} = 0\,, \quad (3.9.22)$$

where

$$r = u + 2(gh)^{\frac{1}{2}} , \qquad (3.9.23)$$

$$s = u - 2(gh)^{\frac{1}{2}} , \qquad (3.9.24)$$

r and s are the Riemann invariants.

The theory developed so far applied only to shallow water. The dispersion effect, which is important in water waves, has not been included. However, this simple theory sometimes gives quite reasonable description of the phenomena of bores and hydraulic jumps.

In section 5.12, when we discuss the stability of a fluid flowing down an inclined plane, we shall again deal with this nonlinear shallow water theory, and show how we may include frictional force due to viscosity.

CHAPTER 4

Waves in Nonhomogeneous Media

4.1 Internal Waves of Layered Fluid Media

Before introducing the main subject of gravity waves in continuously stratified fluid, we consider a very simple case of internal waves. In some deep estuaries, the lighter fresh river water tends to move seawards above the heavier salty sea water. This natural phenomenon provides a good setting for generation of internal waves on the interface regarded as a thin surface separating the heavier sea water of uniform density ρ_2 from the lighter fresh river water of uniform density $\rho_1 < \rho_2$. If the sea surface is relatively calm, the tidal motions will not be strong enough to upset the gravitational stability of the interface. Thus any effects of turbulent diffusion of salty sea water into the overlying fresh river water may be neglected.

We may proceed to investigate the problem in systematic analyses by the same approach as that described in Sections 3.1 and 3.2 when we discussed the water wave problem, and in fact that is what we shall do in Chapter 5 when we discussed the interfacial stability problems. However, we shall here approach the problem from a slightly different approach. In subsequent discussions, we shall limit our discussion to two-dimensional problems, although in many cases, the results can be easily generalized to three-dimensional problems.

If $z = 0$ is the undisturbed position of the interface, then its elevation to $z = \xi(x,t)$ increased the potential energy of the lower fluid by $\frac{1}{2}\rho_2 g \xi^2$ per unit horizontal area. This increase of gravitational potential energy is obtained by integrating $\rho_2 gzdz$ from $z = 0$ to the free surface $z = \xi$, giving

$$\int_0^\xi \rho_2 g z dz = \frac{1}{2}\rho_2 g \xi^2 \ .$$

However, the elevation of the interface to $z = \xi$ decreases the potential energy of the upper fluid by $\frac{1}{2}\rho_1 g \xi^2$ per unit horizontal area. The net gain in gravitational potential energy per unit horizontal area is $\frac{1}{2}\xi^2(\rho_2 - \rho_1)g$. The term

$$(\rho_2 - \rho_1)g \ , \tag{4.1.1}$$

which is the coefficient of $\frac{1}{2}\xi^2$, gives a measure of the stiffness of the interface, i.e. its tendency to oppose by the action of gravity any disturbances in which the heavier fluid may penetrate the lighter fluid, and vice versa. We shall neglect the effects of surface tension in this section. Therefore, the only stiffness we take into consideration is that due to gravity.

Also due to movement of the interface, it generates motion in the lower fluid with kinetic energy $\frac{1}{2}\rho_2 k^{-1}\left(\frac{\partial \xi}{\partial t}\right)^2$ per unit horizontal area where k is the wave number of the monochromatic disturbances. This kinetic energy of fluid ρ_2 can be obtained from a general expression :

$$\frac{1}{2}\rho_2 \int \phi \frac{\partial \phi}{\partial n} dS$$

in fluid flow for which the velocity potential φ satisfies the Laplace's equation $\nabla^2 \phi = 0$. This statement can be proved by using the divergence theorem to express the above surface integral as a volume integral of $\frac{1}{2}\rho_2 \nabla \cdot (\phi \nabla \phi)$ which, because of $\nabla^2 \phi = 0$, is the kinetic energy per unit volume $\frac{1}{2}\rho_2(\nabla\phi)^2$. Since $\frac{\partial\phi}{\partial n}=0$ at the bottom boundary of the lower fluid, the surface integral gives up to the order of square of disturbances, a kinetic energy

$$\frac{1}{2}\rho_2\left[\phi\frac{\partial\phi}{\partial z}\right]_{z=0} \tag{4.1.2}$$

per unit horizontal area.

Since the problem is exactly like that treated in Section 3.2 for the water waves. Thus for sinusoidal waves over a fluid layer of infinite depth, we obtain the kinetic energy expression of $\frac{1}{2}\rho_2 k^{-1}\left(\frac{\partial\xi}{\partial t}\right)^2$. Now movement of the interface generates also an identical but inverted motion in the upper fluid with kinetic energy $\frac{1}{2}\rho_1 k^{-1}\left(\frac{\partial\xi}{\partial t}\right)^2$. Thus the total inertia of the interface, represented by the coefficient of $\frac{1}{2}\left(\frac{\partial\xi}{\partial t}\right)^2$ in the kinetic energy, is therefore

$$(\rho_2+\rho_1)k^{-1} \tag{4.1.3}$$

per unit horizontal area. To obtain the dispersion relation we write ω^2 as the ratio of stiffness (4.1.1) to inertia (4.1.3), i.e.

$$\omega^2 = gk\left[\frac{\rho_2-\rho_1}{\rho_2+\rho_1}\right]. \tag{4.1.4}$$

When $\rho_1=0$, from (4.1.4), $\omega^2=gk$, which is the result for surface gravity waves. Thus for $\rho_1\neq 0$, according to (4.1.4) the internal wave motions for given wavelength $\frac{2\pi}{k}$ at an interface separating a lighter upper fluid from a heavier lower fluid behave exactly as surface gravity waves but with the frequency ω and therefore also the wave speed $c = \frac{\omega}{k}$ reduced by a factor equal to $\left[\frac{(\rho_2 - \rho_1)}{(\rho_2 + \rho_1)}\right]^{\frac{1}{2}}$. That reduction factor is around 11% to 12% for river water superposing on sea water.

Now wave drag due to the motion of a ship usually correlates with the Froude number $\frac{U^2}{c^2}$, where U is the speed of the ship. The waves generated by the ship have preferentially wavelengths of the order of the length of ship. Thus for given length of ship, the corresponding wave speed for the surface wave is faster than that for internal wave. Therefore although a ship may be travelling at a much too slow speed relative to its length to generate any surface waves, it may be travelling fast enough to generate internal waves on such an esturial interface. In other words the ship may experience a significantly enhanced drag as it travels into the estuary, even though the ship excites no observable waves on the free surface.

We shall now consider the case that both fluids will be of finite depth with both top and bottom being rigid, as shown schematically in Fig. 4-1.

Then the stiffness of the interface is again $(\rho_2 - \rho_1)g$. But the kinetic energy will be modified due to finite thickness of the layers. It is easy to see from (3.2.3) and (3.2.9), that the term (4.1.2), i.e., the kinetic energy per unit horizontal area of the lower fluid, becomes

$$\frac{1}{2}\rho_2 k^{-1} \coth kh \left(\frac{\partial \xi}{\partial t}\right)^2 . \qquad (4.1.5)$$

Similarly, the corresponding term for the upper fluid is

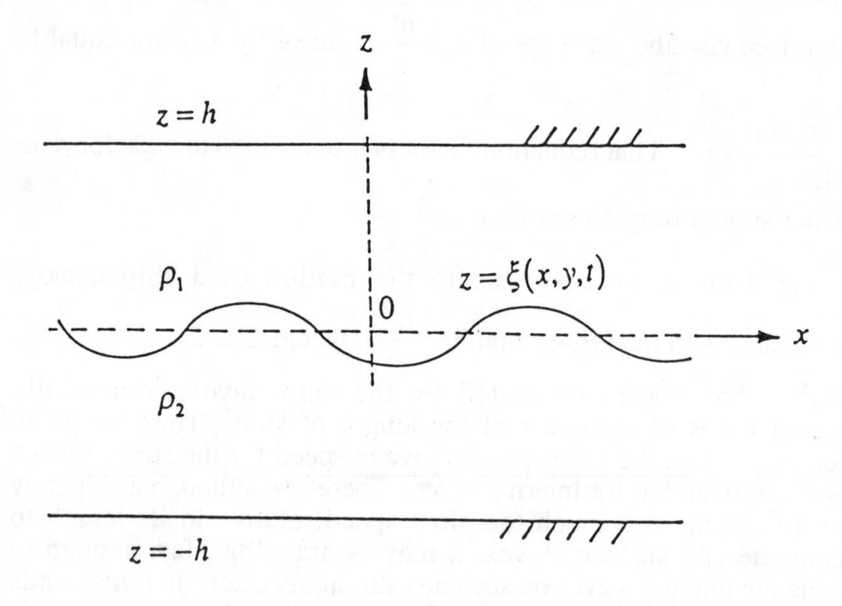

Fig. 4-1 Fluids ρ_1 ,ρ_2 between rigid horizontal planes.

$$\frac{1}{2}\rho_1 k^{-1}\coth kh'\left(\frac{\partial\xi}{\partial t}\right)^2 . \tag{4.1.6}$$

Therefore the total inertia of the interface per unit horizontal area is

$$k^{-1}(\rho_2 \coth kh + \rho_1 \coth kh') . \tag{4.1.7}$$

Thus we obtain again from the ratio of stiffness (4.1.1) and inertia (4.1.7) the dispersion relation :

$$\omega^2 = \frac{gk(\rho_2 - \rho_1)}{\rho_2 \coth kh + \rho_1 \coth kh'} . \tag{4.1.8}$$

When kh and kh' are both very large, i.e., a semi-infinite fluid with ρ_1 , overlying another semi-infinite fluid with ρ_2, the dispersion relation (4.1.8) reduces to (4.1.4). From either (4.1.4) or (4.1.8) the natural frequency of oscillation of the interface of two fluids of nearly equal densities are very small. This explains why the natural period of the interfacial oscillation are very long compared with those of a free surface where $\frac{\rho_1}{\rho_2} << 1$.

From (4.1.8) we may obtain a formula for the speed of an internal wave

$$c^2 = \frac{\omega^2}{k^2} = \frac{g(\rho_2 - \rho_1)}{k[\rho_2 \coth kh + \rho_1 \coth kh']} . \tag{4.1.9}$$

When kh' is large and kh small, (4.1.9) becomes

$$c^2 = \left(1 - \frac{\rho_1}{\rho_2}\right) gh , \quad \text{or} \quad c = \left[\left(1 - \frac{\rho_1}{\rho_2}\right) gh\right]^{\frac{1}{2}} .$$

Thus the internal wave speed now depends only on the depth h of the lower fluid and the ratio of fluid densities $\frac{\rho_1}{\rho_2}$. On the other hand if kh' is small and kh large, then (4.1.9) leads approximately to

$$c = \left[\left(\frac{\rho_2}{\rho_1} - 1\right) gh'\right]^{\frac{1}{2}} .$$

The internal wave speed now depends only on the depth h' of the upper fluid and the ratio $\frac{\rho_2}{\rho_1}$. Thus, the thickness of the thin fluid layer controls the characteristics of the propagation of the internal waves.

Upper Fluid Surface Free

Now suppose that the upper surface of the upper fluid is a free surface and other conditions will be the same as shown in Fig. 4-2. For small disturbances, the complex form of the velocity potentials in the two fluids may be represented by ϕ as follows.

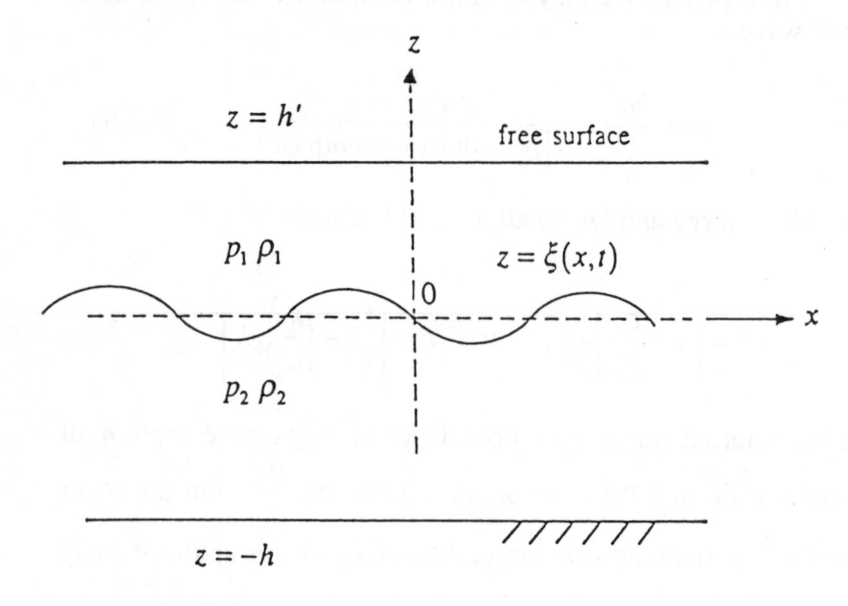

Fig. 4-2 Fluids ρ_1 ,ρ_2 between free surface and rigid plane $z=-h$.

$$\phi=\begin{cases}A\cosh k(z+h)\exp[i(\omega t-kx)]\,, & -h<z<0\,,\\ (B_1\cosh kz+B_2\sinh kz)\exp[i(\omega t-kx)]\,, & 0<z<h'\,,\end{cases}$$

satisfying $\left[\dfrac{\partial\phi}{\partial z}\right]_{z=-h}=0$, and $\xi=C\exp[i(\omega t-kx)]$.

By the kinematical condition at the interface or approximately at its undisturbed position $z=0$, we have, using $\dfrac{\partial\xi}{\partial t}=-\left[\dfrac{\partial\phi}{\partial z}\right]_{z=0}$,

$$-KA\sinh kh=kB_2=i\omega C\,. \tag{4.1.10}$$

The condition for continuity of pressure at the interface, i.e. $p_1 = p_2$, leads to

$$\rho_2\left(\frac{\partial\phi}{\partial t} - g\xi\right)_2 = \rho_1\left(\frac{\partial\phi}{\partial t} - g\xi\right)_1 ,$$

or

$$\rho_2(i\omega A\cosh kh - gC) = \rho_1(i\omega B_1 - gC) . \qquad (4.1.11)$$

To apply the condition for constant pressure at the free surface $z = h'$ of the upper fluid i.e. $\frac{Dp}{Dt} = 0$, or $\frac{\partial^2\phi}{\partial t^2} + g\frac{\partial\phi}{\partial z} = 0$, it requires

$$-\omega^2(B_1\cosh kh' + B_2\sinh kh')$$

$$+ gk(B_1\sinh kh' + B_2\cosh kh') = 0 . \qquad (4.1.12)$$

Eliminating A , B_1 , B_2 and C between equations (4.1.10), (4.1.11), (4.1.12) yields the equation

$$\omega^4(\rho_2\coth kh\coth kh' + \rho_1) - \omega^2\rho_2(\coth kh + \coth kh')gk$$

$$+(\rho_2 - \rho_1)g^2k^2 = 0 . \qquad (4.1.13)$$

Since (4.1.13) is quadratic in ω^2 , there are two possible wave trains of any given wave number k . These two possible systems of waves correspond to two independent modes of oscillation about the undisturbed state. For example, in the case where fluid ρ_2 is much heavier than fluid ρ_1 , i.e. $\frac{\rho_1}{\rho_2}$ is small, one mode may be considered as an oscillation of the upper fluid ρ_1 which behaves almost the same as if the lower fluid ρ_2 were an almost solid material, while the other mode may be regarded as an oscillation of the lower fluid ρ_2 which is almost the same as if the upper surface of fluid ρ_2 (or the interface) were a free surface. To see this, since , for any $kh\ > 0$ and $kh'\ > 0$, coth$kh\ > 1$ and coth$kh'\ > 1$, letting $\frac{\rho_1}{\rho_2} << 1$, we have from (4.1.13)

$$\omega^4 \coth kh \coth kh' - \omega^2(\coth kh + \coth kh')gk + g^2k^2 = 0 \ . \quad (4.1.14)$$

The frequencies of the two independent modes of oscillation about the undisturbed state will be given by the two roots of equation (4.1.14) as a quadratic equation in ω^2 :

$$\omega^2 = gk\tanh kh' \ , \quad \text{or} \quad \omega^2 = gk\tanh kh \ . \quad (4.1.15)$$

Thus when $\omega^2 = gk\tanh kh'$, the upper fluid oscillates in almost the same mode as if the lower fluid were solidified. But when $\omega^2 = gk\tanh kh$, the lower fluid may be described as oscillating in the same mode as if its upper surface were free.

There are other interesting special cases. For instances, when kh is large, i.e. the depth of the lower fluid is great compared with the wavelength. Letting coth$kh\ = 1$, from (4.1.13) we have

$$\omega^4(\rho_2 \coth kh' + \rho_1) - \omega^2\rho_2(1 + \coth kh')gk$$

$$+(\rho_2 - \rho_1)g^2k^2 = 0 \ . \quad (4.1.16)$$

In this case the frequencies of the two possible modes of oscillation are given by

$$\omega^2 = gk \ , \quad \text{or} \quad \omega^2 = \frac{(\rho_2 - \rho_1)gk}{\rho_2 \coth kh' + \rho_1} \ . \quad (4.1.17)$$

The first root of (4.1.17) corresponds exactly to the case of a single fluid of infinite depth. From the second root the wave speed for the other mode of oscillation is

$$c^2 = \frac{\omega^2}{k^2} = \frac{g(\rho_2 - \rho_1)}{k(\rho_2 \coth kh' + \rho_1)} < \frac{g(\rho_2 - \rho_1)}{k(\rho_2 + \rho_1)}, \quad (4.1.18)$$

which is the wave speed at the interface between two semi-infinite fluids, but

$$c^2 = \frac{g(\rho_2 - \rho_1)}{k(\rho_2 + \rho_1)}, \quad \text{as} \quad kh' \to \infty .$$

This shows that the presence of a finite upper fluid has an effect of diminishing the wave speed more than the presence of a semi-infinite upper fluid does. Since $\coth kh'$ increases as kh' decreases, but according to earlier discussion the denominator $k(\rho_2 \coth kh' + \rho_1)$ in (4.1.18) represents the inertia (or kinetic energies) of the disturbed interface, while the numerator represents stiffness (or potential energies), therefore the inertia increases as kh' decreases. Hence wave speed c decreases as kh' decreases. This explains the above statement but seems to be contrary to intuition.

When two fluids of densities ρ_1 and ρ_2 with one overlying the other move parallel to the x-direction with velocities U_1 and U_2 respectively, dispersion relation can again be readily obtained. The problem leads to so-called Kelvin-Helmholtz instability, which we shall discuss in Section 5.5.

4.2 Waves in Continuously Stratified Fluids

We shall now turn to a detailed study of motions of stratified fluids whose densities vary continuously with the depth.

Let a heterogeneous fluid of depth h with the bottom surface be the plane $z = 0$ and the undisturbed upper surface at $z = h$. If at equilibrium the density of the fluid $\rho_0(z)$ is a function of z only, where z is the height above the bottom surface, hence writing the pressure p and density ρ as

$$p = p_0 + p' \ , \qquad \rho = \rho_0 + \rho' \ , \tag{4.2.1}$$

where p_0 and ρ_0 are the equilibrium pressure and density, while p' and ρ' departures from the equilibrium values p_0 and ρ_0 . The equations of motions and of continuity are

$$\rho\left[\frac{\partial u}{\partial t} + u\frac{\partial u}{\partial x} + v\frac{\partial u}{\partial z}\right] = -\frac{\partial p}{\partial x} \ , \tag{4.2.2}$$

$$\rho\left[\frac{\partial v}{\partial t} + u\frac{\partial v}{\partial x} + v\frac{\partial v}{\partial z}\right] = -\frac{\partial p}{\partial z} - \rho g \ , \tag{4.2.3}$$

$$\frac{\partial \rho}{\partial t} + \frac{\partial}{\partial x}(\rho u) + \frac{\partial}{\partial z}(\rho v) = 0 \ . \tag{4.2.4}$$

Take p' , ρ' , u, and v to be small quantities, then after linearization, Equations (4.2.2) - (4.2.4) become respectively :

$$\rho_0 \frac{\partial u}{\partial t} = -\frac{\partial p'}{\partial x} \ , \tag{4.2.5}$$

$$\rho_0 \frac{\partial v}{\partial t} = -\frac{\partial p'}{\partial z} - g\rho' \ , \tag{4.2.6}$$

$$\frac{\partial \rho'}{\partial t} + v\frac{d\rho_0}{dz} = -\rho_0\left(\frac{\partial u}{\partial x} + \frac{\partial v}{\partial z}\right) , \tag{4.2.7}$$

where we have used

$$\frac{dp_0}{dz} + g\rho_0 = 0 \ , \tag{4.2.8}$$

to obtain (4.2.6).

The fluid is assumed to be incompressible, i.e., $\dfrac{D\rho}{Dt} = 0$, thus (4.2.7) becomes

$$\frac{\partial \rho'}{\partial t} + v\frac{d\rho_0}{dz} = 0 \ , \tag{4.2.9}$$

and

$$\frac{\partial u}{\partial x} + \frac{\partial v}{\partial z} = 0 \ . \tag{4.2.10}$$

If ψ denotes the streamfunction, we may write

$$u = -\frac{\partial \psi}{\partial z} \ , \qquad v = \frac{\partial \psi}{\partial x} \ .$$

Eliminating ρ' between (4.2.6) and (4.2.9) yields

$$\rho_0 \frac{\partial^2 v}{\partial t^2} = -\frac{\partial}{\partial z}\left(\frac{\partial p'}{\partial t}\right) + gv\frac{d\rho_0}{dz} \ . \tag{4.2.11}$$

From (4.2.5), we obtain,

$$\rho_0 \frac{\partial^2 u}{\partial t^2} = -\frac{\partial}{\partial x}\left(\frac{\partial p'}{\partial t}\right) \ . \tag{4.2.12}$$

Eliminating p' between (4.2.11) and (4.2.12) leads to

$$\rho_0\left(\frac{\partial^2}{\partial x^2} + \frac{\partial^2}{\partial z^2}\right)\ddot{\psi} + \frac{d\rho_0}{dz}\frac{\partial \ddot{\psi}}{\partial z} = g\frac{d\rho_0}{dz}\frac{\partial^2 \psi}{\partial x^2} \ ,$$

or

$$\nabla^2 \ddot{\psi} + \frac{1}{\rho_0}\frac{d\rho_0}{dz}\left\{\frac{\partial \ddot{\psi}}{\partial z} - g\frac{\partial^2 \psi}{\partial x^2}\right\} = 0 \ ,$$

$$-\infty < x < \infty \ , \qquad 0 \le z < h \ , \qquad (4.2.13)$$

where we have denoted $\ddot{\psi} = \dfrac{\partial^2 \psi}{\partial t^2}$. Since on the free surface $s = 0$, we have $p = 0$, now $\dfrac{Ds}{Dt} = 0$ therefore $\dfrac{Dp}{Dt} = 0$. Hence, after linearization, we have, if the upper surface is a free surface that

$$\frac{\partial p'}{\partial t} + v\frac{\partial p_0}{\partial z} = 0 \ , \qquad \text{on } z = h \ ,$$

or

$$\frac{\partial p'}{\partial t} = g\rho_0 \frac{\partial \psi}{\partial x} \ , \qquad \text{on } z = h \ . \qquad (4.2.14)$$

Eliminating p' between (4.2.5) and (4.2.14) on $z = h$, we obtain

$$\frac{\partial \ddot{\psi}}{\partial z} = g\frac{\partial^2 \psi}{\partial x^2} \ , \qquad -\infty < x < \infty \ , \quad z = h \ . \quad (4.2.15)$$

Equations (4.2.13) and (4.2.15) will enable us to describe wave-motions on a free surface and inside heterogeneous fluids with densities varying continuously with depth, bounded below by a plane $z = 0$, and above by a free surface.

To investigate wave-motions in a heterogeneous fluids, let the wave be travelling in the direction of the x-coordinate, so that the streamfunction is in the form

$$\psi = \Psi(z)e^{i(\omega t - kx)} \ .$$

The equation (4.2.13) becomes

$$\frac{d^2\Psi}{dz^2} - k^2\Psi + \frac{1}{\rho_0}\frac{d\rho_0}{dz}\left(\frac{d\Psi}{dz} - \frac{gk^2}{\omega^2}\Psi\right) = 0 , \quad 0 \le z < h , \quad (4.2.16)$$

while for free upper surface equation (4.2.15) is reduced to

$$\frac{d\Psi}{dz} - \frac{gk^2}{\omega^2}\Psi = 0 , \quad \text{on } z = h . \quad (4.2.17)$$

Thus solution of (4.2.16) and (4.2.17) will be $\Psi = \Psi^{(0)}e^{k(z-h)}$, if

$$\omega^2 = gk , \quad (4.2.18)$$

or

$$c^2 = \frac{g}{k} . \quad (4.2.19)$$

For a heterogeneous fluid of infinite depth the relation between wave number k and frequency ω of a surface wave is then the same as in the case of a homogeneous fluid, and the motion is also irrotational in this case.

The Case that $\rho_0 = \rho^{(0)}e^{-\beta z}$

For further discussion of waves in heterogeneous fluids, consider the special case that the equilibrium value of density $\rho_0 = \rho^{(0)}e^{-\beta z}$. In this case equation (4.2.16) becomes

$$\frac{d^2\Psi}{dz^2} - \beta\frac{d\Psi}{dz} + \left(\frac{g\beta}{\omega^2} - 1\right)k^2\Psi = 0 . \quad (4.2.20)$$

The characteristic equations for equation (4.2.20) is

$$\lambda^2 - \beta\lambda + \left(\frac{g\beta}{\omega^2} - 1\right)k^2 = 0 , \quad (4.2.21)$$

and if λ_1 , λ_2 are the characteristic roots of (4.2.21), then a solution of (4.2.20) and then of (4.2.13) is finally

$$\psi = \left(Ae^{\lambda_1 z} + Be^{\lambda_2 z}\right)e^{i(\omega t - kx)} , \qquad (4.2.22)$$

$$\text{for} \quad -\infty < x < \infty , \quad 0 \le z < h , \quad t > 0 .$$

Rigid Surface on Top

For this case, the boundary conditions to be satisfied are that the normal component of velocity at $z = 0$ and $z = h$ must vanish, i.e., $\dfrac{\partial \psi}{\partial x} = 0$, for $z = 0$ and $z = h$.

Thus we have

$$A + B = 0 ,$$

and

$$Ae^{\lambda_1 h} + Be^{\lambda_2 h} = 0 .$$

In order that not both $A = 0$ and $B = 0$, this requires

$$\begin{vmatrix} 1 & 1 \\ e^{\lambda_1 h} & e^{\lambda_1 h} \end{vmatrix} = 0 ,$$

or

$$e^{(\lambda_1 - \lambda_2)h} = 1 ,$$

or

$$(\lambda_1 - \lambda_2)h = 2\pi i n , \quad (n = 1, 2, \cdots) . \qquad (4.2.23)$$

But according to equation (4.2.21) :

$$\lambda_1 + \lambda_2 = \beta \, , \quad \lambda_1 \lambda_2 = \left(\frac{g\beta}{\omega^2} - 1 \right) k^2 \, .$$

Hence

$$\lambda_1 = \frac{1}{2}\beta + \frac{in\pi}{h} \, , \quad \lambda_2 = \frac{1}{2}\beta - \frac{in\pi}{h} \, , \quad (n = 1, 2, \cdots) \, , \quad (4.2.24)$$

and

$$\left(\frac{g\beta}{\omega^2} - 1 \right) k^2 = \frac{1}{4}\beta^2 + \frac{n^2\pi^2}{h^2} \, . \qquad (4.2.25)$$

Therefore if $\beta > 0$, ω is real; but if $\beta < 0$, ω is imaginary. However according to solution (4.2.22), the equilibrium state if disturbed will be stable if ω is real i.e. if $\beta > 0$, i.e. if the density ρ_0 diminishes vertically upwards. But ψ will grow exponentially or the equilibrium state is unstable when disturbed if ω is imaginary, i.e. if $\beta < 0$, i.e. if the density ρ_0 increases vertically up.

Free Surface on Top

Now consider the case that a heterogeneous fluid of depth h has a free surface. Let the roots of (4.2.21) be complex. Then

$$\lambda_1, \lambda_2 = \frac{1}{2}\beta \pm i\alpha \, ,$$

where

$$\alpha^2 = \left(\frac{g\beta}{\omega^2} - 1 \right) k^2 - \frac{1}{4}\beta^2 \, , \qquad (4.2.26)$$

and solution (4.2.22) takes the form

$$\psi = C \sin \alpha z \exp\left[\frac{1}{2}\beta z + i(\omega t - kx)\right], \qquad (4.2.27)$$

after using the boundary condition at the bottom surface $z = 0$. Because of $\Psi(z) = Ce^{\frac{1}{2}\beta z} \sin \alpha z$ the free surface condition (4.2.17) leads to

$$\frac{1}{2}\beta \sin \alpha h + \alpha \cos \alpha h - \frac{gk^2}{\omega^2} \sin \alpha h = 0 ,$$

or

$$\tan \alpha h = \frac{\alpha}{\dfrac{gk^2}{\omega^2} - \dfrac{1}{2}\beta}$$

or

$$\tan \alpha h = \beta h \frac{\alpha h}{\alpha^2 h^2 + k^2 h^2 - \frac{1}{4}\beta^2 h^2} , \qquad (4.2.28)$$

making use of the relation (4.2.26). Equation (4.2.28) can now be solved graphically for αh , because its solutions are simply the intersections of the two curves

$$z = \tan x , \qquad z = \frac{\mu x}{x^2 + b^2} ,$$

where $\mu = \beta h$, $b^2 = k^2 h^2 - \frac{1}{4}\beta^2 h^2$, $x = \alpha h$.

Consider the case that βh is small. Rewriting the frequency expression (4.2.26) yields

$$\alpha^2 h^2 = \left(\frac{g\beta}{\omega^2} - 1\right)k^2h^2 - \frac{1}{4}\beta^2 h^2 \approx \left(\frac{g\beta}{\omega^2} - 1\right)k^2 h^2 .$$

Now from (4.2.28), $\tan \alpha h \approx 0$, for small βh . Thus

$$\alpha h \approx n\pi , \quad (n = 1, 2, \cdots) , \tag{4.2.29}$$

and hence

$$n^2\pi^2 \approx \frac{g\beta}{\omega^2}k^2h^2 - k^2h^2 ,$$

or

$$\omega^2 \approx g\beta \frac{k^2h^2}{n^2\pi^2 + k^2h^2} . \tag{4.2.30}$$

From (4.2.27) and (4.2.29) it is clear that the vertical motion $\dfrac{\partial \psi}{\partial x}$ is very small at the free surface where $\alpha h \approx n\pi$, while the maximum vertical oscillation takes place at the levels

$$z = \left(n - \frac{1}{2}\right)\frac{\pi}{\alpha} , \quad n = 1, 2, \cdots .$$

Again this is consistent with earlier results for homogeneous stratified fluids that internal waves are dynamically more energetic than waves at the free surface.

The solution (4.2.27) represents wave propagating in the direction $\mathbf{k} = (k, \alpha)$. The dispersion relation (4.2.26) is in fact of the form

$$\omega = \omega(\mathbf{k}) . \tag{4.2.31}$$

Thus the phase velocity

$$\mathbf{c}_p = \frac{\omega}{|\mathbf{k}|^2}\mathbf{k} , \tag{4.2.32}$$

and the group velocity

$$\mathbf{c}_g = \left(\frac{\partial \omega}{\partial k}, \frac{\partial \omega}{\partial \alpha} \right), \tag{4.2.33}$$

are in general not in the same direction. We shall not go into details to discuss these issues, but simply mention here this aspect of anisotropic dispersion.

If the denisty variation $\rho_0(z)$ is not the simple case of exponential variation, then in (4.2.20), we have

$$\beta = -\frac{1}{\rho_0}\frac{d\rho_0}{dz} , \tag{4.2.34}$$

and hence β is a function of z. If β is a slowly varying function of z, then the behavior of the wave propagation is very similar to the case of constant β. The quantity

$$N = (g\beta)^{\frac{1}{2}} , \tag{4.2.35}$$

is known as the Väisäla-Brunt frequency for the incompressible fluids.

4.3 Waves in Compressible Stratified Fluids

In this section we shall investigate the internal and sound waves in nonhomogeneous and compressible fluids as well as the decoupling of the two types of waves. Let $p_0(z)$ and $\rho_0(z)$ be respectively the pressure and density of an equilibrium distribution of a heterogeneous fluid such as an atmosphere or an ocean. If $\rho_0(z)$ is a continuously decreasing function of the height z, then

$p_0(z)$ must also decrease with height in accordance with the following hydrostatic law of force :

$$\frac{dp_0(z)}{dz} = -\rho_0(z)g \; . \tag{4.3.1}$$

We shall first investigate the internal stabilities of a stratified fluid when small disturbances are introduced to an equilibrium distribution of such a fluid, and show how the stabilizing effect of the gravitational restoring force acts on fluid particles in the vicinity of small disturbances. Suppose that a fluid element at height z, with density $\rho_0(z)$, is displaced to a higher height $z+\xi$. This displacement carries it to an evironment where the equilibrium density has a smaller value

$$\rho_0(z)+\rho_0'(z)\xi \; , \tag{4.3.2}$$

with $\rho_0'(z)<0$ because $\rho_0(z)$ is a continuously decreasing function of the height z. The pressure there is likewise diminished to

$$p_0(z)+p_0'(z)\xi = p_0(z)-\rho_0(z)g\xi \; . \tag{4.3.3}$$

Thus the fluid element must experience this pressure drop at the same time. Therefore its density as a result of the pressure change of the surrounding fluid is reduced to

$$\rho_0(z)+\frac{d\rho_0}{dp_0}\cdot dp_0 = \rho_0(z)+\frac{1}{c_0^2(z)}\cdot\frac{dp_0}{dz}\xi$$

$$= \rho_0(z)-\frac{\rho_0(z)g\xi}{c_0^2(z)} \; , \tag{4.3.4}$$

where $c_0^2 = \dfrac{dp_0}{d\rho_0}$. The fluid element's excess density over that of the surrounding fluid is therefore equal to the difference of (4.3.4) from (4.3.3), viz.

$$\rho^*(z) = \left\{ \frac{-\rho_0(z)g}{c_0^2(z)} - \rho_0'(z) \right\} \xi \ . \tag{4.3.5}$$

Now the fluid element's excess of weight $\rho^*(z) \cdot g$ over buoyancy of the surrounding fluid per unit volume, i.e. the gravitational restoring force can be obtained by multiplying (4.3.5) by the gravity constant g :

$$g \left\{ -\frac{\rho_0(z)g}{c_0^2(z)} - \rho_0'(z) \right\} \xi \ . \tag{4.3.6}$$

Thus the restoring force (4.3.6) is positive if

$$\rho_0'(z) < \frac{-\rho_0(z)g}{c_0^2(z)} \ , \tag{4.3.7}$$

or if

$$-\frac{\rho_0'(z)}{\rho_0(z)} > \frac{g}{c_0^2(z)} \ . \tag{4.3.8}$$

In other words a disturbed stratified atmosphere or ocean is stable only if the relative rate of decrease of density with height is greater than $\dfrac{g}{c_0^2(z)}$. When the criterion (4.3.7) or (4.3.8) for stability is satisfied, the restoring force per unit volume (4.3.6) may be written as

$$\rho_0(z)[N(z)]^2 \xi \ ,$$

where

$$N(z) = \left\{ -\frac{\rho_0'(z)}{\rho_0(z)} g - \frac{g^2}{c_0^2(z)} \right\}^{\frac{1}{2}} \ , \tag{4.3.9}$$

defines a positive quantity with the dimensions of frequency, and is often known as the Väisälä-Brunt frequency.

We shall now study the full linearized equations for a stratified compressible fluid, in order to find solutions that will include both internal gravity waves and sound waves. Let p^* and ρ^* be the departures in pressure and density from the equilibrium values, i.e.,

$$p^* = p - p_0 , \quad \rho^* = \rho - \rho_0 .$$

If the gravitational vector $\mathbf{g} = (0,\ 0,\ -g)$ is directed downwards, the hydrostatic law gives

$$\nabla p_0 = \rho_0 \mathbf{g} .$$

Then by means of this hydrostatic condition the equation of momentum in linearized form leads to

$$\rho_0 \frac{\partial \mathbf{v}}{\partial t} + \nabla p^* = \rho^* \mathbf{g} , \tag{4.3.10}$$

where $\mathbf{v}$ denotes the fluid velocity. The equation of continuity also in linearized form becomes

$$\frac{\partial \rho^*}{\partial t} + \nabla \cdot (\rho_0 \mathbf{v}) = 0 . \tag{4.3.11}$$

Taking divergence of (4.3.10) yields

$$\frac{\partial}{\partial t}[\nabla \cdot (\rho_0 \mathbf{v})] + \nabla^2 p^* = \nabla \cdot (\rho^* \mathbf{g}) . \tag{4.3.12}$$

Eliminating $\nabla \cdot (\rho_0 \mathbf{v})$ between (4.3.11) and (4.3.12) leads to

$$\nabla^2 p^* = \frac{\partial^2 \rho^*}{\partial t^2} - g \frac{\partial \rho^*}{\partial z} . \tag{4.3.13}$$

If equation (4.3.13) may be treated as the combined equation of

$$\nabla^2 p^* = \frac{\partial^2 \rho^*}{\partial t^2} \quad \text{and} \quad \nabla^2 p^* = -g\frac{\partial \rho^*}{\partial z} ,$$

then the first equation could be regarded as the value of $\nabla^2 p^*$ for pure sound waves and the second as the value of $\nabla^2 p^*$ for pure internal gravity waves. Now for barotropic fluids, i.e., $p = p(\rho)$ we have

$$\frac{\partial p}{\partial t} + \mathbf{v} \cdot \nabla p = c^2 \left(\frac{\partial \rho}{\partial t} + \mathbf{v} \cdot \nabla \rho \right) , \qquad (4.3.14)$$

where c is the local sound speed. Take $\mathbf{v} = (u, v, w)$, the linearized form of (4.3.14) is

$$\frac{\partial p^*}{\partial t} + w\frac{dp_0}{dz} = c_0^2(z)\left(\frac{\partial \rho^*}{\partial t} + w\frac{d\rho_0}{dz} \right) . \qquad (4.3.15)$$

Denote $q = \rho_0 w$ which is the vertical component of the mass flux, then with the hydrostatic relation :

$$\frac{dp_0}{dz} = -\rho_0 g ,$$

and the definition of the Väisälä-Brunt frequency (4.3.9), we obtain from (4.3.15)

$$\frac{\partial \rho^*}{\partial t} = \frac{1}{c_0^2}\frac{\partial p^*}{\partial t} + \frac{1}{g}N^2(z)q . \qquad (4.3.16)$$

From the linearized equation of momentum (4.3.10), taking the upward z-component gives

$$\frac{\partial q}{\partial t} + \frac{\partial p^*}{\partial z} = -g\rho^* \qquad (4.3.17)$$

Eliminating $\dfrac{\partial \rho^*}{\partial t}$ between equations (4.3.16) and (4.3.17) leads to

$$\frac{\partial^2 q}{\partial t^2} + N^2(z)q = -\frac{\partial^2 p^*}{\partial t \partial z} - \frac{g}{c_0^2}\frac{\partial p^*}{\partial t} , \qquad (4.3.18)$$

while taking derivative $\dfrac{\partial}{\partial t}$ of (4.3.16) and derivative $\dfrac{\partial}{\partial z}$ of (4.3.17) and then substituting these $\dfrac{\partial^2 \rho^*}{\partial t^2}$ and $\dfrac{\partial \rho^*}{\partial z}$ into (4.3.13) lead to

$$\nabla^2 p^* = \frac{1}{c_0^2}\frac{\partial^2 p^*}{\partial t^2} + \frac{1}{g}N^2(z)\frac{\partial q}{\partial t} + \frac{\partial^2 q}{\partial z \partial t} + \frac{\partial^2 p^*}{\partial z^2} ,$$

or

$$\frac{\partial^2 p^*}{\partial x^2} + \frac{\partial^2 p^*}{\partial y^2} - \frac{1}{c_0^2}\frac{\partial^2 p^*}{\partial t^2} = \frac{\partial^2 q}{\partial z \partial t} + \frac{1}{g}N^2(z)\frac{\partial q}{\partial t} . \qquad (4.3.19)$$

Equations (4.3.18) and (4.3.19) for q and p^* govern the combined propagation of internal and sound waves in a stratified fluid, from which one can gain a lot of insight to problems involving wave motions in fluids with continuously varying densities. First define the wave energy flux as

$$\mathbf{F} = (p - p_0)\mathbf{v} = p^* \mathbf{v} . \qquad (4.3.20)$$

Then $\mathbf{F} \cdot \mathbf{n}$ is the component of $\mathbf{F}$ in the direction of a unit vector $\mathbf{n}$. That is the rate at which wave energy is being transported in the direction of $\mathbf{n}$ normal to a unit area of a small plane element, through the rate $p^*(\mathbf{v} \cdot \mathbf{n})$ of doing work by the perturbation in pressure p^*. The associated wave energy per unit volume will consist of

$$W = \frac{1}{2}\rho_0 \mathbf{v} \cdot \mathbf{v} + \frac{1}{2}\frac{p^{*2}}{c_0^2 \rho_0} + \frac{1}{2}\rho_0 N^2(z)\xi^2 , \qquad (4.3.21)$$

which is the sum of the kinetic energy, the acoustic potential energy, and the internal wave potential energy (because the restoring force per unit volume of the perturbed fluid is $\rho_0 N^2(z)\xi$). With the wave energy density W defined by (4.3.21) we are now going to show that the time rate of change of this wave energy density can always be expressed as the negative divergence of the wave energy flux $\mathbf{F}$, i.e.

$$\frac{\partial W}{\partial t} = -\nabla \cdot \mathbf{F} \ . \tag{4.3.22}$$

Since the vertical displacement ξ is given by

$$\xi = \int w dt = \frac{1}{\rho_0} \int q dt \ ,$$

therefore integrating (4.3.16) with respect to time gives

$$N^2(z) = \frac{g}{\rho_0}\left\{\rho * - \frac{p*}{c_0^2}\right\} \ . \tag{4.3.23}$$

Also from the equation of continuity (4.3.11)

$$\frac{\partial \rho *}{\partial t} + w\frac{\partial \rho_0}{\partial z} = -\rho_0 \nabla \cdot \mathbf{v} \ .$$

Substituting this into (4.3.15) and making use of the hydrostatic condition $\dfrac{dp_0}{dz} = -g\rho_0$, we have

$$\frac{\partial p *}{\partial t} = wg\rho_0 - c_0^2 \rho_0 \nabla \cdot \mathbf{v} \ . \tag{4.3.24}$$

Thus the rate of change of wave energy density W defined by (4.3.21) is

$$\frac{\partial W}{\partial t} = \mathbf{v} \cdot \rho_0 \frac{\partial \mathbf{v}}{\partial t} + \frac{1}{c_0^2 \rho_0} p^* \frac{\partial p^*}{\partial t} + \rho_0 N^2(z) \xi w \ ,$$

since $w = \frac{\partial \xi}{\partial t}$. Replacing $\frac{\partial \mathbf{v}}{\partial t}$ by the momentum equation (4.3.10), $\frac{\partial p^*}{\partial t}$ by equation (4.3.24) and $N^2 \xi$ by equation (4.3.23) leads to

$$\frac{\partial W}{\partial t} = \mathbf{v} \cdot (\rho^* \mathbf{g} - \nabla p^*) + \frac{1}{c_0^2 \rho_0} p^* \left(w g \rho_0 - c_0^2 \rho_0 \nabla \cdot \mathbf{v} \right)$$

$$+ w \rho_0 \frac{g}{\rho_0} \left(\rho^* - \frac{p^*}{c_0^2} \right)$$

$$= -wg\rho^* - \mathbf{v} \cdot \nabla p^* + \frac{1}{c_0^2} p^* wg - p^* \nabla \cdot \mathbf{v} + wg\rho^* - \frac{1}{c_0^2} wgp^* \ ,$$

i.e.

$$\frac{\partial W}{\partial t} = -\mathbf{v} \cdot \nabla p^* - p^* \nabla \cdot \mathbf{v} = -\nabla (p^* \mathbf{v}) = -\nabla \cdot \mathbf{F} \ ,$$

which is equation (4.3.22). Thus it has been verified that the linearized equations of momentum (4.3.10), of continuity (4.3.11) and of compressibility (4.3.15) we are using satisfy an energy conservation relation with the wave energy flux defined as $\mathbf{F} = p^* \mathbf{v}$ and with wave energy density W defined by (4.3.21) in which the acoustic potential energy and the internal wave potential energy are merely added together.

Now we shall investigate properties of solutions of the pair of second-order partial differential equations (4.3.18) and (4.3.19) which were derived from the linearized equations of continuity, of momentum and of compressibility by eliminating all variables except p^* and q. In particular we look for conditions under which solutions of (4.3.18) and (4.3.19) are such that sound waves and internal gravity waves are decoupled.

It is not possible so far to obtain exact solutions of the pair of equations (4.3.18) and (4.3.19) for fairly general forms of the Väisälä-Brunt frequency and sound speed $c_0(z)$ as functions of height z. However, one can deduce a lot of important results about wave-like solutions to this pair of equations from the local dispersion relationship.

We consider the following expression for p^* and q

$$p^* = p_1(\mathbf{x})\exp[i(\omega t - \mathbf{k}\cdot\mathbf{x})] , \tag{4.3.25}$$

and

$$q = q_1(\mathbf{x})\exp[i(\omega t - \mathbf{k}\cdot\mathbf{x})] , \tag{4.3.26}$$

where $p_1(\mathbf{x})$ and $q_1(\mathbf{x})$ are slowly varying functions and $\mathbf{k}(k_1, k_2, k_3)$ is the wave number vector. Then a local dispersion relation of the form

$$\omega = \omega(\mathbf{k}; z) , \tag{4.3.27}$$

may be obtained.

If the variation of $p_1(\mathbf{x})$ and $q_1(\mathbf{x})$ may be ignored compared with the variation in the exponential factor, substituting equations (4.3.25) and (4.3.26) into the pair of equations (4.3.18) and (4.3.19) yields respectively

$$\left(\omega^2 - N^2\right)q_1 = \omega\left(k_3 + \frac{ig}{c_0^2}\right)p_1 , \tag{4.3.28}$$

and

$$\left(\frac{\omega^2}{c_0^2} - k_1^2 - k_2^2\right)p_1 = \omega\left(k_3 + i\frac{N^2}{g}\right)q_1 . \tag{4.3.29}$$

When equations (4.3.18) and (4.3.19) have the exact solutions (4.3.25) and (4.3.26) with p_1, q_1, ω, k_1, k_2 and k_3 being constants satisfying (4.3.28) and (4.3.29), these correspond to cases with $c_0(z)$ and $N(z)$ being constant; for instance in case of an isothermal atmosphere of temperature T_0 and constant ratio of specific heats, γ, then $c_0(z) = \left(\dfrac{dp}{d\rho}\right)_0^{\frac{1}{2}} = (\gamma R T_0)^{\frac{1}{2}}$ is constant, and since

$$\frac{\rho_0'(z)}{\rho_0(z)} = \frac{1}{p_0}\frac{dp_0}{dz} = \frac{1}{RT_0\rho_0}(-g\rho_0) = -\frac{g}{RT_0} = -\frac{\gamma g^2}{c_0^2(z)},$$

therefore

$$N(z) = \left\{\frac{\gamma g^2}{c_0^2(z)} - \frac{g^2}{c_0^2(z)}\right\}^{\frac{1}{2}} = \frac{(\gamma-1)^{\frac{1}{2}} g}{c_0^2(z)},$$

is also a constant. In many other cases when $c_0(z)$ and $N(z)$ do vary but only gradually on a scale of wavelengths, the nature of the waves can be determined by a study of the following dispersion relation. Eliminating p_1 and q_1 between equations (4.3.28) and (4.3.29) we have

$$\left(\omega^2 - N^2\right)\left(\frac{\omega^2}{c_0^2} - k_1^2 - k_2^2\right) = \omega^2\left(k_3 + \frac{ig}{c_0^2}\right)\left(k_3 + \frac{iN^2}{g}\right),$$

or

$$\frac{\omega^4}{c_0^2} - \left[k^2 + ik_3\left(\frac{g}{c_0^2} + \frac{N^2}{g}\right)\right]\omega^2 + \left(k_1^2 + k_2^2\right)N^2 = 0\ , \quad (4.3.30)$$

where $k^2 = |\mathbf{k}|^2 = k_1^2 + k_2^2 + k_2^3$. Consider the case that the wave number k is large compared with $-\frac{\rho_0'(z)}{\rho_0(z)}$. Since $-\frac{\rho_0'(z)}{\rho_0(z)} = \frac{g}{c_0^2} + \frac{N^2}{g}$, therefore $k >> \frac{g}{c_0^2}$ and $k >> \frac{N^2}{g}$; and hence $k^2 >> \frac{N^2}{c_0^2}$. Under these conditions the coefficient of ω^2 in (4.3.30) is approximately k^2, and $k^2 >> \frac{1}{c_0^2} N^2(k_1^2 + k_2^2)$, the geometric mean squared of the coefficients of the first and last terms in (4.3.30). For such a quadratic equation for ω^2 (4.3.30) it has two roots : one larger λ_l and one smaller λ_s. From (4.3.30) we have

$$\lambda_l + \lambda_s \approx c_0^2 k^2 \quad \text{and} \quad \lambda_l \lambda_s = c_0^2 N^2 (k_1^2 + k_2^2) .$$

By solving these equations it can be shown approximately that

$$\lambda_l = c_0^2 k^2 \quad \text{or} \quad \omega^2 = c_0^2 (k_1^2 + k_2^2 + k_3^2) , \tag{4.3.31}$$

and

$$\lambda_s = \frac{N^2}{k^2}(k_1^2 + k_2^2) \quad \text{or} \quad \omega^2 = \frac{N^2(k_1^2 + k_2^2)}{(k_1^2 + k_2^2 + k_3^2)} . \tag{4.3.32}$$

Here, (4.3.31) is the ordinary acoustic relationship showing the non-dispersive property of sound waves with their wave speed c_0 independent of wave number, while equation (4.3.32) is the dispersion relation for an internal plane wave with its wave number vector $\mathbf{k}$ normal to the plane wavefront and making an angle θ with the upward vertical such that

$$\sin\theta = \frac{|k_3|}{\sqrt{k_1^2 + k_2^2 + k_3^2}} \ , \qquad \text{and} \qquad \omega = N\cos\theta \ .$$

The latter equation is the law of refraction of waves in stratified fluids with continuously varying density. We have proved, therefore, that under the condition when the wave number k is much larger than the relative rate of change of undisturbed density $-\frac{\rho_0'(z)}{\rho_0(z)}$, the sound wave and the internal gravity wave are completely decoupled : neither is influenced by the presence of the other.

4.4 Dynamic Equations of Bubbly Liquids

A special nonhomogeneous fluid is a liquid containing gas bubbles. Before we discuss waves in bubbly liquids, let us try first to establish the governing equations for flow of such a fluid. We shall try to develop a continuum theory, but retaining certain essential features of the dynamics of the individual bubbles. We shall consider the cases that bubbles are all small, hence they can all be assumed to be spherical. Let us restrict our discussion to the case that there is only one species of bubbles, i.e. the total mass of gas in each bubble m is constant. Let $R(\mathbf{x},t)$ be the average bubble radius. $R(\mathbf{x},t)$ is macroscopic field variable. It is thus implied that the macroscopic characteristic length is much larger than the bubble radius. Let ρ_g be density of the gas phase. Then we have

$$\frac{4\pi}{3}R^3\rho_g = m \ . \qquad (4.4.1)$$

We shall define another macroscopic variable $n(\mathbf{x},t)$ as the average number density of the bubbles, then

$$\beta = \frac{4\pi}{3}R^3 n \ , \qquad (4.4.2)$$

where β is the average volume fraction of the gas phase. Thus by definition the volume fraction of the liquid phase is $(1-\beta)$. We shall use subscript g to designate the gas phase and the subscript f to designate the liquid phase. Then the continuity equations for gas phase and liquid phase are respectively :

$$\frac{\partial}{\partial t}(\beta\rho_g)+\nabla\cdot(\beta\rho_g\mathbf{v}_g)=0\ , \tag{4.4.3}$$

$$\frac{\partial}{\partial t}\left[(1-\beta)\rho_f\right]+\nabla\cdot\left[(1-\beta)\rho_f\mathbf{v}_f\right]=0\ . \tag{4.4.4}$$

From (4.4.1) - (4.4.3), we may also obtain

$$\frac{\partial n}{\partial t}+\nabla\cdot(n\mathbf{v}_g)=0\ , \tag{4.4.5}$$

which states that the number of bubbles is conserved, i.e. there is no coalescence or splitting of bubbles.

The momentum equations for gas phase and liquid phase may be expressed as follows :

$$\beta\rho_g\left[\frac{\partial\mathbf{v}_g}{\partial t}+(\mathbf{v}_g\cdot\nabla)\mathbf{v}_g\right]=-\beta\nabla p_g+\beta\rho_g\mathbf{b}+\mathbf{f}_g+\mathbf{F}\ , \tag{4.4.6}$$

$$(1-\beta)\rho_f\left[\frac{\partial\mathbf{v}_f}{\partial t}+(\mathbf{v}_f\cdot\nabla)\mathbf{v}_f\right]$$

$$=-(1-\beta)\nabla p_f+(1-\beta)\rho_f\mathbf{b}+\mathbf{f}_f+\mathbf{F}\ , \tag{4.4.7}$$

where p_g is the pressure of the gas phase, p_f the pressure of the liquid phase, **b** the external body force, $\mathbf{f}_g$ the self-force of the gas phase, $\mathbf{f}_f$ the self-force of the liquid phase, and **F** the mutual frictional force between the phases.

A Simple Model of Bubbly Liquids

We shall limit our consideration to a relatively simple model of the bubbly liquid. We shall assume that each phase is essentially barotropic, thus

$$p_g = p_g(\rho_g)\,, \tag{4.4.8}$$

and

$$p_f = p_f(\rho_f)\,. \tag{4.4.9}$$

We shall assume that p_g and p_f are interrelated, and they are related through the variable R. Schematically, we can write

$$p_g - p_f = P\{R\}\,. \tag{4.4.10}$$

A particular form of $P\{R\}$ based on the Rayleigh-Plesset equation of the dynamics of a single bubble is given by

$$P\{R\} = \rho_f\left[R\left(\frac{\partial}{\partial t}+\mathbf{v}_g\cdot\nabla\right)^2 R+\frac{3}{2}\left(\frac{\partial R}{\partial t}+\mathbf{v}_g\cdot\nabla R\right)^2\right]+\frac{2\sigma}{R}\,, \tag{4.4.11}$$

where η_f is the viscosity coefficient in the liquid phase.

The mutual friction for $\mathbf{F}$ is assumed to be of the following form :

$$\mathbf{F} = -\frac{9}{2}\frac{\beta\eta_f}{R^2}\left(\mathbf{v}_g-\mathbf{v}_f\right)$$

$$-C_m\beta\rho_f\left[\left(\frac{\partial}{\partial t}+\mathbf{v}_g\cdot\nabla\right)\mathbf{v}_g-\left(\frac{\partial}{\partial t}+\mathbf{v}_f\cdot\nabla\right)\mathbf{v}_f\right], \tag{4.4.12}$$

where the first term represents the mutual friction due to viscous drag, the second term the virtual mass force, and C_m is a coefficient which has been argued to be equal to $\frac{1}{2}$, but its value is still an open question.

For many cases, the mutual friction is more important than self frictional forces. In this simple model, we shall neglect the self-forces $\mathbf{f}_f$ and $\mathbf{f}_g$.

4.5 Waves in Bubbly Liquids

The equations governing the flow of bubbly liquids, described in Section 4.4, are very complex even for the simple model. Furthermore, there are still uncertainties concerning the constitutive relations. In this section, we shall discuss some of the relatively simple cases of wave propagation in bubbly liquids.

Sound Waves

Consider the small disturbances from the equilibrium state. The equilibrium state variables will be designated by the subscript 0, e.g., β_0, R_0, ρ_{g0} , etc. At equilibrium $\mathbf{v}_{g0} = \mathbf{v}_{f0} = 0$. Then we set

$$\beta = \beta_0(1+\beta') \text{ , etc.} \tag{4.5.1}$$

We shall neglect the external body force **b**. The self-forces are also neglected in this simple model. Then the linearized version of the equations (4.4.1), (4.4.3), (4.4.4), (4.4.6) and (4.4.7) become respectively :

$$3R' + \rho_g' = 0 \text{ ,} \tag{4.5.2}$$

$$\frac{\partial}{\partial t}(\beta' + \rho_g') + \nabla \cdot \mathbf{v}_g = 0 \text{ ,} \tag{4.5.3}$$

$$(1-\beta_0)\left(\frac{\partial \rho_f'}{\partial t}+\nabla\cdot\mathbf{v}_f\right)-\beta_0\frac{\partial \beta'}{\partial t}=0\ , \qquad (4.5.4)$$

$$\frac{\partial \mathbf{v}_g}{\partial t}=-c_g^2\nabla\rho_g'+\frac{\mathbf{F}}{\beta_0\rho_{g0}}\ , \qquad (4.5.5)$$

$$\frac{\partial \mathbf{v}_f}{\partial t}=-c_f^2\nabla\rho_f'+\frac{\mathbf{F}}{(1-\beta_0)\rho_{f0}}\ , \qquad (4.5.6)$$

where

$$c_f^2=\left(\frac{dp_f}{d\rho_f}\right)_0\ , \qquad c_g^2=\left(\frac{dp_g}{d\rho_g}\right)_0\ , \qquad (4.5.7)$$

are velocities of sound in liquid and gas phases respectively. Based on the simple model, we obtain from the linearized version of (4.4.10), (4.4.11) and (4.4.12) that

$$\left(\frac{\rho_{g0}}{\rho_{f0}}\right)c_g^2\rho_g'-c_f^2\rho_f'=R_0^2\frac{\partial^2 R'}{\partial t^2}-\frac{2\sigma}{\rho_{f0}R_0}R'\ , \qquad (4.5.8)$$

and

$$\mathbf{F}=\frac{9}{2}\frac{\beta_0\eta_f}{R_0^2}(\mathbf{v}_g-\mathbf{v}_f)-C_m\beta_0\rho_{f0}\frac{\partial}{\partial t}(\mathbf{v}_g-\mathbf{v}_f)\ . \qquad (4.5.9)$$

Let us consider monochromatic waves with variations proportional to $\exp[i(\mathbf{k}\cdot\mathbf{x}-\omega t)]$, and denote :

$$\varepsilon_1=\frac{\rho_{g0}}{\rho_{f0}}\ , \qquad \varepsilon_2=\frac{c_g^2}{c_f^2}\ , \qquad (4.5.10)$$

$$q = (kR_0)^2 , \quad q_\sigma = \frac{2\sigma}{3c_g^2 \rho_{g0} R_0} , \tag{4.5.11}$$

$$\omega_f = kc_f , \quad \omega_0^2 = \frac{3\varepsilon_1 \varepsilon_2 \omega_f^2}{q}(1 - q_\sigma) , \tag{4.5.12}$$

and

$$\alpha_d = \frac{9}{2} \frac{\eta_f}{(1-\beta_0)\rho_{g0} R_0^2} . \tag{4.5.13}$$

Then we obtain from (4.5.2) and (4.5.8) that

$$\rho_f' = BR' , \tag{4.5.14}$$

where

$$B = \frac{q}{\omega_f^2}(\omega^2 - \omega_0^2) . \tag{4.5.15}$$

Using (4.5.2) and (4.5.14), (4.5.3) and (4.5.4) become respectively :

$$\mathbf{k} \cdot \mathbf{v}_g = \omega(\beta' - 3R') , \tag{4.5.16}$$

$$\mathbf{k} \cdot \mathbf{v}_f = \omega\left[BR' - \left(\frac{\beta}{1-\beta_0}\right)\beta'\right] . \tag{4.5.17}$$

Thus from (4.5.9) we obtain

$$\mathbf{k} \cdot \mathbf{F} = \beta_0 \rho_{g0} \omega \left[-\alpha_d + \frac{i\omega C_m}{(1-\beta_0)\varepsilon_1}\right]\left[\beta' - (1-\beta_0)(3+B)R'\right] . \tag{4.5.18}$$

Let equations (4.5.5) and (4.5.6) take dot products with $\mathbf{k}$, then we obtain :

$$\left\{ i\omega\left[1+\frac{C_m}{(1-\beta_0)\varepsilon_1}\right]-\alpha_d \right\}\omega\beta'$$

$$+\left\{ -i\omega^2\left[3+\frac{C_m}{\varepsilon_1}(3+B)\right]+\omega(1-\beta_0)\alpha_d(3+B)+3ik^2c_g^2 \right\}R'=0\,, \tag{4.5.19}$$

$$\left\{ -i\omega\left[1+\frac{C_m}{(1-\beta_0)}\right]+\varepsilon_1\alpha_d \right\}\left(\frac{\beta_0}{1-\beta_0}\right)\omega\beta'$$

$$+\left\{ i\omega^2\left[B+\left(\frac{\beta_0}{1-\beta_0}\right)+C_m(3+B)\right]-\omega\beta_0\varepsilon_1\alpha_d(3+B)-ik^2c_f^2B \right\}R'=0\,. \tag{4.5.20}$$

From (4.5.19) and (4.5.20), we can obtain the characteristic equation which is an equation of degree 5 in terms of ω , but is only a quadratic equation in terms of k. To simplify somewhat the expression, we note that $\varepsilon_1 << 1$, $\varepsilon_2 << 1$, and $\omega_0^2 << \frac{\omega_f^2}{q}$. Thus we shall neglect these small terms wherever there is no other parameters involved. Then we obtain :

$$\left(1+\frac{C_m}{\varepsilon_1}\right)\omega^5-\left[\left(1+\frac{C_m}{(1-\beta_0)\varepsilon_1}\right)+\frac{3\beta_0}{q(1-\beta_0)}\left(1+\frac{C_m}{\varepsilon_1}\right)\right]\omega_f^2\omega^3$$

$$+\left[\left(1+\frac{C_m}{(1-\beta_0)\varepsilon_1}\right)\omega_0^2+\frac{3\beta_0\varepsilon_2}{q(1-\beta_0)}\left(1+\frac{C_m}{(1-\beta_0)}\right)\omega_f^2\right]\omega_f^2\omega$$

$$+i\alpha_d\left\{(1-\beta_0)\omega^4-\left(1+\frac{3\beta_0}{q}\right)\omega_f^2\omega^2+\frac{3\varepsilon_1\varepsilon_2}{(1-\beta_0)q}\left[1-(1-\beta_0)q_\sigma\right]\omega_f^4\right\}=0\,, \tag{4.5.21}$$

or

$$k^2 = \frac{\omega^2}{c_f^2}\left\{\left[\frac{\omega^2 C_m m_1}{(1-\beta_0)\varepsilon_1^2}\left(1+\frac{\varepsilon_1}{C_m}\right)+(1-\beta_0)\alpha_d m_2\right]\right.$$

$$\left.+\frac{i\omega\alpha_d}{\varepsilon_1}\left[\omega_r^2\left(\beta_0\{1-\beta_0\}-\varepsilon_1\{1-(1-\beta_0)q_\sigma\}\right)+\varepsilon_1\beta_0\omega^2\right]\right\}\frac{(\omega_a^2-\omega^2)}{m_3}, \quad (4.5.22)$$

where

$$\omega_r^2 = \frac{3\varepsilon_1\varepsilon_2 c_f^2}{(1-\beta_0)R_0^2}, \quad \omega_a^2 = \frac{3\beta_0 c_f^2}{(1-\beta_0)R_0^2}, \quad (4.5.23)$$

$$m_1 = \omega_r^2\left[1-(1-\beta_0)q_\sigma + \frac{\beta_0(1-\beta_0)}{C_m}\right] - \omega^2\left[1+\frac{(1-\beta_0)\varepsilon_1}{C_m}\right], \quad (4.5.24)$$

$$m_2 = \omega_r^2\left[1-(1-\beta_0)q_\sigma\right]-\omega^2, \quad (4.5.25)$$

$$m_3 = \frac{\omega^2 C_m^2 m_1^2}{(1-\beta_0)^2\varepsilon_1^2} + \alpha_d^2 m_2. \quad (4.5.26)$$

Let us consider the frequency relation (4.5.21) first. We shall discuss two limiting cases, i.e. the case that $\alpha_d >> \omega$ and the case $\alpha_d << \omega$. Before we discuss these limiting cases, a rough idea of the magnitude of the relevant parameter may be helpful. Take water and air, then in cgs units, $\eta_f \approx 10^{-2}$, $\rho_g \approx 10^{-3}$. If we take $R_0 = 10^{-1}$, $\beta = 0.1$, then $\alpha_d \approx 5 \text{ x } 10^3 \sec^{-1}$ In comparison, the bubble resonance frequency $\omega_0 \approx \left[\frac{3\varepsilon_1 c_g^2}{R_0^2}\right]^{\frac{1}{2}} \approx 2 \text{ x } 10^4 \sec^{-1}$.

Thus α_d is comparable with ω_0. Now let us discuss the limiting cases :

(i) $\omega << \alpha_d$. This is the case that the mutual friction is important. The approximate solutions are

$$\omega^2 = \omega_1^2 = \frac{\varepsilon_1 \varepsilon_2 \omega_f^2 \left[1-(1-\beta_0)q_\sigma\right]}{\beta_0(1-\beta_0)\left[1+\left(\frac{q}{3\beta_0}\right)\right]}, \tag{4.5.27}$$

and

$$\omega^2 = (\omega_1')^2 = \frac{3\beta_0\left[1+\left(\frac{q}{3\beta_0}\right)\right]}{1-\beta_0}\frac{c_f^2}{R_0^2}. \tag{4.5.28}$$

Since we usually have $q << 1$, the mode associated with ω_1' is thus not a propagating mode and is not relevant to sound propagation.

Damping coefficients can be estimated by making use of (4.5.21). Let us take $\omega = \omega_1 - i\delta_1$, and since $\omega_1 << \omega_f$, we obtain

$$\delta_1 \approx \frac{\varepsilon_2\left(1+\frac{C_m}{(1-\beta_0)}\right)\omega_f^2}{2(1-\beta_0)\left(1+\frac{q}{3\beta_0}\right)\alpha_d}. \tag{4.5.29}$$

The frequency relation (4.5.27) is the most important one that is relevant to sound propagation in bubbly liquid, since for most cases the mutual friction is important. For $q <<1$, (4.5.27) yields the velocity of propagation

$$c_b = \left[\frac{1-(1-\beta_0)q_\sigma}{\beta_0(1-\beta_0)} \left(\frac{\rho_{g0}}{\rho_{f0}} \right) \right]^{\frac{1}{2}} c_g \ , \qquad (4.5.30)$$

which is usually an order of magnitude smaller than c_g , the sound speed in the gaseous medium.

It may be noted that c_b does not depend on the parameter C_m . Thus when $\omega << \alpha_d$, the effect of virtual mass is negligible on the sound speed in the bubbly liquids. However, the damping of the sound wave does depend on C_m, as can be seen from (4.5.29).

(ii) $\omega >> \alpha_d$, i.e., when the mutual friction is insignificant. The approximate solutions are

$$\omega^2 = \omega_2^2 = \frac{\varepsilon_1 \varepsilon_2 \omega_f^2}{\beta_0(1-\beta_0)} \frac{\left[1-(1-\beta_0)q_\sigma + \dfrac{\beta_0(1-\beta_0)}{C_m} \right]}{\left[1+\dfrac{q}{3\beta_0} + \dfrac{\varepsilon_1}{C_m}\left(1+\dfrac{q(1-\beta_0)}{\beta_0} \right) \right]} \ , \qquad (4.5.31)$$

and

$$\omega^2 = (\omega_2')^2 = \frac{3\beta_0}{(1-\beta_0)} \frac{\left[1+\dfrac{q}{3\beta_0} + \dfrac{\varepsilon_1}{C_m}\left(1+\dfrac{q(1-\beta_0)}{\beta_0} \right) \right]}{\left(1+\dfrac{\varepsilon_1}{C_m} \right)} \frac{c_f^2}{R_0^2} \ . \qquad (4.5.32)$$

Note that for $\varepsilon_1 << 1$, $\omega_2' = \omega_1'$. Thus this mode is again non-propagating. It is interesting to note that for finite C_m ,

$\omega_2 = 0(\omega_1)$; while if we neglect C_m , i.e., letting $C_m \to 0$, then $\frac{\omega_1^2}{\omega_2^2} = 0(\varepsilon_1)$, i.e. ω_2 is much larger than ω_1 .

We may remark here that as seen from (4.5.30), the dynamics of the sound wave associated with the mode ω_1 are that the inertia of the system is largely that of the liquid while the compressibility is mainly due to the gas bubbles. This is to be expected, since the liquid and gas move together when the mutual friction is important. When the mutual friction is insignificant, we expect that the liquid phase and gas phase would move independently of each other. Indeed if $C_m = 0$, we have $\omega_2 \approx c_g k$. But if C_m is finite, the virtual mass effect will couple the motion of liquid and gas together and again yield the sound speed of the same order of c_b .

Dead Zone and Negative Damping

Let us now investigate the characteristic equation in the representation of (4.5.22). It can be seen from (4.5.24) and (4.5.25) that for small ω , both m_1 and m_2 are positive. As ω increases to exceed a value of the order of ω_r , which we shall denote as $\overline{\omega}_r$, m_1 and m_2 become negative, and they stay negative for large ω . Thus from (4.5.22), we can see that

$$\mathrm{Rl}k^2 > 0 , \quad \text{for} \quad \omega < \overline{\omega}_r , \quad \text{or} \quad \omega > \omega_a , \tag{4.5.33}$$

$$\mathrm{Rl}k^2 < 0 , \quad \text{for} \quad \overline{\omega}_r < \omega < \omega_a . \tag{4.5.34}$$

Thus for $\overline{\omega}_r < \omega < \omega_a$, k has a very large imaginary part, and hence the wave is not propagating. We call this frequency range the *Dead Zone* .

$\overline{\omega}_r$ is associated with the resonance of a gas bubble in a liquid. The frequency ω_a, aside from a factor $\dfrac{\beta}{(1-\beta)}$, is the resonance frequency of the liquid outside an enclosure of radius R_0. ω_a is sometimes called the anti-resonance frequency.

We can visualize the whole picture as follows. For low frequencies, or equivalently when the wavelength is long compared with the bubble radius, the mode of propagation is essentially that due to the combined effect of the compressibility of gas and the inertia of liquid. We encounter the Dead Zone as ω reaches the bubble resonance frequency $\overline{\omega}_r$.

For very high frequencies, the propagation mode is essentially that in a pure liquid, since the gas and liquid phase move independently. For this case, the value $q=(kR)^2$ in fact is large. Thus (4.5.32) may yield a propagating mode for large q. As ω decreases towards the anti-resonance frequency ω_a, the Dead Zone is again encountered.

From (4.5.22), we may also notice that

$$\operatorname{Im} k^2 > 0\,, \quad \text{for} \quad \omega < \omega_a\,, \tag{4.5.35}$$

$$\operatorname{Im} k^2 < 0\,, \quad \text{for} \quad \omega > \omega_a\,. \tag{4.5.36}$$

Thus if other dissipation mechanisms, such as thermal damping, viscous damping, etc., are not included as in (4.5.22), there will be negative damping for $\omega > \omega_a$. Whether this negative damping is real is not clear. It may be a revelation that the present model for the bubbly liquids is not adequate for high frequency phenomena.

Experimental investigations on sound propagations in bubbly liquids are few and far apart. The theory and experiment are largely in agreement.

Waves in Locked Bubbly Liquids

From the analysis of small amplitude waves, it is clear that at least for low frequency cases, the dominant mode of propagation is that for which the compressibility is controlled by the gas and the inertia is controlled by the liquid. Therefore for a large class of problems we may make the approximation that ρ_f is constant and $\varepsilon_1 << 1$. Thus (4.4.4) can be simplified to become

$$\frac{\partial \beta}{\partial t} = \nabla \cdot \left[(1-\beta)\mathbf{v}_f\right] . \tag{4.5.37}$$

Using (4.4.1), (4.4.3) can be rewritten as

$$\frac{\partial}{\partial t}\left(\frac{\beta}{R^3}\right) + \nabla \cdot \left(\frac{\beta}{R^3}\mathbf{v}_g\right) = 0 . \tag{4.5.38}$$

Neglecting again the self frictions and the body force, adding (4.4.6) and (4.4.7) leads now approximately to

$$\frac{\partial \mathbf{v}_f}{\partial t} + \left(\mathbf{v}_f \cdot \nabla\right)\mathbf{v}_f = -\frac{1}{\rho_f}\left[\nabla p_f + \frac{\beta}{1-\beta}\nabla p_g\right] . \tag{4.5.39}$$

Using again (4.4.12) for **F**, then the substraction of (4.4.7) from (4.4.6) leads approximately to

$$\nabla p_g = \frac{9}{2}\frac{\eta_f}{R^2}\left(\mathbf{v}_f - \mathbf{v}_g\right)$$

$$+ C_m \rho_f \left[\frac{\partial \mathbf{v}_f}{\partial t} + \left(\mathbf{v}_f \cdot \nabla\right)\mathbf{v}_f - \frac{\partial \mathbf{v}_g}{\partial t} - \left(\mathbf{v}_g \cdot \nabla\right)\mathbf{v}_g\right] . \tag{4.5.40}$$

We shall take a simplified version of (4.4.10), just to retain the essential feature of the mechanisms involved, i.e.,

$$p_g - p_f = \rho_f \left[R\frac{\partial^2 R}{\partial t^2} + \frac{3}{2}\left(\frac{\partial R}{\partial t}\right)^2 \right] + \frac{2\sigma}{R} \,. \qquad (4.5.41)$$

For gas bubbles in liquid, there is evidence to support the isothermal behavior for a wide frequency range. Therefore (4.4.8) becomes

$$p_g = p_{g0}\left(\frac{R_0}{R}\right)^3 \,. \qquad (4.5.42)$$

Equations (4.5.37) - (4.5.42) form a closed system which governs approximately the general motion of a liquid containing gas bubbles.

Using (4.5.37), we can rewrite (4.5.38) as

$$\frac{\partial R}{\partial t} = -(\mathbf{v}_f \cdot \nabla)R + \frac{R}{3\beta}\nabla \cdot \mathbf{v}_f$$

$$+\left[(\mathbf{v}_f - \mathbf{v}_g)\cdot \nabla\right]R - \frac{R}{3\beta}\nabla \cdot \left[\beta(\mathbf{v}_f - \mathbf{v}_g)\right] \,. \qquad (4.5.43)$$

Using (4.5.41) and (4.5.42), (4.5.39) can be rewritten as

$$\frac{\partial \mathbf{v}_f}{\partial t} = -(\mathbf{v}_f \cdot \nabla)\mathbf{v}_f + \nabla\left[\frac{3p_{go}R_0^3}{(1-\beta)\rho_f R^4} - \frac{2\sigma}{\rho_f R^2}\right]\nabla R$$

$$+\nabla\left\{ R\frac{\partial^2 R}{\partial t^2} + \frac{3}{2}\left(\frac{\partial R}{\partial t}\right)^2 \right\} \,. \qquad (4.5.44)$$

Let us now considered the case in which the liquid and gas phases move together more or less. That will be realized if bubbles are small and the pressure gradient is not large. Since (4.5.40) can be rewritten as

$$\left(\mathbf{v}_f-\mathbf{v}_g\right)=\frac{2R^2}{9\eta_f}\left\{\nabla p_g-C_m\rho_f\left[\frac{\partial}{\partial t}\left(\mathbf{v}_f-\mathbf{v}_g\right)+\left(\mathbf{v}_f\cdot\nabla\right)\left(\mathbf{v}_f-\mathbf{v}_g\right)\right]\right\}$$

$$-\frac{2R^2}{9\eta_f}C_m\rho_f\left\{\left[\left(\mathbf{v}_f-\mathbf{v}_g\right)\cdot\nabla\right]\mathbf{v}_f-\left[\left(\mathbf{v}_f-\mathbf{v}_g\right)\cdot\nabla\right]\left(\mathbf{v}_f-\mathbf{v}_g\right)\right\} \qquad (4.5.45)$$

therefore, to a first approximation, we have

$$\left(\mathbf{v}_f-\mathbf{v}_g\right)\approx\frac{2R^2}{9\eta_f}\nabla p_g=-\frac{2p_{go}R_0^3}{3\eta_f R^2}\nabla R\ . \qquad (4.5.46)$$

Substitute (4.5.46) into (4.5.43), we obtain

$$\frac{\partial R}{\partial t}=-\left(\mathbf{v}_f\cdot\nabla\right)R+\frac{R}{3\beta}\nabla\cdot\mathbf{v}_f$$

$$+\frac{2p_{go}R_0^3}{3\eta_f}\left\{\frac{R}{3\beta}\nabla\cdot\left(\frac{\beta}{R^2}\nabla R\right)-\left(\frac{\nabla R}{R}\right)^2\right\}\ . \qquad (4.5.47)$$

Equations (4.5.37), (4.5.44) and (4.5.47) are the basic governing equations for the "locked" bubbly liquids. It is a somewhat more manageable system for discussion on finite amplitude waves. The term in the curly brackets in (4.5.47) are responsible for the dissipation due to the mutual friction or more appropriately the mutual slippage; while the terms in the curly bracket in (4.5.44) will cause dispersion due to bubble oscillation.

We now describe some of the results relating to waves in "locked" bubbly liquids.

(i) Small Amplitude Waves

Denote

$$c_0^2=\frac{p_{go}}{\beta_0\left(1-\beta_0\right)\rho_f}-\frac{2\sigma}{3\beta_0R_0\rho_f}\ , \qquad (4.5.48)$$

where equilibrium properties are designated by the subscript o. Consider the sinusoidal waves with a factor $\exp[i(\mathbf{k}\cdot\mathbf{x}-\omega t)]$. The frequency relation can be readily shown to be :

$$k^2 = \frac{\omega^2}{c_0^2\Gamma} , \tag{4.5.49}$$

where

$$\Gamma = \left(1 - \frac{R_0^2}{3\beta_0 c_0^2}\omega^2 - i\omega\frac{2p_{go}R_0^2}{9\eta}\right) . \tag{4.5.50}$$

The dissipation due to mutual slippage and dispersion due to bubble oscillation is clearly revealed in the expression of Γ .

(ii) Nonlinear Waves

Let us neglect the dissipation and dispersion terms.

Denote

$$c^2 = \frac{p_{go}R_0^3}{\rho_f\beta(1-\beta)R^3} - \frac{2\sigma}{3\rho_f\beta R} . \tag{4.5.51}$$

Then characteristics of the system can be readily determined. For one-dimensional waves, the characteristics C_0 and $C_\pm$ are given by :

$$C_0 : \qquad \frac{dx}{dt} = v_f , \tag{4.5.52}$$

$$C_\pm : \qquad \frac{dx}{dt} = v_f \pm c . \tag{4.5.53}$$

We also obtain that

$$\frac{dR}{dv_f} = \mp \frac{R}{3\beta c}, \quad \text{on} \quad C_\pm . \tag{4.5.54}$$

(iii) Weakly Nonlinear Waves

Denote

$$\xi = x - c_0 t . \tag{4.5.55}$$

We may apply a multiple-scale expansion scheme to the one-dimensional version of the system, treating the amplitude, the dissipation and dispersion terms as small quantities. If we neglect further the surface tension term, then we can obtain the following equation :

$$\frac{\partial v_f}{\partial t} + \left[\frac{3}{2} + \frac{1}{2}\frac{(1-\beta_0)}{\beta_0}\right] v_f \frac{\partial v_f}{\partial \xi}$$

$$= \frac{p_{go} R_0^2}{9\eta} \frac{\partial^2 v_f}{\partial \xi^2} - \frac{c_0 R_0^2}{6\beta_0} \frac{\partial^3 v_f}{\partial \xi^3} . \tag{4.5.56}$$

(4.5.56) is a wave equation of the Burgers-KdV type, which again reveals both the effect of dissipation due to mutual slippage and dispersion due to bubble oscillation.

CHAPTER 5

Stability

When physical problems are formulated in terms of partial differential equations, three mathematical questions are usually asked, i.e., existence, uniqueness and stability of the solutions. From the point of view of the physical scientist, the main goal of the study is to find the solution. Thus the question of existence is naturally resolved if a solution is found. The question of uniqueness is also extremely important for the scientist, because scientific study is largely concerned with predictions. If the solution is not unique, then no definite prediction can be made for the outcome based on known conditions. The question of stability relates to the practical predictability of the outcome. In real world situation, conditions can never be prescribed with perfect precision. Would small deviations from the prescribed conditions result in large deviation of the outcome? If the answer is yes, then the original primary state, which is the exact solution of the problem, is not stable. When the physical system is unstable, prediction is again not possible in practice.

Stability of the system implies that the deviation from the exact solution would be small if the deviation from the prescribed condition is small. Therefore a linearized analysis based on small perturbation about the exact solution state is sufficient to determine whether the primary state is stable. However, if the primary state is unstable, the development of the solution based on the linearized analysis usually will not be the real final outcome. A full nonlinear study is required to get the complete picture. Often the instability predicted by the linear theory may lead to unbounded growth of the physical quantities. The real outcome may be just another different regular state, or a bounded but chaotic or turbulent state, or the collapse of physical assumptions underlying the governing equation, hence the outcome may be a different physical system.

In the following, we shall discuss various stability problems in fluid flow. The emphasis is to understand the essential mechanisms of the instability, without going into technical details of variations and interplays of many different parameters.

5.1 Interfacial Conditions

Before we discuss any specific stability problems, let us first discuss the interfacial conditions between two fluids, since we need to use these interfacial conditions for the study of some stability problems.

The governing equations of the fluid motion can all be expressed in the general form of a transport equation :

$$\frac{\partial \psi}{\partial t} + \nabla \cdot \mathbf{f} = S \ , \tag{5.1.1}$$

where ψ is the density of the physical quantity in concern, $\mathbf{f}$ the flux of the physical quantity and S the source term, i.e. the rate at which the physical quantity is created per unit volume. Equation (5.1.1) is actually derived from the more general integral version :

$$\frac{d}{dt}\int_V \psi(\mathbf{x},t) d^3\mathbf{x} + \int_A \mathbf{f} \cdot \mathbf{n} d^2\mathbf{x} = \int_V S(\mathbf{x},t) d^3\mathbf{x} \ , \tag{5.1.2}$$

where A is the boundary of any region V, and $\mathbf{n}$ is the outward unit normal vector. Equation (5.1.1) can be derived from (5.1.2) by using Gauss theorem if the physical quantities are continuously differentiable functions. In connection with the governing equations, we can identify ψ, $\mathbf{f}$ and s as follows :

Equation of Continuity (1.1.1) :

$$\psi = \rho \ , \quad \mathbf{f} = \rho \mathbf{v} \ , \quad S = 0 \ . \tag{5.1.3}$$

Equation of Motion (1.1.3) :

$$\psi = \rho v_i \ , \quad f_j = \rho v_i v_j - \sigma_{ij} \ , \quad S = \rho b_i \ . \tag{5.1.4}$$

Equation of Energy (1.1.13) :

$$\psi = \rho U' + \frac{\rho v^2}{2} \ ,$$

$$f_i = (\rho U + \frac{\rho v^2}{2}) v_i - \sigma_{ij} v_j + q_i \ , \quad S = \rho Q \qquad (5.1.5)$$

where Q represents the heat source term, and $\mathbf{q}$ is the heat flux and $\mathbf{q} = -\kappa \nabla T$ if Fourier's law of heat conduction is adopted.

We shall now derive the general interfacial condition based on (5.1.2).

Let the interface between two fluids be defined by

$$F(\mathbf{x}, t) = 0 \ . \qquad (5.1.6)$$

Specifically, let fluid (1) occupy the region $F(\mathbf{x}, t) > 0$, and fluid (2) the region $F(\mathbf{x}, t) < 0$. Consider a small right cylinder situated astride the interface (See Fig. 5-1). The axis of the cylinder is normal to the interface. The height of the cylinder is l and the base area B. The cylinder is fixed in space. The part of the cylinder in fluid (1) has height l_1 . Now although the height l and base area B are constant since the cylinder is fixed, l_1 is changing since the interface is in general moving. We shall take the region inside the cylinder to be V when we apply (5.1.2). Then we shall let both l and B approach zero. Thus we obtain from (5.1.2)

$$\frac{d}{dt}\left[\psi^{(1)} l_1 B + \psi^{(2)} (l - l_1) B \right] + \left[\mathbf{f}^{(1)} \cdot \frac{\nabla F}{|\nabla F|} - \mathbf{f}^{(2)} \cdot \frac{\nabla F}{|\nabla F|} \right] B$$

$$= SlB + 0(l) \ , \qquad (5.1.7)$$

where we have included in the term $0(l)$ the flux across the side wall of the cylinder. The term SlB is also $0(l)$ if S refers only to the volumetric sources like those in (5.1.3), (5.1.4) and (5.1.5).

However, the source usually consists of two parts, the volumetric source S_V and the interfacial surface source S_S. Therefore, we have in fact

$$Sl = S_V l + S_S . \tag{5.1.8}$$

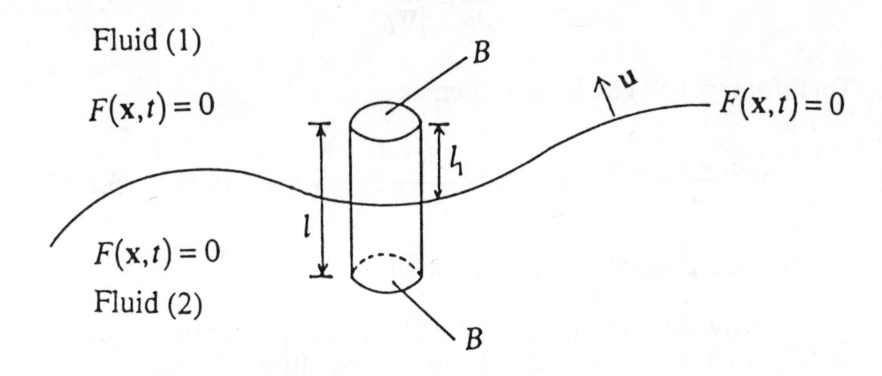

Fig. 5-1 The interface between fluids (1) and (2).

Substitute (5.1.8) into (5.1.7), and since l and B are constants, we obtain as $l \to 0$:

$$\left(\psi^{(1)} - \psi^{(2)}\right)\frac{dl_1}{dt} + \left(\mathbf{f}^{(1)} - \mathbf{f}^{(2)}\right)\cdot\frac{\nabla F}{|\nabla F|} = S_S . \tag{5.1.9}$$

As we may see from Fig. 5-1, on the interface $F(\mathbf{x},t) = 0$:

$$\frac{dl_1}{dt} = -\mathbf{n}_s \cdot \frac{d\mathbf{x}}{dt}$$

$$= -\frac{\nabla F}{|\nabla F|}\cdot\frac{d\mathbf{x}}{dt} . \tag{5.1.10}$$

Now

$$dF = \frac{\partial F}{\partial t} dt + \nabla F \cdot d\mathbf{x} = 0 , \quad \text{on } F(\mathbf{x},t) = 0 . \quad (5.1.11)$$

Combining (5.1.10) and (5.1.11), we have

$$\frac{dl_1}{dt} = \frac{\frac{\partial F}{\partial t}}{|\nabla F|} . \quad (5.1.12)$$

Therefore (5.1.9) can be rewritten as

$$\psi^{(1)} \frac{\partial F}{\partial t} + \mathbf{f}^{(1)} \cdot \nabla F = \psi^{(2)} \frac{\partial F}{\partial t} + \mathbf{f}^{(2)} \cdot \nabla F + S_S |\nabla F| , \quad (5.1.13)$$

which is the general interfacial condition.

Now let us apply (5.1.13) to the governing equations (5.1.3), (5.1.4) and (5.1.5). From the equation of continuity, we thus obtain the following interfacial condition :

$$\rho^{(1)} \left(\frac{\partial F}{\partial t} + \mathbf{v}^{(1)} \cdot \nabla F \right) = \rho^{(2)} \left(\frac{\partial F}{\partial t} + \mathbf{v}^{(2)} \cdot \nabla F \right) , \quad (5.1.14)$$

since $S_S = 0$ for this case. Let

$$w_n = \frac{\nabla F}{|\nabla F|} \cdot \frac{d\mathbf{x}}{dt} = -\frac{\frac{\partial F}{\partial t}}{|\nabla F|} . \quad (5.1.15)$$

w_n is the normal component of velocity of the interface $F(\mathbf{x},t) = 0$. Then (5.1.14) can be written as

$$\rho^{(1)} \left(v_n^{(1)} - w_n^{(1)} \right) = \rho^{(2)} \left(v_n^{(2)} - w_n^{(2)} \right) . \quad (5.1.16)$$

As $\left(v_n^{(1)} - w_n^{(1)} \right)$ and $\left(v_n^{(2)} - w_n^{(2)} \right)$ are respectively the normal components of the velocities of fluids (1) and (2) relative to the interface, the quantities on the left and right sides of (5.1.16) represent the masses of these two fluids flowing across the interface. Since there is no interfacial material source, they naturally should be equal; thus reflecting the conservation of mass. In this way, the physical meaning of (5.1.14) is very clear.

If fluid (1) and fluid (2) are two immiscible fluids, then there is no flow of mass across the interface, and both sides of (5.1.14) should be equal to zero. Hence

$$\frac{\partial F}{\partial t}+\mathbf{v}^{(1)}\cdot\nabla F=0\ , \tag{5.1.17}$$

$$\frac{\partial F}{\partial t}+\mathbf{v}^{(2)}\cdot\nabla F=0\ . \tag{5.1.18}$$

Equations (5.1.17) and (5.1.18) state that the normal velocity of the fluid particle on the interface relative to the interface is zero. In other words, fluid particles stay on the interface once they are on the interface. We have used this argument to derive the free surface kinematics condition when we discussed the problem of water waves. This is also the argument usually adopted in the literature. The present derivation seems to be physically more transparent.

Now let us apply (5.1.13) to the equation of motion (5.1.4). For this case we need to introduce an interfacial source term S_S due to the surface tension. Let the surface tension coefficient be σ, then it can be shown (See Appendix 2) that

$$S_S=-\mathbf{n}[\nabla\cdot(\sigma\mathbf{n})]+\nabla\sigma\ .$$

Thus we obtain another interfacial condition by applying (5.1.13) :

$$\rho^{(1)}v_i^{(1)}\left(\frac{\partial F}{\partial t}+v_j^{(1)}\frac{\partial F}{\partial x_j}\right)-\sigma_{ij}^{(1)}\frac{\partial F}{\partial x_j}=\rho^{(2)}v_i^{(2)}\left(\frac{\partial F}{\partial t}+v_j^{(2)}\frac{\partial F}{\partial x_j}\right)$$

$$-\sigma_{ij}^{(2)}\frac{\partial F}{\partial x_j}-\left[\nabla\cdot\left(\frac{\sigma\nabla F}{|\nabla F|}\right)\right]\frac{\partial F}{\partial x_i}+|\nabla F|\frac{\partial \sigma}{\partial x_i}. \qquad (5.1.19)$$

The surface tension coefficient σ will vary over the interface if the interface is not homogeneous due to impurity and contamination. However σ is usually taken to be constant for homogeneous interface. Now, according to differential geometry, we can write

$$\nabla\cdot\left(\frac{\nabla F}{|\nabla F|}\right)=\frac{1}{R_1}+\frac{1}{R_2},$$

where R_1 and R_2 are the principal radii of curvature of the interface, and positive value is taken if the center of curvature is in region (2). Then (5.1.19) becomes

$$\rho^{(1)}v_i^{(1)}\left(\frac{\partial F}{\partial t}+v_j^{(1)}\frac{\partial F}{\partial x_j}\right)-\sigma_{ij}^{(1)}\frac{\partial F}{\partial x_j}=\rho^{(2)}v_i^{(2)}\left(\frac{\partial F}{\partial t}+v_j^{(2)}\frac{\partial F}{\partial x_j}\right)$$

$$-\sigma_{ij}^{(2)}\frac{\partial F}{\partial x_j}-\sigma\left(\frac{1}{R_1}+\frac{1}{R_2}\right)\frac{\partial F}{\partial x_i}, \quad i=1,\ 2,\ 3\ . \qquad (5.1.20)$$

When the fluids (1) and (2) are immiscible, then using (5.1.17) and (5.1.18), (5.1.20) becomes

$$\sigma_{ij}^{(1)}\frac{\partial F}{\partial x_j}=\sigma_{ij}^{(2)}\frac{\partial F}{\partial x_j}+\sigma\left(\frac{1}{R_1}+\frac{1}{R_2}\right)\frac{\partial F}{\partial x_i}, \quad i=1,2,3\ . \qquad (5.1.21)$$

For the particular case of inviscid fluids, since

$$\sigma_{ij}^{(k)}=-p^{(k)}\delta_{ij}, \quad k=1,2, \qquad (5.1.22)$$

(5.1.21) becomes

$$p^{(2)} - p^{(1)} = \sigma\left(\frac{1}{R_1} + \frac{1}{R_2}\right), \tag{5.1.23}$$

which states that the pressure difference on the interface is balanced by the surface tension, and it has again been derived before when we discussed the water waves.

Finally let us apply (5.1.13) to the energy equation (5.1.5). Any increase of the interfacial energy will be a drain from the kinetic and internal energy. Thus with ρQ_s given by (A2.6), the interfacial energy condition is

$$\rho^{(1)}\left(U^{(1)} + \frac{v^{(1)^2}}{2}\right)\left(\frac{\partial F}{\partial t} + \mathbf{v}^{(1)} \cdot \nabla F\right) - \sigma_{ij}^{(1)} v_i^{(1)} \frac{\partial F}{\partial x_j} + \mathbf{q}^{(1)} \cdot \nabla F$$

$$= \rho^{(2)}\left(U^{(2)} + \frac{v^{(2)^2}}{2}\right)\left(\frac{\partial F}{\partial t} + \mathbf{v}^{(2)} \cdot \nabla F\right) - \sigma_{ij}^{(2)} v_i^{(2)} \frac{\partial F}{\partial x_j}$$

$$+ \mathbf{q}^{(2)} \cdot \nabla F - \sigma \frac{\partial F}{\partial t}\left(\frac{1}{R_1} + \frac{1}{R_2}\right). \tag{5.1.24}$$

Besides these interfacial conditions which are derivable from the fundamental conservation laws, there are other interfacial condition which may need to be imposed. We shall discuss them when we come to those situations.

5.2 The Classical Rayleigh-Taylor Stability

Consider two immiscible, incompressible and inviscid fluids separated by an interface. Let the interface be defined by

$$F(\mathbf{x},t) = y - \eta(x,z,t) = 0 . \tag{5.2.1}$$

We shall use superscripts (2) and (1) to denote the upper and lower fluids. The fluid (2) occupies the region $\eta < y < h^{(2)}$, while the fluid (1) occupies the region $-h^{(1)} < y < \eta$. The only external force is the gravitational force g in the direction of $(-y)$. The configuration is shown in Fig. 5-2.

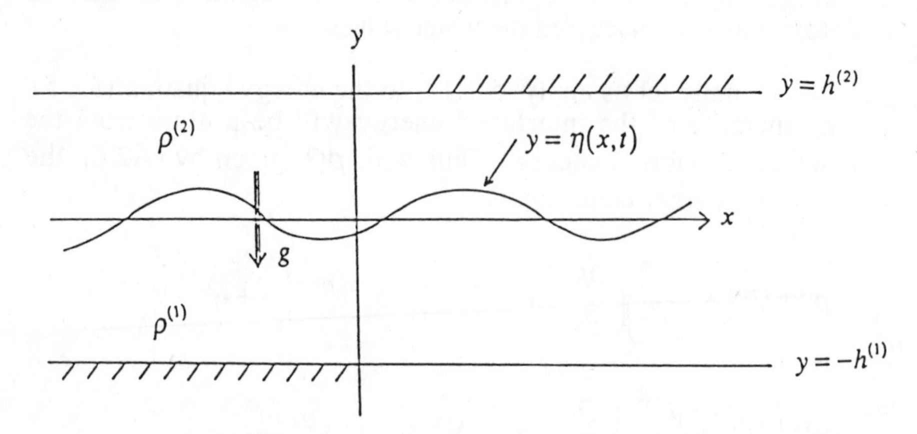

Fig. 5-2 Configuration of two superposed fluids in gravitational field.

For incompressible fluids, since the densities are constant, the continuity equation (1.1.1) becomes now :

$$\nabla \cdot \mathbf{v}^{(i)} = 0 , \quad i = 1,2 . \tag{5.2.2}$$

As the fluids are inviscid, let us introduce the velocity potentials :

$$\mathbf{v}^{(i)} = \nabla \varphi^{(i)} , \quad i = 1,2 . \tag{5.2.3}$$

Then (5.2.2) becomes

$$\nabla^2 \varphi^{(i)} = 0 , \quad i = 1,2 , \tag{5.2.4}$$

in their respective regions.

For potential flows, the equations of motion can be integrated to yield the Bernoulli equations :

$$\frac{p^{(i)}}{\rho^{(i)}}+\frac{1}{2}\left(\nabla\varphi^{(i)}\right)^2+gy+\frac{\partial\varphi^{(i)}}{\partial t}=f^{(i)}(t)\,,\qquad i=1,2\,. \quad (5.2.5)$$

On the interface $y=\eta(x,z,t)$, we have the interfacial conditions (5.1.17), (5.1.18) and (5.1.23) :

$$\frac{\partial\eta}{\partial t}-\frac{\partial\varphi^{(i)}}{\partial y}+\left(\nabla\varphi^{(i)}\right)\cdot(\nabla\eta)=0\,,\qquad i=1,2\,, \quad (5.2.6)$$

and

$$p^{(2)}-p^{(1)}=\sigma\left(\frac{1}{R_1}+\frac{1}{R_2}\right). \quad (5.2.7)$$

On the fixed walls $y=h^{(2)}$ and $y=-h^{(1)}$, the normal components of the fluid velocities are zero, thus

$$\frac{\partial\varphi^{(1)}}{\partial y}=0\,,\qquad y=-h^{(1)}\,, \quad (5.2.8)$$

and

$$\frac{\partial\varphi^{(2)}}{\partial y}=0\,,\qquad y=h^{(2)}\,. \quad (5.2.9)$$

Equations (5.2.4) - (5.2.9) are the basic equations for the stability problem under consideration.

It is evident that the following primary state satisfy the basic equations :

$$\eta_0=0\,,\quad \varphi_0^{(1)}=\varphi_0^{(2)}=0\,,\quad f^{(1)}=f^{(2)}=0\,,$$

$$p_0^{(1)}=-\rho^{(1)}gy\,,\quad p_0^{(2)}=-\rho^{(2)}gy\,. \quad (5.2.10)$$

This primary state represents an equilibrium state with a plane interface.

To analyze the stability of the system, a small perturbation is introduced to the primary state. Let

$$\eta = \eta_0 + \eta_1 , \quad \varphi^{(i)} = \varphi_0^{(i)} + \varphi_1^{(i)} , \quad p^{(i)} = p_0^{(i)} + p_1^{(i)} ,$$

$$i = 1,\ 2 . \tag{5.2.11}$$

Substitute (5.2.11) into (5.2.4) - (5.2.9) and retain only terms up to first order of the perturbed quantities. We thus obtain

$$\nabla^2 \varphi_1^{(1)} = 0 , \qquad -h^{(1)} < y < 0 , \tag{5.2.12}$$

$$\nabla^2 \varphi_1^{(2)} = 0 , \qquad 0 < y < h^{(2)} , \tag{5.2.13}$$

$$\frac{\partial \eta_1}{\partial t} - \frac{\partial \varphi_1^{(1)}}{\partial y} = 0 , \qquad on\ y = 0 , \tag{5.2.14}$$

$$\frac{\partial \eta_1}{\partial t} - \frac{\partial \varphi_1^{(2)}}{\partial y} = 0 , \qquad on\ y = 0 , \tag{5.2.15}$$

$$-\rho^{(2)}\left(g\eta_1 + \frac{\partial \varphi_1^{(2)}}{\partial t} \right) + \rho^{(1)}\left(g\eta_1 + \frac{\partial \varphi_1^{(1)}}{\partial t} \right)$$

$$= \sigma\left(\frac{\partial^2 \eta_1}{\partial x^2} + \frac{\partial^2 \eta_1}{\partial z^2} \right), \qquad on\ y = 0 , \tag{5.2.16}$$

$$\frac{\partial \varphi_1^{(1)}}{\partial y} = 0 , \qquad on\ y = -h^{(1)} , \tag{5.2.17}$$

$$\frac{\partial \varphi_1^{(2)}}{\partial y} = 0 , \qquad on\ y = h^{(2)} . \tag{5.2.18}$$

In deriving (5.2.16), we have made use of (5.2.5) and (5.2.7) and also the fact that to the first order of η_1, $\left(\frac{1}{R_1}+\frac{1}{R_2}\right)=\left(\frac{\partial^2\eta_1}{\partial x^2}+\frac{\partial^2\eta_1}{\partial z^2}\right)$.

Since the system (5.2.12) - (5.2.18) is linear, there is no loss of generality to consider just a single Fourier mode. Let

$$\eta_1 = Ae^{i(k_x x+k_z z)+nt}, \qquad (5.2.19)$$

$$\varphi_1^{(1)} = \Phi^{(1)}(y)e^{i(k_x x+k_z z)+nt}, \qquad (5.2.20)$$

$$\varphi_1^{(2)} = \Phi^{(2)}(y)e^{i(k_x x+k_z z)+nt}. \qquad (5.2.21)$$

Substitute (5.2.20) and (5.2.21) into (5.2.12) and (5.2.13) and using the boundary conditions (5.2.17) and (5.2.18) respectively, we obtain :

$$\varphi_1^{(1)} = B\cosh k(y+h^{(1)})e^{i(k_x x+k_z z)+nt}, \qquad (5.2.22)$$

$$\varphi_1^{(2)} = C\cosh k(y-h^{(2)})e^{i(k_x x+k_z z)+nt}. \qquad (5.2.23)$$

Substitute (5.2.19), (5.2.22) and (5.2.23) into (5.2.14), (5.2.15) and (5.2.16) respectively, we then obtain :

$$nA - kB\sinh kh^{(1)} = 0, \qquad (5.2.24)$$

$$nA + kC\sinh kh^{(2)} = 0, \qquad (5.2.25)$$

$$n\left(\rho^{(1)}B\cosh kh^{(1)} - \rho^{(2)}C\cosh kh^{(2)}\right)$$

$$+\left[g\left(\rho^{(1)}-\rho^{(2)}\right)+\sigma k^2\right]A = 0, \qquad (5.2.26)$$

where

$$k^2 = k_x^2 + k_z^2. \qquad (5.2.27)$$

Eliminate B and C from (5.2.24) - (5.2.26), we obtain

$$\frac{n^2}{k}\left(\rho^{(1)}\coth kh^{(1)}+\rho^{(2)}\coth kh^{(2)}\right)$$

$$+\left[g\left(\rho^{(1)}-\rho^{(2)}\right)+\sigma k^2\right]=0\ . \qquad (5.2.28)$$

Or

$$n^2=-\frac{\left(\rho^{(1)}-\rho^{(2)}\right)gk+\sigma k^3}{\rho^{(1)}\coth kh^{(1)}+\rho^{(2)}\coth kh^{(2)}}\ . \qquad (5.2.29)$$

If n^2 is positive, then there is a positive real n as the solution. Thus the perturbed quantities will grow exponentially with t, hence the primary state is unstable. On the other hand, if n^2 is negative, then n will be purely imaginary. Thus the perturbed quantities will remain small if their initial values are small. The primary state is thus stable.

It is clear that if $\rho^{(1)}>\rho^{(2)}$, i.e. if the heavy fluid is underneath the lighter fluid, then n^2 is always negative. Thus the equilibrium state is stable. On the other hand if $\rho^{(1)}<\rho^{(2)}$, i.e. if the heavy fluid is on top of the lighter fluid, then a critical wave number k_c can be defined :

$$k_c=\left[\frac{g}{\sigma}\left(\rho^{(2)}-\rho^{(1)}\right)\right]^{\frac{1}{2}}\ . \qquad (5.2.30)$$

The equilibrium is stable if $k>k_c$, and unstable if $k<k_c$. We can appreciate from these analyses that the surface tension plays a stabilizing role.

This type of stability is known as Rayleigh-Taylor stability. The essence of Rayleigh-Taylor stability is that the equilibrium state is stable if the heavy fluid is underneath the lighter fluid, which is consistent with our daily life experience.

Take the example of air and water. We have $\rho^{(1)} \approx 10^{-3}$, $\rho^{(2)} \approx 1$ and $\sigma \approx 73$ in cgs units. Then $k_c \approx 3.7\ \text{cm}^{-1}$, or the critical wavelength λ_c is

$$\lambda_c = \frac{2\pi}{k_c} \approx 1.7\ \text{cm}\ .$$

Thus for a small tube with diameter less than $\frac{\lambda_c}{2} \approx 0.85$ cm , the air pressure can hold up the water in the tube, even though the opening of the tube is pointing downward.

5.3 Rayleigh-Taylor Stability with Mass and Heat Transfer Across the Interface

There are many generalizations and extensions of the classical Rayleigh-Taylor stability problem, e.g., the inclusion of viscosity in fluids, and the influence of magnetic fields on conducting fluids. We shall discuss here the case when there is mass and heat transfer across the interface. The problem is relevant, for instance, to the study of boiling heat transfer. Thus we need to deal with thermally conducting fluids. Moreover, the superposed fluids are no longer immiscible. They can transform from one to the other. Most typically, the superposed fluids are a liquid and its vapor phase. The motion of interface is thus usually accompanied by some phase transformation at the interface with the simultaneous release or absorption of latent heat of the phase transformation.

Consider now two incompressible, inviscid and thermally conducting fluids separated by the interface given by (5.2.1). We are treating even the vapor as incompressible since the Mach number is expected to be small. The configuration of the system is again as shown in Fig. 5-2. Equations (5.2.4), (5.2.5), (5.2.8) and (5.2.9) are again valid. However since there is mass transfer across the interface, the two interfacial conditions (5.2.6) should be replaced by the single condition (5.1.14); or

$$\rho^{(1)}\left[\frac{\partial\eta}{\partial t}-\frac{\partial\varphi^{(1)}}{\partial y}+\left(\nabla\varphi^{(1)}\right)\cdot\left(\nabla\eta\right)\right]$$

$$=\rho^{(2)}\left[\frac{\partial\eta}{\partial t}-\frac{\partial\varphi^{(2)}}{\partial y}+\left(\nabla\varphi^{(2)}\right)\cdot\left(\nabla\eta\right)\right]. \qquad (5.3.1)$$

Also the interfacial condition (5.2.7), modified according to (5.1.20), becomes now

$$\rho^{(1)}v_n^{(1)}\left(v_n^{(1)}-w_n\right)+p^{(1)}$$

$$=\rho^{(2)}v_n^{(2)}\left(v_n^{(2)}-w_n\right)+p^{(2)}-\sigma\left(\frac{1}{R_1}+\frac{1}{R_2}\right). \qquad (5.3.2)$$

Since the fluids are thermally conducting, we have the energy equations (1.1.16) for both fluids, i.e.,

$$\frac{\partial T^{(i)}}{\partial t}+\left(\nabla\varphi^{(i)}\right)\cdot\left(\nabla T^{(i)}\right)=D_T^{(i)}\nabla^2T^{(i)}. \quad i=1,2. \quad (5.3.3)$$

With the introduction of (5.3.3), we need additional boundary and interfacial conditions. At the walls $y=-h^{(1)}$ and $y=h^{(2)}$, we may prescribe either temperatures or heat fluxes. We shall consider the case that constant temperatures are maintained on both walls, thus

$$T^{(1)}=T_I, \qquad \text{on } y=-h^{(1)}, \qquad (5.3.4)$$

$$T^{(2)}=T_{II}, \qquad \text{on } y=h^{(2)}. \qquad (5.3.5)$$

On the interface, we assume that there is local thermal equilibrium, thus

$$T^{(1)} = T^{(2)}\,, \quad on\; y = \eta(x,z,t)\,. \tag{5.3.6}$$

Moreover, we assume that there is also local phase equilibrium, thus we have

$$p^{(2)} = p_v\left(T^{(2)}\right), \quad on\; y = \eta(x,z,t)\,, \tag{5.3.7}$$

where p_v is the equilibrium vapor pressure which could be given for instance, by the Clausius-Clapeyron equation. Here we have also assigned the fluid (2) to be the vapor. Referring to the configuration shown in 5-2, we need only to take the sign of g negative to consider the case that vapor is underneath liquid in the gravitation field.

Let the heat fluxes be given by the Fourier's law, i.e. $\mathbf{q} = -\kappa \nabla T$. Then the interfacial condition (5.1.24) can be rewritten now as :

$$\rho^{(1)}\left(v_n^{(1)} - w_n\right)\left[\frac{1}{2}\left(\mathbf{v}^{(1)} - \mathbf{w}\right)^2 - \frac{1}{2}\left(\mathbf{v}^{(2)} - \mathbf{w}\right)^2 - L\right]$$

$$-\kappa^{(1)}\frac{\partial T^{(1)}}{\partial n} = -\kappa^{(2)}\frac{\partial T^{(2)}}{\partial n}\,, \tag{5.3.8}$$

where we have written

$$\mathbf{w} = -\frac{\dfrac{\partial F}{\partial t}\nabla F}{|\nabla F|^2}\,, \tag{5.3.9}$$

so that w_n is given by (5.1.15), and

$$L = U^{(2)} + \frac{p^{(2)}}{\rho^{(2)}} - U^{(1)} - \frac{p^{(1)}}{\rho^{(1)}} \tag{5.3.10}$$

is the latent heat of the phase transformation.

Equations (5.2.4), (5.2.5), (5.2.8), (5.2.9) and (5.3.1) - (5.3.8) are basic equations for this stability problem which takes into consideration of the heat and mass transfer across the interface.

Again let us start with the equilibrium state which satisfies the above basic equations exactly :

$$\eta_0 = 0\ , \quad \varphi_0^{(1)} = \varphi_0^{(2)} = 0\ , \quad f^{(1)} = f^{(2)} = 0\ ,$$

$$p_0^{(1)} = -\rho^{(1)} g y + p_0\ , \quad p_0^{(2)} = -\rho^{(2)} g y + p_0\ ,$$

$$T_0^{(1)} = G_1 y + T_b\ , \quad T_0^{(2)} = G_2 y + T_b\ , \qquad (5.3.11)$$

$$G_1 = \frac{T_b - T_I}{h^{(1)}}\ , \quad G_2 = \frac{T_{II} - T_b}{h^{(2)}}\ , \quad \kappa_1 G_1 = \kappa_2 G_2\ ,$$

$$p_0 = p_v(T_b)\ .$$

Let us again discuss the stability of the system with respect to small perturbations. From the analysis of the classical Rayleigh-Taylor stability, it is clear that a two dimensional analysis is sufficient to bring out the essential features of the problem. Let

$$\eta_1 = \zeta e^{ikx+nt}\ , \qquad (5.3.12)$$

and let other quantities all have the same *(x, t)* variation, then just like (5.2.22) and (5.2.23), we obtain, without writing the factor e^{ikx+nt} from now on, that

$$\varphi_1^{(1)} = A_1 \cosh k\left(y + h^{(1)}\right)\ , \qquad (5.3.13)$$

$$\varphi_1^{(2)} = A_2 \cosh k\left(y - h^{(2)}\right)\ . \qquad (5.3.14)$$

From (5.3.3) and (5.3.11), we have for the perturbed temperature field in fluid (1) :

$$\frac{d^2T_1^{(1)}}{dy^2} - \left(k^2 + \frac{n}{D_T^{(1)}}\right)T_1^{(1)} = A_1 \frac{G_1 k}{D_T^{(1)}} \sinh k\left(y + h^{(1)}\right) . \quad (5.3.15)$$

Using the boundary condition (5.3.4), we thus obtain

$$T_1^{(1)} = B_1 \sinh\left[\left(k^2 + \frac{n}{D_T^{(1)}}\right)^{1/2} \left(y + h^{(1)}\right)\right]$$

$$- \frac{G_1 k}{n} A_1 \sinh k\left(y + h^{(1)}\right) . \quad (5.3.16)$$

Similarly, we have for fluid (2) :

$$T_1^{(2)} = B_2 \sinh\left[\left(k^2 + \frac{n}{D_T^{(2)}}\right)^{1/2} \left(y - h^{(2)}\right)\right]$$

$$- \frac{G_2 k}{n} A_2 \sinh k\left(y - h^{(2)}\right) . \quad (5.3.17)$$

From (5.2.5), we have, on the interface, that

$$p_1^{(1)} = -n\rho^{(1)} A_1 \cosh kh^{(1)} - \rho^{(1)} g \zeta , \quad (5.3.18)$$

$$p_1^{(2)} = -n\rho^{(2)} A_2 \cosh kh^{(2)} - \rho^{(2)} g \zeta . \quad (5.3.19)$$

Thus the interfacial conditions (5.3.1), (5.3.2), (5.3.6), (5.3.7) and (5.3.8) for the small perturbations become respectively:

$$\rho^{(1)}\left(n\zeta - A_1 k \sinh kh^{(1)}\right) = \rho^{(2)}\left(n\zeta + A_2 k \sinh kh^{(2)}\right) , \quad (5.3.20)$$

$$\rho^{(1)}\left(nA_1 \cosh kh^{(1)} + g\zeta\right)$$

$$= \rho^{(2)}\left(nA_2 \cosh kh^{(2)} + g\zeta\right) - \sigma k^2 \zeta , \quad (5.3.21)$$

$$B_1 \sinh l^{(1)} h^{(1)} - \frac{G_1 k}{n} A_1 \sinh k h^{(1)} + G_1 \zeta$$

$$= -B_2 \sinh l^{(2)} h^{(2)} + \frac{G_2 k}{n} A_2 \sinh k h^{(2)} + G_2 \zeta, \quad (5.3.22)$$

$$-n\rho^{(2)} A_2 \cosh k h^{(2)} - \rho^{(2)} g \zeta$$

$$= \beta(-B_2 \sinh l^{(2)} h^{(2)} + \frac{G_2 k}{n} A_2 \sinh k h^{(2)} + G_2 \zeta), \quad (5.3.23)$$

$$L\rho^{(1)}\left(A_1 k \sinh k h^{(1)} - n\zeta\right) - \kappa^{(1)}\left(B_1 l^{(1)} \cosh l^{(1)} h^{(1)} - \frac{G_1 k^2}{n} A_1 \cosh k h^{(1)}\right)$$

$$= -\kappa^{(2)}\left(B_2 l^{(2)} \cosh l^{(2)} h^{(2)} - \frac{G_2 k^2}{n} A_2 \cosh k h^{(2)}\right), \quad (5.3.24)$$

where

$$l^{(i)} = \left(k^2 + \frac{n}{D_T^{(i)}}\right)^{1/2}, \quad (5.3.25)$$

and

$$\beta = \left(\frac{dp_v}{dT}\right)_{T=T_b}. \quad (5.3.26)$$

Eliminating A_1, A_2, B_1, B_2 and ζ from (5.3.20) - (5.3.24), we obtain the following dispersion relation :

$$\left[L + \frac{G_1 \kappa^{(1)}}{n}\left(\frac{l^{(1)}}{\rho^{(1)}} \coth l^{(1)} h^{(1)} + \frac{l^{(2)}}{\rho^{(2)}} \coth l^{(2)} h^{(2)}\right)\right] F_1$$

$$= F_2\left[\frac{G_1\kappa^{(1)}k\left(\rho^{(1)}-\rho^{(2)}\right)}{n\rho^{(1)}\rho^{(2)}} - \frac{n\left(\kappa^{(1)}l^{(1)}\coth l^{(1)}h^{(1)} + \kappa^{(2)}l^{(2)}\coth l^{(2)}h^{(2)}\right)}{\beta k}\right], \tag{5.3.27}$$

where

$$F_1 = n^2\left(\rho^{(1)}\cosh kh^{(1)}\sinh kh^{(2)} + \rho^{(2)}\cosh kh^{(2)}\sinh kh^{(1)}\right)$$
$$+\left[g\left(\rho^{(1)}-\rho^{(2)}\right)+\sigma k^2\right]k\sinh kh^{(1)}\sinh kh^{(2)}\ , \tag{5.3.28}$$

$$F_2 = n^2\left(\rho^{(1)}-\rho^{(2)}\right)\cosh kh^{(1)}\cosh kh^{(2)}$$
$$+k\left[\left(g\rho^{(1)}+\sigma k^2\right)\sinh kh^{(1)}\cosh kh^{(2)} + g\rho^{(2)}\sinh kh^{(2)}\cosh kh^{(1)}\right]. \tag{5.3.29}$$

The dispersion relation (5.3.27) is a complicated expression involving many parameters. We shall not go into the details of its analysis. For many physical systems the latent heat L is very large. Take water, in cgs units, the latent heat of vaporization ranges from 2.5×10^{10} to 2.257×10^{10} between 0°C and 100°C. For comparison, $\kappa^{(1)} \approx 6 \times 10^4$. Thus, to a first approximation, we obtain from (5.3.27) that

$$F_1 = 0\ , \tag{5.3.30}$$

which is the same as the dispersion relation for the classical case (5.2.28). The effect of heat and mass transfer will just serve as some corrections. Without going into details, we can distinguish two sources of the correction, one due to the temperature gradient G_1 and the other due to the effect of changing vapor pressure β.

Consider first the effect of changing vapor pressure. The pressure at the crest of the disturbance wave, i.e., the side protruding against the gravity, tends to be smaller than the equilibrium value. If the vapor is on top of the liquid, then the pressure of the vapor at the crest tends to fall below the equilibrium vapor pressure, and evaporation will take place and diminish the amplitude of the disturbance wave. If the liquid is on top of the vapor, then the additional evaporation will increase the amplitude of the disturbance wave. Therefore the effect of changing vapor pressure tends to damp the stable interfacial waves and enhance the growth rate when the system is unstable.

The effect of temperature gradient is independent of whether the system is stable or not. The system tends to be more stable or damped if the vapor is hotter than the liquid, because the protruding of the liquid to a hotter region will result in evaporation and thus diminishing the protrusion. If the liquid is hotter than the vapor, then the system tends to be destabilized. The growth rate of the instability can be greatly reduced if the thermal gradient is large, although the criterion of instability is still governed by (5.3.30) according to the linear analysis (Hsieh 1972).

Linear stability of viscous fluids with mass and heat transfer has also been carried out (Ho 1980).

A Simplified Version

Although the explicit dispersion relation (5.3.27) has been found for the Rayleigh-Taylor stability with mass and heat transfer, the expression is very complicated and it is difficult to grasp its essential feature. Moreover, for certain class of problems of application, e.g. the problem of boiling heat transfer, a nonlinear analysis may be required to really understand the physical mechanisms. Therefore it is desirable that a simplified version of the problem, which incorporates the essential effects of mass and heat transfer, can be established and explored first.

The simplified version to be discussed is based on the assumption that the amount of released latent heat depends mainly on the instantaneous position of the interface. More specifically we impose an interfacial condition for energy transfer as follows :

$$-L\rho^{(1)}\left(v_n^{(1)} - w_n\right) = H(\eta) , \tag{5.3.31}$$

where

$$H(y) = \frac{\kappa^{(2)}(T_b - T_{II})}{h^{(2)} - y} - \frac{\kappa^{(1)}(T_I - T_b)}{h^{(1)} + y} . \qquad (5.3.32)$$

Comparing (5.3.31) with (5.3.8), we see the simplification results from two aspects : the neglect of kinetic energy terms which are usually small in comparison with the latent heat, and the replacement of heat flux terms by their pseudo-equilibrium counterparts. However, if we adopt this simplification assumption, then we can forego the complication of the coupled energy equation. Specifically, we no longer need to consider equations (5.3.3) - (5.3.8). The basic system now consists of (5.2.4), (5.2.5), (5.2.8), (5.2.9), (5.3.1), (5.3.2) and (5.3.31).

The linear stability analysis can be readily carried out. The perturbed quantities are again (5.3.12) - (5.3.14). The interfacial conditions (5.3.1) and (5.3.2) lead again to (5.3.20) and (5.3.21). To apply the interfacial condition (5.3.31), we note that in the neighborhood of $\eta = 0$, we have

$$H(\eta) = H'(0)\eta + \frac{1}{2}H''(0)\eta^2 + \cdots ,$$

since $H(0) = 0$ from the equilibrium condition. From (5.3.32), we obtain

$$H'(0) = H_0\left(\frac{1}{h^{(2)}} + \frac{1}{h^{(1)}}\right) , \qquad (5.3.33)$$

and

$$H_0 = \frac{\kappa^{(2)}(T_b - T_{II})}{h^{(2)}} = \frac{\kappa^{(1)}(T_I - T_b)}{h^{(1)}} , \qquad (5.3.34)$$

is the equilibrium heat flux from the plane $y = -h^{(1)}$ to $y = h^{(2)}$. Therefore, the linearized version of the interfacial condition (5.3.31) becomes

$$\rho^{(1)}\left(A_1 k \sinh kh^{(1)} - n\zeta\right) = \alpha\zeta \ , \tag{5.3.35}$$

where

$$\alpha = -\frac{H_0}{L}\left(\frac{1}{h^{(2)}} + \frac{1}{h^{(1)}}\right) . \tag{5.3.36}$$

Consider the liquid-vapor system. Usually the vapor phase is hotter than the liquid phase; then α is always positive. Because if fluid (2) is vapor and fluid (1) is liquid, then L is positive while H_0 is negative; if fluid (1) is vapor and fluid (2) is liquid, then L is negative while H_0 is positive.

Eliminating ζ, A_1 and A_2 from (5.3.20), (5.3.21) and (5.3.35), we thus obtain the dispersion relation :

$$n^2 + 2an + b = 0 \ , \tag{5.3.37}$$

where

$$a = \frac{\alpha}{2c} \sinh k\left(h^{(1)} + h^{(2)}\right) , \tag{5.3.38}$$

$$b = \frac{1}{c}\left[gk\left(\rho^{(1)} - \rho^{(2)}\right) + \sigma k^3\right] \sinh kh^{(1)} \sinh kh^{(2)} \ , \tag{5.3.39}$$

$$c = \rho^{(1)} \cosh kh^{(1)} \sinh kh^{(2)} + \rho^{(2)} \cosh kh^{(2)} \sinh kh^{(1)} \ . \tag{5.3.40}$$

If $\alpha = 0$, i.e., when there is no heat transfer, then we recover the classical Rayleigh-Taylor dispersion relation. From (5.3.37), we thus obtain

$$n = -a \pm \left(a^2 - b\right)^{1/2} . \tag{5.3.41}$$

If $b < 0$, i.e., the classically unstable case, then the system is also unstable, since there is a positive real root for n. However the growth rate is diminished from the classical value for this case that the vapor is hotter than the liquid. If $b > 0$, i.e., the classically stable case, then the system is also stable. But, in contrast with the classical case, there is no permanent periodic wave state, and the system will settle down to an asymptotic equilibrium because of the evaporation effect.

If the liquid is hotter than the vapor, then α and a will be negative, the evaporation effect will then tend to destablize the system. These findings are consistent with those found from the previous analysis of the general case.

Nonlinear Stability Based on Simplified Version

From the previous linear analysis, it is found that when the vapor region is hotter than the liquid region, the effect of mass and heat transfer tends to inhibit the growth of the instability. However, the stability criterion remains the same as that of the classical theory. The question we may ask now is that whether nonlinear effect would change the stability criterion. We shall consider the simplified version first, and make use of the multi-scale expansion to analyze the situation near the critical state. We shall outline only the main ideas, and refer the details to original works (Hsieh, 1978, 1979).

Let us begin with the dispersion relation (5.3.37). The critical state is defined by $n = 0$, or $b = 0$, i.e., when the disturbed wave neither grows nor decays. Thus the critical wave number is

$$k_c = \left[\frac{g}{\sigma}\left(\rho^{(2)} - \rho^{(1)}\right)\right]^{1/2} . \tag{5.3.42}$$

Notice that we are now considering the case that the heavier fluid is on top of the lighter fluid, or the liquid is on top of the vapor in the liquid-vapor system, so $\rho^{(2)}$ is larger than $\rho^{(1)}$.

Introducing a small parameter ε , we assume the following expansion of variables:

$$\eta = \sum_{n=1}^{3} \varepsilon^n \eta_n(x, t_0, t_1, t_2) + 0(\varepsilon^4) , \qquad (5.3.43)$$

$$\varphi^{(i)} = \sum_{n=1}^{3} \varepsilon^n \varphi_n^{(i)}(x, y, t_0, t_1, t_2) + 0(\varepsilon^4) , \quad i = 1,2 , \quad (5.3.44)$$

where

$$t_n = \varepsilon^n t . \qquad (5.3.45)$$

Substitute (5.3.43) and (5.3.44) into (5.2.4), (5.2.5), (5.2.8), (5.2.9), (5.3.1), (5.3.2) and (5.3.31), and equating like powers of ε, we can obtain for n = 1, 2, 3, three sets of equations for η_n and $\varphi_n^{(i)}$.

To solve these equations in the neighborhood of the linear critical wave number k_c, we assume that the critical wave number, because of the nonlinear effect, will shift to

$$k = k_c + \varepsilon^2 \mu . \qquad (5.3.46)$$

The first order equations and solutions are just what we obtained from the previous linear analysis. Thus for the critical case we can write :

$$\eta_1 = A(t_1, t_2) e^{ikx} + A^*(t_1, t_2) e^{-ikx} , \qquad (5.3.47)$$

$$\varphi_1^{(1)} = \frac{\alpha}{\rho^{(1)} k} \frac{\cosh k\left(y + h^{(1)}\right)}{\sinh k h^{(1)}} (A e^{ikx} + A^* e^{-ikx}) , \quad (5.3.48)$$

$$\varphi_1^{(2)} = -\frac{\alpha}{\rho^{(2)}k}\frac{\cosh k\left(y-h^{(2)}\right)}{\sinh kh^{(2)}}(Ae^{ikx}+A^*e^{-ikx}) , \quad (5.3.49)$$

where the asterisk denotes the complex conjugate quantities.

In order to find $A(t_1,t_2)$, we need to go to higher order equations. By making use of the condition to insure the uniformity of expansion, it is found from the second order analysis that

$$\frac{\partial A}{\partial t_1} = 0 . \quad (5.3.50)$$

A third order analysis is needed to find $A(t_2)$. After carrying out a tedious but straightforward analysis, it can be shown that

$$\frac{dA}{dt_2} + \left(a_1 + a_2 A^2\right)A = 0 , \quad (5.3.51)$$

where

$$a_1 = 2\sigma\mu k k_c \tau , \quad (5.3.52)$$

$$a_2 = \left(\upsilon - \frac{3}{2}\sigma k_c^4\right)k\tau , \quad (5.3.53)$$

$$\tau = \left[\alpha\left(\coth kh^{(1)} + \coth kh^{(2)}\right)\right]^{-1} , \quad (5.3.54)$$

where υ is a parameter depending on α and other parameters.

Denoting $A(0)$ by A_0, we obtain

$$A^2(t) = a_1 A_0^2 \exp(-2a_1 t)\left[a_1 + a_2 A_0^2 - a_2 A_0^2 \exp(-2a_1 t)\right]^{-1} . \quad (5.3.55)$$

With a finite initial value A_0, A may become infinite when the denominator in (5.3.55) vanishes. Otherwise, A will be asymptotically bounded. The situation can be summarized as follows :

(I) $a_2 > 0$: stable.

(i) $a_1 > 0$: $A^2 \to 0$, as $t \to \infty$.

(ii) $a_1 < 0$ $A^2 \to \left(-\frac{a_1}{a_2}\right)$, as $t \to \infty$.

(II) $a_2 < 0$:

(i) $a_1 < 0$: unstable.

(ii) $a_1 > 0$, and $A_0^2 > \left(-\frac{a_1}{a_2}\right)$: unstable.

(iii) $a_1 > 0$, and $A_0^2 < \left(-\frac{a_1}{a_2}\right)$: stable and

$A^2 \to 0$, as $t \to \infty$.

When $a_2 = 0$, the system is stable if $a_1 > 0$ and vice versa. This is the result which can be obtained from the linear stability analysis. As k_c is the wave number for the critical state, $a_1 > 0$, i.e. $\mu > 0$, implies $k > k_c$, and this indeed is the stable wave number regime. With nonlinear effects, the controlling factor shifts from a_1 to a_2 , and a sufficient condition for stability is $a_2 > 0$. When $a_2 > 0$, even if $a_1 < 0$, which corresponds to the linearly unstable regime, the system can be stabilized. On the other hand, if $a_2 < 0$, even though $a_1 > 0$, the system can be unstable if the initial amplitude is large enough.

The stability condition $a_2 > 0$ is equivalent to

$$\upsilon > \frac{3}{2}\sigma k_c^4 . \tag{5.3.56}$$

Roughly speaking, the condition (5.3.56) amounts to requiring that a non-dimensional parameter δ , defined as

$$\delta = \frac{\alpha^2}{\rho^{(1)}\sigma k_c^3} , \tag{5.3.57}$$

is sufficiently large. Since α is proportional to the heat flux of the system, when the heat flux is sufficiently intense, the system can be stabilized. Moreover, with the same heat flux, the thinner the fluid layers, the easier the system can be stabilized.

A nonlinear analysis has also been carried out for the general case without using the simplied quasi-equilibrium assumption (Hsieh and Ho, 1981). The previous analysis based on the simplified version can capture some of the important facets of the problem.

5.4 Stability of Spherical Bubble in Motion

The interfacial stability analysis of the Rayleigh-Taylor type can also be applied to non-planar interface. One such case is the stability of a spherical bubble in motion. Let us present first the simplest case of bubble in motion, then we shall study its stability.

A general bubble is defined by a closed interface $F(\mathbf{x},t)=0$ which separates two fluids. For bubbles, the fluid in the interior is the lighter fluid usually in gaseous state while the fluid outside is usually the liquid. The simplest case is the study of the spherically symmetric motion of a spherical bubble in a liquid of infinite extent. Then the interface, i.e., the bubble wall is given in terms of the spherical polar coordinates by :

$$F(\mathbf{x},t)=r-R(t)=0\ . \qquad (5.4.1)$$

As shown in Fig. 5-3, outside the bubble, i.e., for $r>R(t)$, is the fluid with denisty ρ and pressure p. Inside the bubble, i.e., for $r<R(t)$, is the fluid with density ρ' and pressure p'. The pressure at infinity will be denoted by p_∞.

Let us take the fluid outside to be incompressible and inviscid, and denote the velocity potential by φ. Then the continuity equation and the Bernoulli's equation lead to :

$$\nabla^2\varphi=0\ , \quad \text{for } r>R(t)\ , \qquad (5.4.2)$$

and

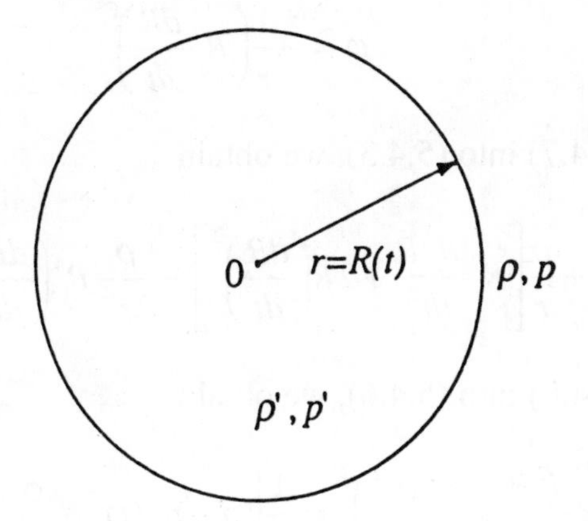

Fig. 5-3 The spherical bubble with radius R.

$$\frac{p}{\rho}+\frac{1}{2}(\nabla\varphi)^2+\frac{\partial\varphi}{\partial t}=f(t)\ , \quad \text{for} \quad r>R(t)\ . \qquad (5.4.3)$$

If the fluids inside and outside the bubble are immiscible, then the interfacial conditions (5.1.16) and (5.1.23) lead, for this spherically symmetric case, to :

$$\frac{\partial\varphi}{\partial r}=\frac{dR}{dt}\ , \quad \text{on} \quad r=R(t)\ , \qquad (5.4.4)$$

$$p'-p=\frac{2\sigma}{R}\ , \quad \text{on} \quad r=R(t)\ . \qquad (5.4.5)$$

For this spherically symmetric case, the solution of (5.4.2) is simply

$$\varphi=\frac{A}{r}+B\ . \qquad (5.4.6)$$

The term B can be absorbed in the pressure term. Thus, using (5.4.4), we obtain

$$\varphi = -\frac{1}{r}\left(R^2 \frac{dR}{dt}\right). \tag{5.4.7}$$

Substitute (5.4.7) into (5.4.3), we obtain

$$p = p_\infty + \frac{\rho}{r}\left[R^2 \frac{d^2R}{dt^2} + 2R\left(\frac{dR}{dt}\right)^2\right] - \frac{\rho}{2r^4} R^4 \left(\frac{dR}{dt}\right)^2. \tag{5.4.8}$$

Substitute (5.4.8) into (5.4.6), we obtain

$$R\frac{d^2R}{dt^2} + \frac{3}{2}\left(\frac{dR}{dt}\right)^2 = \frac{1}{\rho}\left[p' - p_\infty(t) - \frac{2\sigma}{R}\right], \tag{5.4.9}$$

where $p' = p'(R,t)$ is the pressure at the bubble wall of the fluid inside the bubble, and $p_\infty(t)$ is the externally applied pressure. The equation (5.4.9) is known as the Rayleigh-Plesset equation. It is just a second order ordinary differential equation if p' is known. In the general case, however, p' has to be obtained from the solution of the interior flow problem, which can be very involved.

Still, for the simplest case that both p' and p_∞ are constants, a first integral can be readily obtained. (5.4.9) can be rewritten as

$$\frac{1}{2R^2}\frac{d}{dR}\left[R^3\left(\frac{dR}{dt}\right)^2\right] = \frac{1}{\rho}\left[p' - p_\infty - \frac{2\sigma}{R}\right]. \tag{5.4.10}$$

Integrating once, we obtain

$$\left(\frac{dR}{dt}\right)^2 = \frac{2}{3}\left(\frac{p' - p_\infty}{\rho}\right)\left[1 - \left(\frac{R_0}{R}\right)^3\right] - \frac{2\sigma}{\rho R}\left[1 - \left(\frac{R_0}{R}\right)^2\right], \tag{5.4.11}$$

where R_0 is the value of R when $\frac{dR}{dt}$ is zero.

From (5.4.11), we see that if $(p' - p_\infty)$ is positive, i.e., the internal pressure is larger than the external pressure, then the bubble tends to grow, and asymptotically we have

$$\left(\frac{dR}{dt}\right) \approx \left[\frac{2}{3}\frac{(p' - p_\infty)}{\rho}\right]^{1/2}, \quad \text{as } t \to \infty . \quad (5.4.12)$$

On the other hand if $(p' - p_\infty)$ is negative, i.e., the external pressure is larger than the internal pressure, then the bubble will collapse, and asymptotically we have

$$\left(\frac{dR}{dt}\right) \approx -\left[\frac{2}{3}\frac{(p_\infty - p')R_0^3}{\rho} + \frac{2\sigma R_0^2}{\rho}\right]^{\frac{1}{2}} R^{-3/2}, \quad \text{as } R \to 0 . \quad (5.4.13)$$

With these simple background, let us now discuss the stability problem.

We would like to ask the question that when the bubble undergoes the spherically symmetric motion, whether expansion, collapse or oscillation, can the spherical shape of the bubble be always maintained. In other words, let us take the interface equation be

$$r = r_s \equiv R(t) + \sum_{l=1}^{\infty} \sum_{m=-l}^{l} a_{lm}(t) Y_{lm}(\theta, \psi) , \quad (5.4.14)$$

where $R(t)$ is the solution of the spherically symmetric motion. If $a_{lm}(t)$ remains small with small initial conditions, then the spherical motion is stable. Otherwise, the assumption of spherical symmetry needs to be revised. With the bubble wall given by (5.4.14), the flow field is also no longer spherically symmetric, and thus instead of (5.4.6), the general solutions outside the bubble becomes :

$$\varphi = \frac{A(t)}{r} + \sum_{l=1}^{\infty}\sum_{m=-l}^{l} \frac{b_{lm}(t)}{r^{l+1}} Y_{lm}(\theta, \psi)\,, \quad \text{for } r > r_s\,. \quad (5.4.15)$$

What about the flow inside the bubble? It depends on the real physical problem we are dealing with. In order to cover a general class of problems, and with the expectation that perturbations near the interface play the most important roles, we shall adopt the following model for the internal flow. We shall assume that the fluid inside the bubble is also incompressible and inviscid. But a source term is added to accommodate the motion of the bubble wall. Thus we have inside the bubble the velocity potential φ' :

$$\varphi' = \frac{A'(t)}{r} + \sum_{l=1}^{\infty}\sum_{m=-l}^{l} b'_{lm}(t) r^l Y_{lm}(\theta, \psi)\,, \quad \text{for } r < r_s\,. \quad (5.4.16)$$

With φ and φ' given by (5.4.15) and (5.4.16), we can now compute the pressure p and p' by using the Bernoulli's equation (5.4.3) and its counterpart inside the bubble. Then we apply the interfacial conditions (5.1.16), (5.1.17) and (5.1.23). We shall limit our discussion only on linear stability analysis. Thus a_{lm} , b_{lm} and b'_{lm} are all small, and only the first order of these small quantities are retained in those equations to be obtained. From (5.1.16) and (5.1.17), we then have to the first order that

$$\frac{\partial \varphi}{\partial r} = \frac{\partial \varphi'}{\partial r} = \frac{dR}{dt} + \sum_{l=1}^{\infty}\sum_{m=-l}^{l} \frac{da_{lm}}{dt} Y_{lm}(\theta, \psi)\,, \quad \text{on } r = r_s\,. \quad (5.4.17)$$

Substitute (5.4.15) and (5.4.16) into (5.4.17) and make use of the orthonormality of Y_{lm} , we obtain

$$\varphi = -\frac{1}{r}\left(R^2 \frac{dR}{dt}\right) + \sum_{l=1}^{\infty}\sum_{m=-l}^{l} \frac{R^{l+2}}{(l+1) r^{l+1}} \left(\frac{da_{lm}}{dt} + \frac{2a_{lm}}{R}\frac{dR}{dt}\right) Y_{lm}(\theta, \psi)\,,$$

$$r > r_s\,, \quad (5.4.18)$$

and

$$\varphi = -\frac{1}{r}\left(R^2 \frac{dR}{dt}\right) + \sum_{l=1}^{\infty}\sum_{m=-l}^{l} \frac{r^l}{lR^{l-1}}\left(\frac{da_{lm}}{dt} + \frac{2a_{lm}}{R}\frac{dR}{dt}\right) Y_{lm}(\theta, \psi) ,$$

$$r < r_s . \quad (5.4.19)$$

Now we use the Bernoulli's equations to compute the pressure. Specifically we have

$$p = p_\infty(t) - \rho\left[\frac{\partial \varphi}{\partial t} + \frac{1}{2}(\nabla \varphi)^2\right], \quad r > r_s , \quad (5.4.20)$$

and

$$p' = p_i(t) - \rho'\left[\frac{\partial \varphi'}{\partial t} + \frac{1}{2}(\nabla \varphi')^2\right], \quad r < r_s , \quad (5.4.21)$$

where p_∞ and p_i are pressures outside and inside the bubble respectively when $\varphi = \varphi' = 0$, i.e., when the system is in equilibrium. To apply the interfacial condition (5.1.23), we need to compute the values of $\left(\frac{\partial \varphi}{\partial t}\right)$, $\left(\frac{\partial \varphi'}{\partial t}\right)$, $(\nabla \varphi)^2$ and $(\nabla \varphi')^2$ on $r = r_s$. To the first order of the small quantity a_{lm}, we have on $r = r_s$:

$$(\nabla \varphi)^2 = (\nabla \varphi')^2$$

$$= \left(\frac{dR}{dt}\right)^2 + \sum_{l=1}^{\infty}\sum_{m=-l}^{l} 2\left(\frac{dR}{dt}\right)\left(\frac{da_{lm}}{dt}\right) Y_{lm}(\theta, \psi) , \quad (5.4.22)$$

$$\frac{\partial\varphi}{\partial t}=-\frac{1}{R}\frac{d}{dt}\left(R^2\frac{dR}{dt}\right)+\sum_{l=1}^{\infty}\sum_{m=-l}^{l}\frac{a_{lm}}{R^2}\frac{d}{dt}\left(R^2\frac{dR}{dt}\right)Y_{lm}(\theta,\psi)$$

$$-\sum_{l=1}^{\infty}\sum_{m=-l}^{l}\left\{\frac{R}{l+1}\frac{d^2a_{lm}}{dt^2}+\left(\frac{l+4}{l+1}\right)\left(\frac{dR}{dt}\right)\frac{da_{lm}}{dt}\right\}Y_{lm}(\theta,\psi)$$

$$-\sum_{l=1}^{\infty}\sum_{m=-l}^{l}\left\{\left[\frac{2}{l+1}\frac{d^2R}{dt^2}+\frac{2}{R}\left(\frac{dR}{dt}\right)^2\right]a_{lm}\right\}Y_{lm}(\theta,\psi)\ , \tag{5.4.23}$$

$$\frac{\partial\varphi'}{\partial t}=-\frac{1}{R}\frac{d}{dt}\left(R^2\frac{dR}{dt}\right)+\sum_{l=1}^{\infty}\sum_{m=-l}^{l}\frac{a_{lm}}{R}\frac{d}{dt}\left(R^2\frac{dR}{dt}\right)Y_{lm}(\theta,\psi)$$

$$+\sum_{l=1}^{\infty}\sum_{m=-l}^{l}\left\{\frac{R}{l}\frac{d^2a_{lm}}{dt^2}+\left(\frac{l-3}{l}\right)\left(\frac{dR}{dt}\right)\frac{da_{lm}}{dt}\right\}Y_{lm}(\theta,\psi)$$

$$+\sum_{l=1}^{\infty}\sum_{m=-l}^{l}\left[\frac{2}{l}\frac{d^2R}{dt^2}-\frac{2}{R}\left(\frac{dR}{dt}\right)^2\right]a_{lm}Y_{lm}(\theta,\psi)\ . \tag{5.4.24}$$

Also, to the first order of a_{lm}, we have

$$\frac{1}{R_1}+\frac{1}{R_2}=\frac{2}{R}+\sum_{l=1}^{\infty}\sum_{m=-l}^{l}\frac{1}{R^2}(l-1)(l+2)a_{lm}Y_{lm}(\theta,\psi)\ . \tag{5.4.25}$$

With these preliminary calculations, the interfacial condition (5.1.23) yields after making use of the orthonormality of Y_{lm} :

$$R\frac{d^2R}{dt^2}+\frac{3}{2}\left(\frac{dR}{dt}\right)^2=\frac{1}{\rho-\rho'}\left[p_i(t)-p_\infty(t)-\frac{2\sigma}{R}\right], \tag{5.4.26}$$

and

$$\frac{d^2a_{lm}}{dt^2}+\frac{3}{R}\left(\frac{dR}{dt}\right)\frac{da_{lm}}{dt}-A_l a_{lm}=0\ ,\quad l\geq 1\ ,\quad -l\leq m\leq l\ , \tag{5.4.27}$$

where

$$A_l = \frac{[l(l-1)\rho-(l+1)(l+2)\rho']\dfrac{d^2R}{dt^2}-(l-1)l(l+1)(l+2)\dfrac{\sigma}{R^2}}{R[l\rho+(l+1)\rho']} . \tag{5.4.28}$$

(5.4.26) is essentially the same as (5.4.9). The cases for $l = 1$ represent the translation of the bubble as a whole. Only for cases $l \geq 2$, the perturbations represent the distortion of the spherical shape of the bubble. For most cases we have $\rho' << \rho$. Then (5.4.28) becomes

$$A_l = \frac{l-1}{R}\frac{d^2R}{dt^2}-(l-1)(l+1)(l+2)\frac{\sigma}{\rho R^3} . \tag{5.4.29}$$

Let

$$a_{lm} = \left(\frac{R_0}{R}\right)^{3/2} c_{lm} , \tag{5.4.30}$$

where R_0 is some reference value of R, then the first derivative term in (5.4.27) can be eliminated, and (5.4.27) becomes

$$\frac{d^2c_{lm}}{dt^2}+C_lc_{lm} = 0 , \tag{5.4.31}$$

$$C_l = (l-1)(l+1)(l+2)\left(\frac{\sigma}{\rho R^3}\right)$$

$$-\frac{1}{R}\left(l+\frac{1}{2}\right)\frac{d^2R}{dt^2}-\frac{3}{4}\left(\frac{1}{R}\frac{dR}{dt}\right)^2 . \tag{5.4.32}$$

Equation (5.4.27), or (5.4.31) is the equation governing the stability of the spherical shape of the bubble.

Stability of Expanding and Collapsing Bubbles

Let us apply the stability equation (5.4.31) to the expanding and collapsing spherical bubbles. For expanding bubbles, as seen from (5.4.12), $\left(\frac{dR}{dt}\right)$ is constant as $t \to \infty$. Therefore, R will be linearly growing with t, as $t \to \infty$. It is clear from (5.4.32), for such $R(t)$, the dominant term will be $\left[-\frac{3}{4}\left(\frac{1}{R}\frac{dR}{dt}\right)^2\right]$. Thus we have

$$C_l \approx -\frac{3}{4}t^{-2}, \quad as\ t \to \infty. \tag{5.4.33}$$

Substituting (5.4.33) into (5.4.31), we thus obtain

$$c_{lm} \propto t^{\frac{3}{2}}, \quad as\ t \to \infty. \tag{5.4.34}$$

From (5.4.30) we thus obtain

$$a_{lm} \to \text{constant}, \quad \text{as } t \to \infty. \tag{5.4.35}$$

Since $R \propto t$, therefore,

$$\left(\frac{a_{lm}}{R}\right) \to 0, \quad \text{as } t \to \infty. \tag{5.4.36}$$

Hence the spherical shape of the bubble in expansion is stable.

To deal with the collapsing bubble, since from (5.4.13), we can write :

$$\left(\frac{dR}{dt}\right) \approx -\beta R^{-\frac{3}{2}}, \quad \text{as } R \to 0, \tag{5.4.37}$$

where β is a positive real constant. The term C_l in (5.4.31) becomes very large as $R \to 0$. We shall use the WKB-method to evaluate c_{lm}. According to the WKB method, when C_l is large, the asymptotic solutions of (5.4.31) are given by :

$$c_{lm} \approx \frac{1}{C_l^{1/4}} \exp\left[\pm i \int^t C_l^{1/2}(t)dt \right]. \qquad (5.4.38)$$

Now with $\left(\frac{dR}{dt}\right)$ given by (5.4.37), the dominant terms are the last two terms in (5.4.32) as $R \to 0$; and we have

$$C_l \approx l\beta^2 R^{-5}, \qquad \text{as } R \to 0. \qquad (5.4.39)$$

Thus, from (5.4.38), we obtain

$$c_{lm} \approx R^{\frac{5}{4}} \exp\left\{ \pm i\beta l^{\frac{1}{2}} \int^t R^{-\frac{5}{2}}(t)dt \right\}, \qquad \text{as } R \to 0. \quad (5.4.40)$$

Hence, using (5.4.30), we have

$$a_{lm} \approx R^{-\frac{1}{4}} \exp\left\{ \pm i\beta l^{\frac{1}{2}} \int^t R^{-\frac{5}{2}}(t)dt \right\}, \qquad \text{as } R \to 0. \quad (5.4.41)$$

Therefore, the mean value of a_{lm} will grow like $R^{-\frac{1}{4}}$ as $R \to 0$, while the oscillation frequency is growing even more rapidly. Thus the spherical shape of a collapsing bubble is not stable. The assumption of spherical symmetry of the system, on which so much simplification is based, need to be modified when we deal with the later phase of the collapse of a bubble.

5.5 The Classical Kelvin-Helmholtz Stability

Referring back to the configuration of the classical Rayleigh-Taylor stability, shown in Fig. 5-2, we have two immiscible, incompressible and inviscid fluids separated by an interface in the gravitation field. If these two fluids are in relative motion in the primary state, then we are dealing with the Kelvin-Helmholtz stability. Let U_1 and U_2 be the velocity in x-direction of the fluids (1) and (2) respectively. There is no loss of generality to consider the case that both fluids are moving in the same direction, since what distinguishes the Kelvin-Helmholtz stability from the Rayleigh-Taylor stability is the relative velocity between these two fluids.

The governing equations for the Kelvin-Helmholtz stability problem are again (5.2.4) - (5.2.9). With fluids (1) and (2) moving with U_1 and U_2 in the primary state, the primary state of motion is now given by :

$$\eta_0 = 0 \,, \tag{5.5.1}$$

$$\varphi_0^{(1)} = U_1 x \,, \qquad \varphi_0^{(2)} = U_2 x \,, \tag{5.5.2}$$

$$p_0^{(1)} = -\rho^{(1)} g y \,, \qquad p_0^{(2)} = -\rho^{(2)} g y \,, \tag{5.5.3}$$

$$f^{(1)} = \frac{1}{2} U_1^2 \,, \qquad f^{(2)} = \frac{1}{2} U_2^2 \,. \tag{5.5.4}$$

Following the same procedure for the analysis of Rayleigh-Taylor stability, we obtain the following linear set of equations for the perturbed quantities :

$$\nabla^2 \varphi_1^{(1)} = 0 \,, \quad \text{for} \; -h^{(1)} < y < 0 \,, \tag{5.5.5}$$

$$\nabla^2 \varphi_1^{(2)} = 0 \,, \quad \text{for} \; 0 < y < h^{(2)} \,, \tag{5.5.6}$$

$$\frac{\partial \eta_1}{\partial t} - \frac{\partial \varphi_1^{(1)}}{\partial y} + U_1 \frac{\partial \eta_1}{\partial x} = 0 \,, \quad \text{on } y = 0 \,, \tag{5.5.7}$$

$$\frac{\partial \eta_1}{\partial t} - \frac{\partial \varphi_1^{(2)}}{\partial y} + U_2 \frac{\partial \eta_1}{\partial x} = 0 , \quad \text{on } y = 0 , \qquad (5.5.8)$$

$$-\rho^{(2)}\left(g\eta_1 + \frac{\partial \varphi_1^{(2)}}{\partial t} + U_2 \frac{\partial \varphi_1^{(2)}}{\partial x} \right) + \rho^{(1)}\left(g\eta_1 + \frac{\partial \varphi_1^{(1)}}{\partial t} + U_1 \frac{\partial \varphi_1^{(1)}}{\partial x} \right)$$

$$= \sigma\left(\frac{\partial^2 \eta_1}{\partial x^2} + \frac{\partial^2 \eta_1}{\partial z^2} \right) , \qquad \text{on } y = 0 , \qquad (5.5.9)$$

$$\frac{\partial \varphi_1^{(1)}}{\partial y} = 0 , \quad \text{on } y = -h^{(1)} , \qquad (5.5.10)$$

$$\frac{\partial \varphi_1^{(2)}}{\partial y} = 0 , \quad \text{on } y = h^{(2)} . \qquad (5.5.11)$$

Again, let us consider a single Fourier mode. Making use of the boundary conditions (5.5.10) and (5.5.11), the perturbed solutions are again the same as (5.2.19), (5.2.22) and (5.2.23) :

$$\eta_1 = A e^{i(k_x x + k_z z) + nt} , \qquad (5.5.12)$$

$$\varphi_1^{(1)} = B \cosh k\left(y + h^{(1)}\right) e^{i(k_x x + k_z z) + nt} , \qquad (5.5.13)$$

$$\varphi_1^{(2)} = C \cosh k\left(y - h^{(2)}\right) e^{i(k_x x + k_z z) + nt} . \qquad (5.5.14)$$

Substitute (5.5.12) - (5.5.14) into (5.5.7) - (5.5.9) respectively, we obtain :

$$\left(n + ik_x U_1\right) A - kB \sinh kh^{(1)} = 0 , \qquad (5.5.15)$$

$$\left(n + ik_x U_2\right) A + kC \sinh kh^{(2)} = 0 , \qquad (5.5.16)$$

$$\rho^{(1)}(n+ik_xU_1)B\cosh kh^{(1)} - \rho^{(2)}(n+ik_xU_2)C\cosh kh^{(2)}$$

$$+\left[g\left(\rho^{(1)}-\rho^{(2)}\right)+\sigma k^2\right]A = 0 \,. \tag{5.5.17}$$

Eliminating A, B, C from (5.5.15) - (5.5.17), we obtain :

$$\rho^{(1)}(n+ik_xU_1)^2\coth kh^{(1)} + \rho^{(2)}(n+ik_xU_2)^2\coth kh^{(2)}$$

$$+\left[\left(\rho^{(1)}-\rho^{(2)}\right)gk+\sigma k^3\right] = 0 \,, \tag{5.5.18}$$

or

$$\left(s^{(1)}+s^{(2)}\right)n^2 + 2ik_x\left(U_1s^{(1)}+U_2s^{(2)}\right)n - k_x^2\left(s^{(1)}U_1^2+s^{(2)}U_2^2\right)$$

$$+\left[\left(\rho^{(1)}-\rho^{(2)}\right)gk+\sigma k^3\right]\sinh kh^{(1)}\sinh kh^{(2)} = 0 \,, \tag{5.5.19}$$

where

$$s^{(1)} = \rho^{(1)}\cosh kh^{(1)}\sinh kh^{(2)} \,,$$
$$s^{(2)} = \rho^{(2)}\cosh kh^{(2)}\sinh kh^{(1)} \,. \tag{5.5.20}$$

(5.5.19) is the dispersion relation for the Kelvin-Helmholtz stability. If $U_1 = U_2 = 0$, then it is the same as (5.2.28), the dispersion relation for the Rayleigh-Taylor stability. Solving (5.5.19) for n, we obtain

$$n = \frac{1}{s^{(1)}+s^{(2)}}\left[-ik_x\left(U_1s^{(1)}+U_2s^{(2)}\right)\pm T^{\frac{1}{2}}\right], \tag{5.5.21}$$

where

$$T = s^{(1)}s^{(2)}(U_1 - U_2)^2 k_x^2$$

$$-\left(s^{(1)} + s^{(2)}\right)\left[\left(\rho^{(1)} - \rho^{(2)}\right)gk + \sigma k^3\right]\sinh kh^{(1)} \sinh kh^{(2)} \ . \quad (5.5.22)$$

Therefore the primary state is stable only when

$$T < 0 \ . \quad (5.5.23)$$

It is clear that if the stability condition for the Rayleigh-Taylor stability is not met, the primary state is definitely unstable. Even if the state is Rayleigh-Taylor stable, the state can be unstable if the relative velocity is large enough. Specifically, the primary state is stable only if

$$(U_1 - U_2)^2 < \frac{s^{(1)} + s^{(2)}}{s^{(1)}s^{(2)}} \cdot \frac{\left[\left(\rho^{(1)} - \rho^{(2)}\right)gk + \sigma k^3\right]}{k_x^2} \cdot \sinh kh^{(1)} \sinh kh^{(2)} \ . \quad (5.5.24)$$

We may recall that $k^2 = k_x^2 + k_z^2$. Therefore the right hand side assumes the smallest value when $k_x^2 = k^2$, i.e., for the mode $\eta_1 = Ae^{ikx+nt}$, the mode that is independent of z. Thus the most critical modes are disturbances varying in the direction of relative velocity. Therefore, taking into consideration of all possibilities, the primary state is stable only if

$$(U_1 - U_2)^2 < \frac{s^{(1)} + s^{(2)}}{s^{(1)}s^{(2)}} \cdot \left[\left(\rho^{(1)} - \rho^{(2)}\right)\frac{g}{k} + \sigma k\right]\sinh kh^{(1)} \sinh kh^{(2)} \ . \quad (5.5.25)$$

Let us consider a few special cases.

The Case $h^{(1)} \to \infty$ *and* $h^{(2)} \to \infty$.

We are dealing the case of two semi-infinite fluids. Then (5.5.25) becomes

$$(U_1 - U_2)^2 < \left(\frac{\rho^{(1)} + \rho^{(2)}}{\rho^{(1)}\rho^{(2)}}\right)\left[\left(\rho^{(1)} - \rho^{(2)}\right)\frac{g}{k} + \sigma k\right] = G(k) \,. \quad (5.5.26)$$

For $\rho^{(1)} > \rho^{(2)}$, $G(k)$ is schematically shown in Fig. 5-4. $G(k)$ has a minimum at $k = k_c$, and

$$k_c = \left[\frac{g}{\sigma}\left(\rho^{(1)} - \rho^{(2)}\right)\right]^{\frac{1}{2}}. \quad (5.5.27)$$

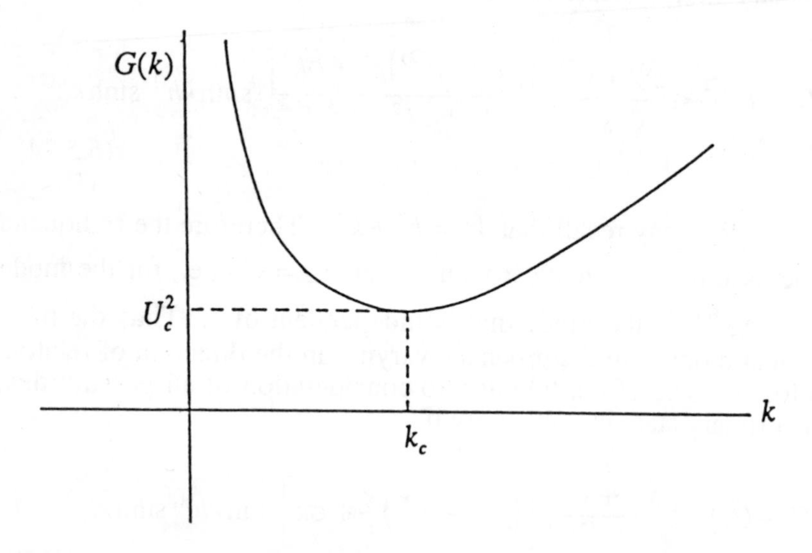

Fig. 5-4 The curve $G(k)$ for $\rho^{(1)} > \rho^{(2)}$.

Denote

$$U_c^2 = G(k_c) = \frac{2\left(\rho^{(1)} + \rho^{(2)}\right)}{\rho^{(1)}\rho^{(2)}}\left[\sigma g\left(\rho^{(1)} - \rho^{(2)}\right)\right]^{\frac{1}{2}} \,. \quad (5.5.28)$$

Then, for $(U_1 - U_2)^2 < U_c^2$, the primary state is stable for all k. However, if $(U_1 - U_2)^2 > U_c^2$, then there is a range of k, for which the system is Kelvin-Helmholtz unstable.

Take the case of air over water, we have in cgs units, $\rho^{(1)} \approx 1$, $\rho^{(2)} \approx 0.0012$, $\sigma \approx 73$, $g \approx 980$. Thus we obtain

$$U_c \approx 668 \text{ cm / sec} .$$

The corresponding k_c is

$$k_c \approx 3.66 \text{ cm}^{-1} ,$$

and the critical wavelength is

$$\lambda_c \approx 1.72 \text{ cm} .$$

It has been observed that when the velocity of wind over the ocean exceeds the value around U_c, the number of white caps suddenly increases and so does the rate of evaporation. It may be due to the onset of Kelvin-Helmholtz instability.

The Case $h^{(1)} \to 0$ *and* $h^{(2)} \to \infty$.

For this case, we have

$$s^{(1)} \approx \frac{1}{2}\rho^{(1)} e^{kh^{(2)}} , \quad s^{(2)} \approx \frac{1}{2}\rho^{(2)} kh^{(1)} e^{kh^{(2)}} .$$

Thus (5.5.25) becomes :

$$(U_1 - U_2)^2 < \frac{1}{\rho^{(2)}}\left[\left(\rho^{(1)} - \rho^{(2)}\right)\frac{g}{k} + \sigma k\right] . \qquad (5.5.29)$$

Hence we have again for the critical case that

$$k_c = \left[\frac{g}{\sigma}\left(\rho^{(1)} - \rho^{(2)}\right)\right]^{\frac{1}{2}} . \tag{5.5.30}$$

But the critical velocity is now

$$U_c^2 = \frac{2}{\rho^{(2)}}\left[\sigma g\left(\rho^{(1)} - \rho^{(2)}\right)\right]^{\frac{1}{2}} . \tag{5.5.31}$$

For the case $\rho^{(1)} >> \rho^{(2)}$, (5.5.31) is essentially the same as (5.5.28).

The Case $h^{(1)} \to \infty$ and $h^{(2)} \to 0$.

For this case, we have

$$s^{(1)} \approx \frac{1}{2}\rho^{(1)} k h^{(2)} e^{kh^{(1)}} , \quad s^{(2)} \approx \frac{1}{2}\rho^{(2)} e^{kh^{(1)}} .$$

Thus (5.5.25) becomes

$$\left(U_1 - U_2\right)^2 < \frac{\rho^{(1)} k h^{(2)} + \rho^{(2)}}{\rho^{(1)}\rho^{(2)}}\left[\left(\rho^{(1)} - \rho^{(2)}\right)\frac{g}{k} + \sigma k\right] . \tag{5.5.32}$$

If $\rho^{(2)} = 0\left(\rho^{(1)}\right)$, then (5.5.32) becomes

$$\left(U_1 - U_2\right)^2 = \frac{1}{\rho^{(1)}}\left[\left(\rho^{(1)} - \rho^{(2)}\right)\frac{g}{k} + \sigma k\right] . \tag{5.5.33}$$

Thus k_c is again given by (5.5.27), while

$$U_c^2 = \frac{2}{\rho^{(1)}}\left[\sigma g\left(\rho^{(1)} - \rho^{(2)}\right)\right]^{\frac{1}{2}} , \tag{5.5.34}$$

which is smaller than the value given by (5.5.28). When $\rho^{(1)}$ is large compared with $\rho^{(2)}$, this critical velocity U_c is essentially the phase velocity of the surface wave at $k \approx k_c$.

If $\rho^{(1)}$ is so large that $\rho^{(1)}kh^{(2)} >> \rho^{(2)}$, then (5.5.32) becomes

$$(U_1 - U_2)^2 < \frac{h^{(2)}}{\rho^{(2)}}\left[\left(\rho^{(1)} - \rho^{(2)}\right)g + \sigma k^2\right]. \quad (5.5.35)$$

Thus, for the critical case, we have

$$k_c = 0 , \quad (5.5.36)$$

and

$$U_c^2 = \frac{\rho^{(1)}}{\rho^{(2)}} g h^{(2)} . \quad (5.5.37)$$

Therefore U_c is the velocity of the shallow gravity wave for the fluid layer (2) increased by a factor of $\left(\dfrac{\rho^{(1)}}{\rho^{(2)}}\right)^{\frac{1}{2}}$.

The Case $h^{(1)} \to 0$ *and* $h^{(2)} \to 0$.

For this case, (5.5.25) becomes

$$(U_1 - U_2)^2 < \left(\frac{h^{(1)}}{\rho^{(1)}} + \frac{h^{(2)}}{\rho^{(2)}}\right)\left[\left(\rho^{(1)} - \rho^{(2)}\right)g + \sigma k^2\right]. \quad (5.5.38)$$

Therefore, we have for the critical case again $k_c = 0$, and

$$U_c^2 = g\left(\rho^{(1)} - \rho^{(2)}\right)\left(\frac{h^{(1)}}{\rho^{(1)}} + \frac{h^{(2)}}{\rho^{(2)}}\right). \qquad (5.5.39)$$

when $\rho^{(1)} >> \rho^{(2)}$, (5.5.39) is the same as (5.5.37).

5.6 Variational Method and Kelvin-Helmholtz Stability

We introduce in this section the variational method to deal with the Kelvin-Helmholtz stability problem. The variational method has been applied extensively to oscillation and wave problems, especially water wave problems. It can also be applied to the study of stability problems. We shall first deal with the linear Kelvin-Helmholtz stability, and then extended to the study of nonlinear problem.

The general problem formulated in terms of the governing equations (5.2.4) - (5.2.9) can be formulated in terms of the variational principle that the following functional J is stationary (Hsieh 1977) :

$$J = \int_{t_1}^{t_2} dt(\mathcal{J}_1 + \mathcal{J}_2 + \mathcal{J}_a) , \qquad (5.6.1)$$

where

$$\mathcal{J}_i = \rho^{(i)} \int_{V_i} d^3\mathbf{x}\left[\frac{\partial \varphi^{(i)}}{\partial t} + \frac{1}{2}\left(\nabla \varphi^{(i)}\right)^2 + \Omega^{(i)}\right], \quad i = 1,2 , \qquad (5.6.2)$$

$$\mathcal{J}_a = \sigma \int_F d^2\mathbf{x} . \qquad (5.6.3)$$

In (5.6.2) we have designate $\Omega^{(i)}$ as the potential of the external force, V_i the region occupied by the fluid (i). In (5.6.3), the integral represents the area of the interface $F(\mathbf{x},t) = 0$. The variations with respect to $\varphi^{(i)}$ lead to (5.2.4) and (5.2.5) as well as the boundary conditions corresponding to (5.2.8) and (5.2.9); while the variation with respect to the interface leads to (5.2.7).

To study the Kelvin-Helmholtz stability, we shall consider the case of two superposed semi-infinite fluids for simplicity. This would corresponds to the case that $h^{(1)} \to \infty$ and $h^{(2)} \to \infty$ in (5.2.8) and (5.2.9) respectively. From the previous analyses, it is clear that there is no loss of generality to consider the problem in two spatial dimensions. Thus all the physical quantities are independent of z. As the only external force is the gravitational force pointing in the direction of $(-y)$, thus

$$\Omega^{(i)} = gy \; . \tag{5.6.4}$$

Let us express the interface as

$$y = \eta(x,t) \equiv A(x,t)\sin\left[S(x,t)\right] . \tag{5.6.5}$$

We can always express $\eta(x,t)$ in the form of (5.6.5), since we have not specified the functions A and S yet. Then we have the following expressions for $\mathcal{J}_i$ and $\mathcal{J}_a$ corresponding to the Kelvin-Helmholtz problem :

$$\mathcal{J}_1 = \rho^{(1)} \int_{-\infty}^{\infty} dx \int_{-\infty}^{\eta} \left[\frac{\partial \varphi^{(1)}}{\partial t} + \frac{1}{2}\left(\frac{\partial \varphi^{(1)}}{\partial x}\right)^2 + \frac{1}{2}\left(\frac{\partial \varphi^{(1)}}{\partial y}\right)^2 + gy \right] dy, \tag{5.6.6}$$

$$\mathcal{J}_2 = \rho^{(2)} \int_{-\infty}^{\infty} dx \int_{\eta}^{\infty} \left[\frac{\partial \varphi^{(2)}}{\partial t} + \frac{1}{2}\left(\frac{\partial \varphi^{(2)}}{\partial x}\right)^2 + \frac{1}{2}\left(\frac{\partial \varphi^{(2)}}{\partial y}\right)^2 + gy \right] dy \tag{5.6.7}$$

and

$$\mathcal{J}_a = \sigma \int_{-\infty}^{\infty} dx \left[1 + \left(\frac{\partial \eta}{\partial x}\right)^2 \right]^{1/2} . \tag{5.6.8}$$

To apply the variational method to the flow problem, we seek trial solutions with certain undetermined parameters, which are consistent with the boundary conditions and are as close as possible to the real solutions that we can anticipate. As suggested by the previous linear analyses, we shall take

$$\varphi^{(1)} = U_1 x + B(x,t)\cos S e^{ky} , \tag{5.6.9}$$

$$\varphi^{(2)} = U_2 x + C(x,t)\cos S e^{-ky} , \tag{5.6.10}$$

where

$$k = \frac{\partial S}{\partial x} > 0 . \tag{5.6.11}$$

In a more complete theory, indeed, the spatial and temporal variations of A, B, C and S should be taken into account in order to study modulation of nonlinear behavior. We shall, however, only consider the case that A, B, C, k and $\omega = \dfrac{\partial S}{\partial t}$ are constants.

Substitute (5.6.5), (5.6.9) and (5.6.10) into (5.6.6) - (5.6.8) and then carry out the integration. The way we carry out the integration can be illustrated by the following example.

$$\int_{-\infty}^{\infty} dx \int_{-\infty}^{\eta} \left(\frac{\partial \varphi^{(1)}}{\partial t} \right) dy = \int_{-\infty}^{\infty} dx \int_{-\infty}^{\eta} -B\omega \sin S e^{ky} dy$$

$$= -\int_{-\infty}^{\infty} dx B \left(\frac{\omega}{k} \right) \sin S e^{k\eta}$$

$$= -\int_{-\infty}^{\infty} dx B \left(\frac{\omega}{k} \right) \sin S \left[1 + kA \sin S + \frac{1}{2} k^2 A^2 \sin^2 S + \frac{1}{6} k^3 A^3 \sin^3 S \right]$$

$$+ 0(A^5 B)$$

$$\approx -\int_{-\infty}^{\infty} dx\omega B\left(\frac{1}{2}A + \frac{1}{16}k^2A^3\right) + 0(A^5B) . \qquad (5.6.12)$$

In the last step of (5.6.12), we have made use of the fact that

$$\int_P dx \sin^2 S = \frac{1}{2}\int_P dx ,$$

$$\int_P dx \sin^3 S = 0 ,$$

$$\int_P dx \sin^4 S = \frac{3}{8}\int_P dx ,$$

where P is the period of the sine function. Thus if A, B, ω and k are slowly varying over the period, we obtain approximately the result expressed in (5.6.12). With similar operations on other terms, then we find up to the order of $0(A^4)$, or $0(B^4)$ or $0(C^4)$;

$$\frac{\mathcal{J}_1}{\rho^{(1)}} = \int_{-\infty}^{\infty} dx\left\{-\frac{1}{2}(\omega + kU_1)BA\left[1+\frac{1}{8}(kA)^2\right] + \frac{k}{4}B^2\left[1+(kA)^2\right] + \frac{g}{4}A^2\right\} , \qquad (5.6.13)$$

$$\frac{\mathcal{J}_2}{\rho^{(2)}} = \int_{-\infty}^{\infty} dx\left\{\frac{1}{2}(\omega + kU_2)CA\left[1+\frac{1}{8}(kA)^2\right] + \frac{k}{4}C^2\left[1+(kA)^2\right] - \frac{g}{4}A^2\right\} , \qquad (5.6.14)$$

$$\mathcal{J}_a = \sigma\int_{-\infty}^{\infty} dx\left[1+\frac{1}{4}(kA)^2 - \frac{3}{64}(kA)^4\right] . \qquad (5.6.15)$$

The variations with respect to A, B, and C thus lead to :

$$\delta A::\quad \rho^{(1)}\left\{-\frac{1}{2}(\omega+kU_1)B\left[1+\frac{3}{8}(kA)^2\right]+\frac{k^3}{2}AB^2+\frac{g}{2}A\right\}$$

$$+\rho^{(2)}\left\{\frac{1}{2}(\omega+kU_2)C\left[1+\frac{3}{8}(kA)^2\right]+\frac{k^3}{2}AC^2-\frac{g}{2}A\right\}$$

$$+\frac{\sigma k^2}{2}A\left[1-\frac{3}{8}(kA)^2\right]=0\ , \tag{5.6.16}$$

$$\delta B::\quad -\frac{1}{2}(\omega+kU_1)A\left[1+\frac{1}{8}(kA)^2\right]+\frac{k}{2}B\left[1+(kA)^2\right]=0\ , \tag{5.6.17}$$

$$\delta C::\quad \frac{1}{2}(\omega+kU_2)A\left[1+\frac{1}{8}(kA)^2\right]+\frac{k}{2}C\left[1+(kA)^2\right]=0\ . \tag{5.6.18}$$

Denote

$$G=\frac{1+\frac{1}{8}(kA)^2}{1+(kA)^2}\ , \tag{5.6.19}$$

then we have

$$B=\frac{1}{k}(\omega+kU_1)GA\ , \tag{5.6.20}$$

$$C=-\frac{1}{k}(\omega+kU_2)GA\ . \tag{5.6.21}$$

Substitute (5.6.20) and (5.6.21) into (5.6.16), we obtain

$$\left[\left(\rho^{(1)}+\rho^{(2)}\right)\omega^2+2k\left(\rho^{(1)}U_1+\rho^{(2)}U_2\right)\omega+\left(\rho^{(1)}U_1^2+\rho^{(2)}U_2^2\right)k^2\right]$$

$$\times G\left[1+\left(\frac{3}{8}-G\right)k^2A^2\right]-\left[\left(\rho^{(1)}+\rho^{(2)}\right)gk+\sigma k^3\left(1-\frac{3}{8}k^2A^2\right)\right]=0\,. \tag{5.6.22}$$

Thus,

$$\omega=\frac{1}{\left(\rho^{(1)}+\rho^{(2)}\right)}\left\{-\left(\rho^{(1)}U_1+\rho^{(2)}U_2\right)k\pm\left[W\left(k,U_1-U_2,A\right)\right]^{\frac{1}{2}}\right\}, \tag{5.6.23}$$

where

$$W\left(k,U_1-U_2,A\right)=\frac{\left(\rho^{(1)}+\rho^{(2)}\right)\left[\left(\rho^{(1)}-\rho^{(2)}\right)gk+\sigma k^3\left(1-\frac{3}{8}k^2A^2\right)\right]}{G\left[1+\left(\frac{3}{8}-G\right)k^2A^2\right]}$$

$$-\rho^{(1)}\rho^{(2)}k^2\left(U_1-U_2\right)^2\,. \tag{5.6.24}$$

Since we have assumed that the trial solutions behaves like $\cos(kx+\omega t)$ or $\sin(kx+\omega t)$, the system is stable only when ω is real or when $W\geq 0$. As $A\to 0$, we recover the results of linear theory, i.e., the corresponding result from (5.5.19).

The critical state is given by

$$W=0\,, \tag{5.6.25}$$

which gives some nonlinear correction to the linear result. But this nonlinear correction is somewhat defective. We have so far computed everything up to only orders of $(kA)^2$. Thus a perturbation based on the smallness of (kA) is implied. As we have seen from the multiple scale nonlinear analysis of the

Rayleigh-Taylor stability, the inclusion of second harmonics is needed to obtain the full nonlinear correction. Therefore, we should include in our trial solutions the second harmonic terms, so that the full nonlinear correction to the order of $(kA)^2$ can be obtained.

From another perspective, we may assume that there may indeed be nonlinearly stable states represented approximately by the trial solutions (5.6.5), (5.6.9) and (5.6.10). The problem is to find the nonlinear dispersion relation. Then we should try to find the full nonlinear relation not just limited to the order of $(kA)^2$. We shall now briefly deal with these questions.

Nonlinear Analysis

To include second harmonics, we shall now take the trial solutions :

$$\eta = A\sin S + A_2 \cos 2S \,, \tag{5.6.26}$$

$$\varphi^{(1)} = U_1 x + B\cos S e^{ky} + B_2 \sin 2S e^{2ky} \,, \tag{5.6.27}$$

$$\varphi^{(2)} = U_2 x + C\cos S e^{-ky} + C_2 \sin 2S e^{-2ky} \,. \tag{5.6.28}$$

It may be remarked that the $\cos 2S$ term is chosen in (5.6.26) out of experience with nonlinear analyses. Then it is expected that the second harmonics in (5.6.27) and (5.6.28) should be $\sin 2S$. We could assume more general expressions, but then the analysis will be more complicated. Substitute (5.6.26) - (5.6.28) into (5.6.6) - (5.6.8) and then carry out the same integration as before, we obtain up to $0\left(A^4, A_2^4\right)$:

$$\frac{\mathcal{J}_1}{\rho^{(1)}} = -\frac{BA}{2}(\omega + kU_1)\left[1 - \frac{1}{2}kA_2 + \frac{k^2}{8}\left(A^2 + 2A_2^2\right)\right]$$

$$-B_2(\omega + kU_1)\left[-A_2 + \frac{1}{2}kA^2 - k^2 A_2\left(A^2 + \frac{1}{2}A_2^2\right)\right]$$

$$+\frac{k^2}{4}B^2\left[1 + k^2\left(A^2 + A_2^2\right)\right] + \frac{k}{2}B_2^2\left[1 + 4k^2\left(A^2 + A_2^2\right)\right]$$

$$+kBB_2\left[kA + \frac{3}{2}k^2 AA_2 + \frac{9}{8}k^3 A\left(A^2 + A_2^2\right)\right] + \frac{g}{4}\left(A^2 + A_2^2\right) . \quad (5.6.29)$$

Let us denote

$$\mathcal{J}_1 = \Phi(\omega, k, \rho^{(1)}, U_1, g, A, A_2, B, B_2),$$

then

$$\mathcal{J}_2 = \Phi(\omega, k, \rho^{(2)}, U_2 - g, -A, -A_2, C, C_2) , \quad (5.6.30)$$

and

$$\mathcal{J}_a = \sigma\left[1 + \frac{k^2}{4}\left(A^2 + 4A_2^2\right) - \frac{3}{64}k^4\left(A^4 + 16A^2A_2^2 + 16A_2^4\right)\right] . \quad (5.6.31)$$

Now we vary with respect to A, B, C, A_2, B_2 and C_2 . We shall limit ourselves to the case that $(kA)^2$ is small, and also assume that $A_2, B_2, C_2 = 0(A^2)$. Then after some calculations, it is found that the dispersion relation (5.6.22) is now modified up to $0\left(k^2A^2\right)$ to :

$$\left(\rho^{(1)} + \rho^{(2)}\right)\omega^2 + 2\left(\rho^{(1)}U_1 + \rho^{(2)}U_2\right)k\omega + \left(\rho^{(1)}U_1^2 + \rho^{(2)}U_2^2\right)k^2$$

$$-gk\left(\rho^{(1)} - \rho^{(2)}\right) - \sigma k^3 - \alpha(kA)^2 = 0 , \quad (5.6.32)$$

where

$$\alpha = \frac{1}{2} gk\left(\rho^{(1)} - \rho^{(2)}\right) + \frac{\sigma}{8} k^3$$
$$+ \frac{\left[\rho^{(1)}(\omega + kU_1)^2 - \rho^{(2)}(\omega + kU_2)^2\right]^2}{2k\left[g\left(\rho^{(1)} - \rho^{(2)}\right) - 2\sigma k^2\right]} . \tag{5.6.33}$$

It is clear that the nonlinearity will inhibit the linear instability if $\alpha > 0$, and enhance the instability if $\alpha < 0$.

Up to order $0\left(k^2 A^2\right)$, we then obtain

$$\omega = \frac{\left\{-\left(\rho^{(1)}U_1 + \rho^{(2)}U_2\right)k \pm \left[W(k, U_1 - U_2, 0) + \alpha\left(\rho^{(1)} + \rho^{(2)}\right)(kA)^2\right]^{\frac{1}{2}}\right\}}{\left(\rho^{(1)} + \rho^{(2)}\right)}, \tag{5.6.34}$$

where W is given by (5.6.24) and in α the value of ω is that given by the linear relation, i.e., the expression (5.6.34) with A=0.

Now we know the system is linearly unstable if W<0. This instability can now be arrested by the nonlinear effect if $\alpha > 0$. The arrested state will be a wave with

$$\omega = -\frac{k\left(\rho^{(1)}U_1 + \rho^{(2)}U_2\right)}{\left(\rho^{(1)} + \rho^{(2)}\right)} , \tag{5.6.35}$$

and

$$(kA)^2 = -\frac{W(k, U_1 - U_2, 0)}{\alpha\left(\rho^{(1)} + \rho^{(2)}\right)} . \tag{5.6.36}$$

The expression α , using (5.6.35) can be put in the form

$$\alpha = -\frac{3}{8}\sigma k^3 + \frac{1}{2}\frac{\rho^{(1)}\rho^{(2)}k^2}{\left(\rho^{(1)}+\rho^{(2)}\right)}(U_1 - U_2)^2$$

$$+\frac{1}{2}\frac{\left(\rho^{(1)}-\rho^{(2)}\right)\left[\rho^{(1)}\rho^{(2)}k^2(U_1-U_2)^2\right]^2}{\left[gk\left(\rho^{(1)}-\rho^{(2)}\right)-2\sigma k^3\right]\left(\rho^{(1)}+\rho^{(2)}\right)^4} . \qquad (5.6.37)$$

This result agrees with that obtained from a multiple-scale analysis.

Fully Nonlinear One-Mode Analysis

It turns out that if we take the one-mode representation of the trial solutions, a fully nonlinear analysis can be carried out. We shall take the trial solutions of the following form :

$$\eta(x,t) = A(t)\sin[kx + \psi(t)] , \qquad (5.6.38)$$

$$\varphi^{(1)}(x,t) = U_1 x + B(t)\cos[kx + \psi(t) + \psi_1(t)]e^{ky} , \quad (5.6.39)$$

$$\varphi^{(2)}(x,t) = U_2 x - C(t)\cos[kx + \psi(t) + \psi_2(t)]e^{-ky} . \quad (5.6.40)$$

These trial solutions contain the previous case of (5.6.5), (5.6.9) and (5.6.10) when we take $S = kx + \omega t$ and *A, B, C* to be constants. Here we allow $A(t), B(t), C(t), \psi(t), \psi_1(t)$ and $\psi_2(t)$ to be functions of *t*, not necessarily slowly varying. After substituting (5.6.38) - (5.6.40) into (5.6.6) - (5.6.8), we found that the integral over *y* will lead to terms with the factor of $e^{kA\sin(kx+\psi)}$. Now the integrals over *x* can be replaced by integrals over a period $\dfrac{2\pi}{k}$. As the modified Bessel function of the first kind is given by

$$I_0(r) = \frac{1}{2\pi}\int_0^{2\pi} e^{r\sin X} dX , \qquad (5.6.41)$$

and the elliptical integral is given by

$$E(m) = \int_0^{\frac{\pi}{2}} \left(1 - m\sin^2 z\right)^{\frac{1}{2}} dz \ , \qquad (5.6.42)$$

$\mathcal{J}_1$, $\mathcal{J}_2$ and $\mathcal{J}_a$ can be explicitly computed in terms of these modified Bessel functions and elliptical integrals. From the variations with respect to A, B, C, ψ, ψ_1 and ψ_2, we can obtain a set of six evolution equations which are fully nonlinear.

The limitation of taking only one dominant spatial Fourier mode is obvious. But the variational approach has the implication that the solutions are the best approximations if the asymptotic states can indeed be characterized by a dominant mode. In this sense, the one-mode solutions may have incorporated into its development some average effects of other modes not represented explicitly in the formulation. The most noteworthy aspect of this analysis, of course, is that it is not based on perturbation expansion like most nonlinear stability analyses. Therefore, it is not restricted to weakly nonlinear case. Some analyses of the evolution equations have been carried out (Hsieh and Chen 1985). But much still need to be explored.

5.7 Kelvin-Helmholtz Stability for Compressible Fluids

We shall now consider the Kelvin-Helmholtz stability for two compressible, barotropic, inviscid, immiscible fluids. The flow configuration is the same as the classical Kelvin-Helmholtz stability discussed in Section 5.5, except that the fluids are now compressible. We shall neglect gravity and surface tension, and consider the flow in two spatial dimensions for simplicity. The interface is thus as before :

$$y = \eta(x,t) \ . \qquad (5.7.1)$$

The fluid (1), as before, occupies the region $-h^{(1)} < y < \eta$, while the fluid (2) occupies region $\eta < y < h^{(2)}$. Since the fluids are compressible, the densities of the fluids are no longer constant. The continuity equations for respective regions are, according to (1.1.1), thus :

$$\frac{\partial \rho^{(i)}}{\partial t} + \nabla \cdot \left(\rho^{(i)} \mathbf{v}^{(i)} \right) = 0 \,, \quad i = 1,\, 2 \,, \qquad (5.7.2)$$

while the momentum equations, according to (1.1.6), for inviscid fluids with no external force are :

$$\frac{\partial \mathbf{v}^{(i)}}{\partial t} + \left(\mathbf{v}^{(i)} \cdot \nabla \right) \mathbf{v}^{(i)} = -\frac{1}{\rho^{(i)}} \nabla p^{(i)} \,, \quad i = 1,\, 2 \,. \quad (5.7.3)$$

Since the fluids are inviscid, we shall again consider irrotational flows. Thus the velocities $\mathbf{v}^{(i)}$ can be expressed in terms of velocity potentials $\varphi^{(i)}$:

$$\mathbf{v}^{(i)} = \nabla \varphi^{(i)} \,. \qquad (5.7.4)$$

Integrate the momentum equations (5.7.3), we obtain the Bernoulli equations for compressible fluids :

$$H^{(i)} + \frac{1}{2} \left(\nabla \varphi^{(i)} \right)^2 + \frac{\partial \varphi^{(i)}}{\partial t} = f_i(t) \,, \qquad (5.7.5)$$

where

$$H^{(i)}\left(\rho^{(i)} \right) = \int^{\rho^{(i)}} \frac{dp^{(i)}}{\rho^{(i)}} \,, \qquad (5.7.6)$$

is the enthalpy.

The interfacial conditions (5.1.16), (5.1.17) and (5.1.23), on $y = \eta(x,t)$ are

$$\frac{\partial \eta}{\partial t} - \frac{\partial \varphi^{(i)}}{\partial y} + \frac{\partial \varphi^{(i)}}{\partial x} \cdot \frac{\partial \eta}{\partial x} = 0 , \quad i = 1, 2 , \tag{5.7.7}$$

$$p^{(1)} = p^{(2)} . \tag{5.7.8}$$

In (5.7.8), we have neglected the surface tension. The boundary conditions at the surface $y = -h^{(1)}$ and $y = h^{(2)}$ are the same as (5.2.8) and (5.2.9) :

$$\frac{\partial \varphi^{(1)}}{\partial y} = 0 , \quad \text{on } y = -h^{(1)} , \tag{5.7.9}$$

$$\frac{\partial \varphi^{(2)}}{\partial y} = 0 , \quad \text{on } y = h^{(2)} . \tag{5.7.10}$$

We now consider the primary state defined as follows :

$$\eta_0 = 0 , \tag{5.7.11}$$

$$\varphi_0^{(1)} = U_1 x , \qquad \varphi_0^{(2)} = U_2 x , \tag{5.7.12}$$

$$\rho_0^{(1)} = \rho_0^{(1)} , \qquad \rho_0^{(2)} = \rho_0^{(2)} , \tag{5.7.13}$$

$$p_0^{(1)} = p^{(1)}\left(\rho_0^{(1)}\right) , \qquad p_0^{(2)} = p^{(2)}\left(\rho_0^{(2)}\right) , \quad p_0^{(1)} = p_0^{(2)} = p_0 , \tag{5.7.14}$$

$$H_0^{(1)} = H^{(1)}\left(\rho_0^{(1)}\right) , \quad H_0^{(2)} = H^{(2)}\left(\rho_0^{(2)}\right) , \tag{5.7.15}$$

$$f^{(1)} = \frac{1}{2}U_1^2 + H_0^{(1)} , \quad f^{(2)} = \frac{1}{2}U_2^2 + H_0^{(2)} , \tag{5.7.16}$$

where $\rho_0^{(1)}$ and $\rho_0^{(2)}$ are constants. It is evident that the primary state satisfies the equations (5.7.1) - (5.7.10). Now let the interface be perturbed to become

$$y = \eta_1(x,t) = Ae^{i(kx+\omega t)} . \tag{5.7.17}$$

For linear stability analysis, we expect other first order perturbed variables will have same (x,t) variation. Thus we obtain

$$\varphi_1^{(i)} = \psi^{(i)}(y)e^{i(kx+\omega t)} , \tag{5.7.18}$$

$$\rho_1^{(i)} = \tau^{(i)}(y)e^{i(kx+\omega t)} , \tag{5.7.19}$$

$$p_1^{(i)} = a_i^2 \tau^{(i)}(y)e^{i(kx+\omega t)} , \tag{5.7.20}$$

$$H_1^{(i)} = \frac{a_i^2}{\rho_0^{(i)}} \tau^{(i)}(y)e^{i(kx+\omega t)} , \tag{5.7.21}$$

where

$$a_i^2 = \left(\frac{dp^{(i)}}{d\rho^{(i)}} \right)_{\rho^{(i)}=\rho_0^{(i)}} . \tag{5.7.22}$$

Thus a_i is the sound speed of the unperturbed fluid *(i)*.

Using the primary state and the perturbed expressions (5.7.11) - (5.7.21), the linearized version of (5.7.2) becomes now :

$$\frac{d^2\psi^{(i)}}{dy^2} - k^2\psi^{(i)} + \frac{i(\omega + kU_i)\tau^{(i)}}{\rho_0^{(i)}} = 0 , \quad i = 1,\ 2 , \tag{5.7.23}$$

while (5.7.5) becomes

$$\frac{a_i^2}{\rho_0^{(i)}} \tau^{(i)} + i(\omega + kU_i)\psi^{(i)} = 0 , \quad i = 1,\ 2 . \tag{5.7.24}$$

From (5.7.23) and (5.7.24), we thus obtain

$$\frac{d^2\psi^{(i)}}{dy^2} - \left[k^2 - \frac{1}{a_i^2}(\omega + kU_i)^2 \right] \psi^{(i)} = 0 , \quad i = 1,\ 2 . \tag{5.7.25}$$

Using the boundary conditions (5.7.9) and (5.7.10), we thus obtain

$$\psi^{(1)}(y) = B_1 \cosh\left\{\left[k^2 - \frac{1}{a_1^2}(\omega + kU_1)^2\right]^{\frac{1}{2}}\left(y + h^{(1)}\right)\right\}, \quad (5.7.26)$$

and

$$\psi^{(2)}(y) = B_2 \cosh\left\{\left[k^2 - \frac{1}{a_2^2}(\omega + kU_2)^2\right]^{\frac{1}{2}}\left(y - h^{(2)}\right)\right\}. \quad (5.7.27)$$

The interfacial conditions (5.7.7), after linearization, become

$$i(\omega + kU_1)A$$

$$-\left[k^2 - \frac{1}{a_1^2}(\omega + kU_1)^2\right]^{\frac{1}{2}} B_1 \sinh\left[k^2 - \frac{1}{a_1^2}(\omega + kU_1)^2\right]^{\frac{1}{2}} h^{(1)} = 0\,, \quad (5.7.28)$$

and

$$i(\omega + kU_2)A$$

$$+\left[k^2 - \frac{1}{a_2^2}(\omega + kU_2)^2\right]^{\frac{1}{2}} B_2 \sinh\left[k^2 - \frac{1}{a_2^2}(\omega + kU_2)^2\right]^{\frac{1}{2}} h^{(2)} = 0\,. \quad (5.7.29)$$

The condition (5.7.8), after using (5.7.24) and linearization, becomes

$$\rho_0^{(1)}(\omega+kU_1)B_1\cosh\left[k^2-\frac{1}{a_1^2}(\omega+kU_1)^2\right]^{\frac{1}{2}}h^{(1)}$$

$$=\rho_0^{(2)}(\omega+kU_2)B_2\cosh\left[k^2-\frac{1}{a_2^2}(\omega+kU_2)^2\right]^{\frac{1}{2}}h^{(2)}\,. \quad (5.7.30)$$

Eliminating A, B_1 and B_2 from (5.7.28) - (5.7.30), we obtain the following dispersion relation :

$$\frac{\rho_0^{(1)}(\omega+kU_1)^2\cosh\left[k^2-\frac{1}{a_1^2}(\omega+kU_1)^2\right]^{\frac{1}{2}}h^{(1)}}{\left[k^2-\frac{1}{a_1^2}(\omega+kU_1)^2\right]^{\frac{1}{2}}\sinh\left[k^2-\frac{1}{a_1^2}(\omega+kU_1)^2\right]^{\frac{1}{2}}h^{(1)}}=-\{1\rightarrow 2\}\,, \quad (5.7.31)$$

where $\{1\rightarrow 2\}$ denotes the same expression as the one on the left hand side except that the subscripts and superscripts 1 are changed to 2.

The Case of Semi-infinite Fluid Layers

If we let $h^{(1)}\rightarrow\infty$ and $h^{(2)}\rightarrow\infty$, then the dispersion relation becomes

$$\frac{\rho_0^{(1)}(\omega+kU_1)^2}{\left[k^2-\frac{1}{a_1^2}(\omega+kU_1)^2\right]^{\frac{1}{2}}}=-\{1\rightarrow 2\}\,, \quad (5.7.32)$$

which is the dispersion relation first obtained by Landau (Landau 1944).

The dispersion relation (5.7.32) may be rewritten as

$$\frac{\rho_0^{(1)}\omega_1^2\left[k^2-\frac{1}{a_2^2}(\omega_1+kU)^2\right]^{\frac{1}{2}}}{\rho_0^{(2)}(\omega_1+kU)^2\left[k^2-\frac{\omega_1^2}{a_1^2}\right]^{\frac{1}{2}}}=-1\,, \qquad (5.7.33)$$

where

$$\omega_1=\omega+kU_1\,, \qquad (5.7.34)$$

$$U=U_2-U_1\,. \qquad (5.7.35)$$

If ω_1 is real, then ω is also real, and the flow state is stable. If ω_1 is complex and has negative imaginary part, then so is ω, and the flow state is unstable.

When $\frac{U^2}{a_1^2}<<1$ and $\frac{U^2}{a_2^2}<<1$, since $\frac{\omega_1}{k}=0(U)$, (5.733) becomes approximately

$$\left(\rho_0^{(1)}+\rho_0^{(2)}\right)\omega_1^2+2\rho_0^{(2)}kU\omega_1+\rho_0^{(2)}k^2U^2=0\,. \qquad (5.7.36)$$

Thus

$$\omega_1=\frac{1}{\rho_0^{(1)}+\rho_0^{(2)}}\left[-\rho_0^{(2)}kU\pm ikU\left(\rho_0^{(1)}\rho_0^{(2)}\right)^{\frac{1}{2}}\right], \qquad (5.7.37)$$

which is actually the same as the classical result for incompressible fluids (5.5.21) for semi-infinite fluid layers when gravitation and surface tension are neglected. Thus the flow state is always unstable since there are no stabilizing gravitational and surface tension effects.

However if U is very large, it can be shown that all the roots for ω_1 are real, and the flow state becomes stable. To show this, let us rewrite (5.7.33) as

$$\frac{\omega_1^2 - k^2 a_1^2}{\left(\rho_0^{(1)} a_1\right)^2 \omega_1^4} = \frac{\left(\omega_1 + kU\right)^2 - k^2 a_2^2}{\left(\rho_0^{(2)} a_2\right)^2 \left(\omega_1 + kU\right)^4} . \qquad (5.7.38)$$

Consider a curve

$$y = f(x) = \frac{\alpha}{x^4}\left(x^2 - k^2 a_1^2\right) . \qquad (5.7.39)$$

The real roots of the equation (5.7.38) can be considered as the intersection points of the curve $y = f(x)$ and a similar curve with the origin shifted by an amount kU along the x-axis. The curve $y = f(x)$ is shown in Fig. 5-5. It is clear that when $|U|$ is sufficiently large, one branch of the curve can intersect at 5 points of the other curve, and there are in total 6 intersection points. In other words, all the six roots of the equation (5.7.38) are real. Thus the flow state is stable.

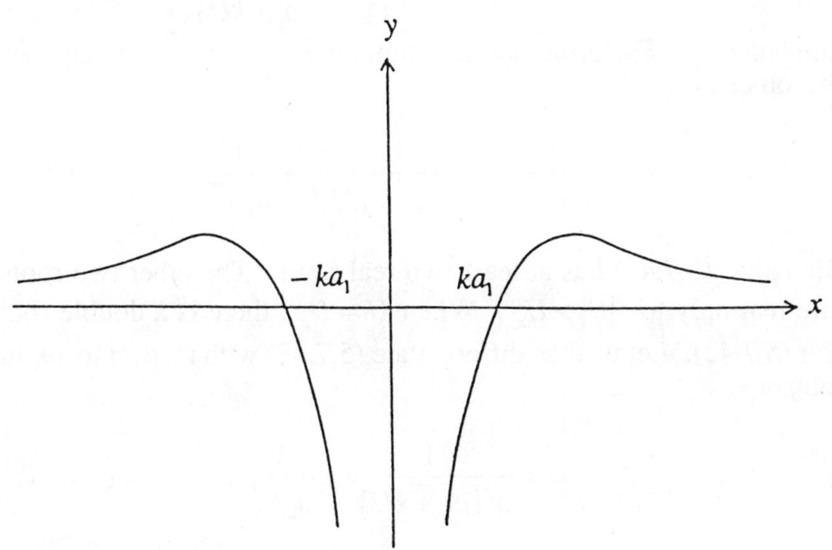

Fig. 5-5. Schematic representation of $y = \frac{\alpha}{x^4}\left(x^2 - k^2 a_1^2\right)$.

A particular case of interest is as follows :

$$p^{(1)}\left(\rho^{(1)}\right)=b_1\left(\rho^{(1)}\right)^{\gamma}, \quad p^{(2)}\left(\rho^{(2)}\right)=b_2\left(\rho^{(2)}\right)^{\gamma}. \quad (5.7.40)$$

For this case, fluids (1) and (2), although having the same γ , can still be different. For instance, if both are monatomic gases, then $\gamma=\frac{5}{3}$. For this particular case, then

$$a_1^2\rho_0^{(1)} = \gamma p_0^{(1)} = \gamma p_0^{(2)} = a_2^2\rho_0^{(2)} . \quad (5.7.41)$$

Using (5.7.41), then (5.7.38) can be rewritten as

$$\frac{a_1^2}{\omega_1^2}-\frac{a_2^2}{(\omega_1+kU)^2}=k^2\left[\frac{a_1^4}{\omega_1^4}-\frac{a_2^4}{(\omega_2+kU)^2}\right]. \quad (5.7.42)$$

A common factor in (5.7.42) is $\left[\frac{a_1^2}{\omega_1^2}-\frac{a_2^2}{(\omega_1+kU)^2}\right]$ whose roots are both real. Factoring out this common factor, the remaining part becomes :

$$\frac{1}{a_1^2a_2^2k^2}=\frac{1}{a_1^2(\omega_1+kU)^2}+\frac{1}{a_2^2\omega_1^2} . \quad (5.7.43)$$

Equation (5.7.43) has at least two real roots. The other two roots are real only for $|U|>U_c$. When $U=U_c$, there is a double root for (5.7.43). Let us thus differentiate (5.7.43) with respect to ω_1 to obtain

$$0=\frac{1}{a_1^2(\omega_1+kU)^3}+\frac{1}{a_2^2\omega_1^3} . \quad (5.7.44)$$

Eliminating ω_1 from (5.7.43) and (5.7.44), we obtain U_c . From (5.7.44), we have

$$(\omega_1 + kU) = -\left(\frac{a_2}{a_1}\right)^{\frac{2}{3}} \omega_1 \,. \tag{5.7.45}$$

Substitute (5.7.45) into (5.7.43), we obtain

$$\omega_1^2 = \left(a_1^2 + a_1^{\frac{4}{3}} a_2^{\frac{2}{3}}\right) k^2 \,. \tag{5.7.46}$$

Substituting (5.7.46) into (5.7.45), and we obtain

$$U_c^{\frac{2}{3}} = a_1^{\frac{2}{3}} + a_2^{\frac{2}{3}} \,. \tag{5.7.47}$$

If $a_1 = a_2 = a$, then

$$U_c = 2\sqrt{2}a \,. \tag{5.7.48}$$

Relationship with Two Phase Flows

The dispersion relation, in the limiting case $h^{(1)} \to 0$ and $h^{(2)} \to 0$, becomes

$$\rho_0^{(1)} h^{(2)} (\omega + kU_1)^2 \left[k^2 - \frac{1}{a_2^2}(\omega + kU_2)^2\right]$$

$$= -\rho_0^{(2)} h^{(1)} (\omega + kU_2)^2 \left[k^2 - \frac{1}{a_1^2}(\omega + kU_1)^2\right] . \tag{5.7.49}$$

We shall show that a similar dispersion relation can be obtained from two phase flow problems.

Consider the flow of a mixture of two compressible, inviscid fluids, which will be designated by the subscripts 1 and 2. We shall consider one dimensional flow problems. The simplest or basic model of such two phase flow has the following governing equations :

$$\frac{\partial}{\partial t}(\alpha_i \rho_i) + \frac{\partial}{\partial x}(\alpha_i \rho_i u_i) = 0 \, , \quad i = 1, \, 2 \, , \qquad (5.7.50)$$

$$\frac{\partial u_i}{\partial t} + u_i \frac{\partial u_i}{\partial x} = -\frac{1}{\rho_i}\frac{\partial p}{\partial x}$$

$$= -\frac{c_i^2}{\rho_i}\frac{\partial \rho_i}{\partial x} \, , \quad i = 1, \, 2 \, , \qquad (5.7.51)$$

where α_i is the volume fraction of the i-th phase, and

$$c_i = \left[\frac{dp_i}{d\rho_i}\right]^{\frac{1}{2}} , \qquad (5.7.52)$$

is the sound speed of the i-th phase. (5.7.50) are the continuity equations, and (5.7.51) the momentum equations. In this basic model we have taken $p_1 = p_2 = p$, and assume the fluids to be barotropic, thus

$$c_1^2 d\rho_1 = c_2^2 d\rho_2 = dp \, . \qquad (5.7.53)$$

It is by definition that

$$\alpha_1 + \alpha_2 = 1 \, . \qquad (5.7.54)$$

The stability of this nonlinear system may be investigated by the characteristics of the partial differential equations (5.7.50) and (5.7.51). If all the characteristics are real, then the system is hyperbolic, or we say the initial value problem is well-posed. In other words, the system is stable. Otherwise, the problem is ill-posed or the system is unstable. For this two-phase flow problem, so far as the stability criterion is concerned, identical results can be obtained from the linear analysis. Let us now take the primary flow state to be that these two phases are moving uniformly relative to each other. Thus in the primary state, we have

$$u_i = U_i \,, \qquad \rho_i = \rho_{i0} \,, \qquad i = 1,\ 2 \,, \tag{5.7.55}$$

$$\alpha_i = \alpha_{i0} \,, \qquad p = p_0 \,, \qquad i = 1,\ 2 \,. \tag{5.7.56}$$

For small perturbations, let us now take

$$u_i = U_i + u_i' \, e^{i(kx+\omega t)} \,, \qquad i = 1,\ 2 \,, \tag{5.7.57}$$

$$\alpha_i = \alpha_{i0} + \alpha_i' \, e^{i(kx+\omega t)} \,, \qquad i = 1,\ 2 \,, \tag{5.7.58}$$

$$\rho_i = \rho_{i0} + \rho_i' \, e^{i(kx+\omega t)} \,, \qquad i = 1,\ 2 \,, \tag{5.7.59}$$

and

$$p = p_0 + p' \, e^{i(kx+\omega t)} \,, \qquad i = 1,\ 2 \,. \tag{5.7.60}$$

Substituting (5.7.57) - (5.7.60) into (5.7.50) and (5.7.51), using (5.7.53) and (5.7.54) and neglecting second order and higher order small quantities, we obtain the following linearized relations :

$$\rho_{10}(\omega + kU_1)\alpha_1' + \frac{\alpha_{10}}{c_1^2}(\omega + kU_1)p' + k\alpha_{10}\rho_{10}u_1' = 0 \,, \tag{5.7.61}$$

$$-\rho_{20}(\omega + kU_2)\alpha_1' + \frac{\alpha_{20}}{c_2^2}(\omega + kU_2)p' + k\alpha_{20}\rho_{20}u_2' = 0 \,, \tag{5.7.62}$$

$$(\omega + kU_1)u_1' + \frac{k}{\rho_{10}}p' = 0 \,, \tag{5.7.63}$$

$$(\omega + kU_2)u_2' + \frac{k}{\rho_{20}}p' = 0 \,. \tag{5.7.64}$$

Eliminating α_1' , u_1' , u_2' and p' from (5.7.61) - (5.7.64), we obtain the dispersion relation :

$$\left[\frac{\alpha_{10}\rho_{20}}{c_1^2}+\frac{\alpha_{20}\rho_{10}}{c_2^2}\right](\omega+kU_1)^2(\omega+kU_2)^2$$

$$-\alpha_{20}\rho_{10}k^2(\omega+kU_1)^2-\alpha_{10}\rho_{20}k^2(\omega+kU_2)^2=0\,. \quad (5.7.65)$$

If we make the identification of α_{i0} with $h^{(i)}$, and c_i with a_i , equation (5.7.65) is exactly the same as (5.7.49). Thus we have demonstrated that the underlying physical mechanism of the "ill-posedness" of the two-phase flow problem is the Kelvin-Helmholtz instability.

From (5.7.65), it can also be shown that all four roots for $\left(\frac{\omega}{k}\right)$ are real if $(U_1-U_2)^2$ is large enough. In fact, denote

$$U_c^2=\left[(\alpha_{10}\rho_{20})^{\frac{1}{3}}+(\alpha_{20}\rho_{10})^{\frac{1}{3}}\right]\left[\frac{\alpha_{10}\rho_{20}}{c_1^2}+\frac{\alpha_{20}\rho_{10}}{c_2^2}\right]^{-1}\,. \quad (5.7.66)$$

It can be shown that all four roots for $\left(\frac{\omega}{k}\right)$ are real if

$$(U_1-U_2)^2\geq U_c^2\,, \quad (5.7.67)$$

and there are two complex roots for $\left(\frac{\omega}{k}\right)$, if

$$0<(U_1-U_2)^2<U_c^2\,. \quad (5.7.68)$$

U_c is clearly of the order of c_1 or c_2 . Thus the flow system can be stabilized if the relative speed is sufficiently supersonic.

The assumption that $h^{(1)} \to 0$ and $h^{(2)} \to 0$, or equivalently $kh^{(1)} << 1$, or $kh^{(2)} << 1$, implies that we are considering the long wavelength problem. The basic model of the two-phase flow has also implied that the characteristic length of the problem is long compared with the dimension of bubbles if we are dealing with bubbly liquids. For the study of this class of problems, a long wavelength Kelvin-Helmholtz problem can be formulated from the outsets. Then the gravitational and viscous effects can also be readily incorporated into the study. (Hsieh 1989)

5.8 The Faraday Problem

Move a glass of water up and down, and that is the Faraday problem. When we move the glass very slowly, the water surface will remain plane, as in the equilibrium case, if the effect of surface tension is neglected. However, when we increase the frequency and amplitude of the up-and-down motion, the water surface will no longer remain plane, i.e., the plane surface becomes unstable. Surface waves start to form, and eventually water spouts out of the glass. Faraday problem is the study of this type of instability.

Although the general Faraday problem deals with flow in a container of any shape moving in a general manner, we shall concentrate on the case of the cylindrical container moving vertically up and down, as illustrated in Fig. 5-6.

The radius of the cylindrical container is R. At equilibrium, the bottom of the container is given by Z=0, while the water surface is given by *Z=H*. Now the container will be moving up and down. Thus the bottom is given by *Z=a(t)*, and the free surface will be given by

$$Z = H + a(t) + \eta(r,\theta,t) \ . \qquad (5.8.1)$$

Except for the up and down motion, the general formulation of the problem is exactly the same as the water wave problem. Thus we have

$$\nabla^2 \varphi = 0\,, \quad \text{for} \quad a(t) < Z < H + a(t) + \eta(r,\theta,t)\,, \quad r < R\,, \tag{5.8.2}$$

with the following boundary and free surface conditions :

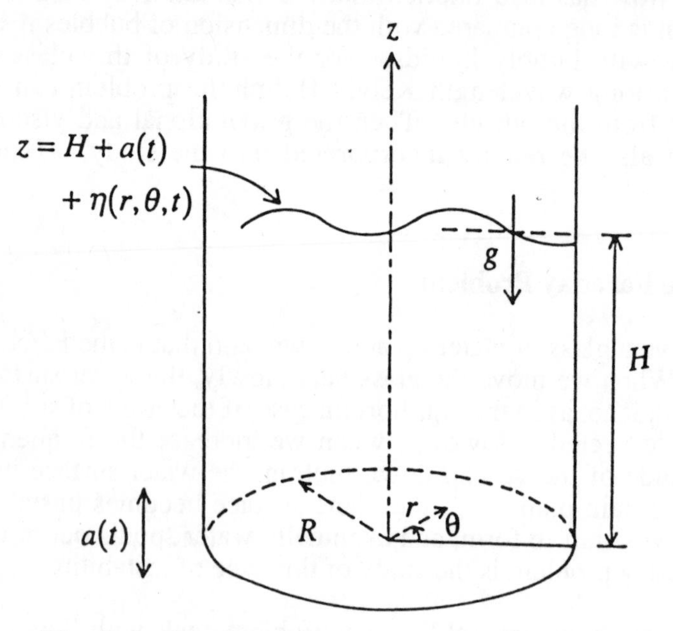

Fig. 5-6 The Schematic Diagram for Faraday Problem.

$$\frac{\partial \varphi}{\partial Z} = 0\,, \quad \text{on} \quad Z = a(t)\,, \tag{5.8.3}$$

$$\frac{\partial \varphi}{\partial r} = 0\,, \quad \text{on} \quad r = R\,, \tag{5.8.4}$$

$$\frac{\partial \eta}{\partial t} - \frac{\partial \varphi}{\partial Z} + (\nabla \varphi)\cdot(\nabla \eta) + \frac{da}{dt} = 0\,,$$

$$\text{on} \qquad Z = H + a(t) + \eta(r,\theta,t)\,, \tag{5.8.5}$$

$$\frac{\partial \varphi}{\partial t} + gZ + \frac{1}{2}(\nabla \varphi)^2 = f(t) \, ,$$

$$\text{on} \qquad Z = H + a(t) + \eta(r,\theta,t) \, . \tag{5.8.6}$$

We have used the Bernoulli's equation and the condition that p=0 on the free surface to arrive at (5.8.6). We have also neglected the surface tension and viscosity.

A primary state which satisfies (5.8.1) - (5.8.6) is given by

$$\eta = 0 \, , \tag{5.8.7}$$

$$\varphi = \varphi_0 \equiv Z\frac{da}{dt} \, , \tag{5.8.8}$$

and

$$[H + a(t)]\frac{d^2 a}{dt^2} + g[H + a(t)] + \frac{1}{2}\left(\frac{da}{dt}\right)^2 = f(t) \, . \tag{5.8.9}$$

Equation (5.8.9) defines $f(t)$ when $a(t)$ is specified.

The primary state represents the motion of the whole system moving up and down as a rigid body, with the free surface remaining plane. To investigate the stability of the primary state, we expect that the free surface may be distorted to give rise to a non-vanishing $\eta(r,\theta,t)$, which can be represented as follows :

$$\eta(r,\theta,t) = \sum_{m=0}^{\infty}\sum_{n=0}^{\infty} \alpha_{mn}(t) J_m(k_{mn} r)\cos m\theta \, , \tag{5.8.10}$$

where k_{mn} are chosen so that $k_{m,\,n+1} > k_{mn}$, and

$$J_m{}'(k_{mn}R) = 0 \, . \tag{5.8.11}$$

The velocity potential φ will also be modified to become

$$\varphi = \varphi_0 + \varphi_1 , \tag{5.8.12}$$

where

$$\varphi_1 = \sum_{m=0}^{\infty}\sum_{n=0}^{\infty} A_{mn}(t)\cosh k_{mn}[Z - a(t)]J_m(k_{mn}r)\cos m\theta . \tag{5.8.13}$$

It can be easily verified that $\nabla^2\varphi_1 = 0$, and the boundary conditions (5.8.3) and (5.8.4) are also satisfied. To solve the general problem, we need to substitute (5.8.10) and (5.8.12) into (5.8.5) and (5.8.6), and it is clear that it is very difficult to find $\alpha_{mn}(t)$ and $A_{mn}(t)$ in general. However, for linear stability analysis, i.e., if α_{mn} and A_{mn} are small, then it can be shown that the disturbance modes are decoupled, and individual equations governing α_{mn} and A_{mn} can be obtained. We shall now consider the linear case.

Linear Stability

When α_{mn} and A_{mn} are small, by neglecting terms nonlinear in α_{mn} and A_{mn} , the free surface conditions (5.8.5) and (5.8.6) become

$$\frac{\partial\eta}{\partial t} - \frac{\partial\varphi_1}{\partial Z} = 0 , \quad \text{on} \qquad Z = H + a(t) , \tag{5.8.14}$$

and

$$\frac{\partial\varphi_1}{\partial t} + \eta\frac{d^2a}{dt^2} + g\eta + (\nabla\varphi_0)\cdot(\nabla\varphi_1) = 0 , \text{ on } Z = H + a(t) . \tag{5.8.15}$$

Because of the orthogonality of the functions $J_m(k_{mn}r)\cos m\theta$, the modes are decoupled, thus we obtain from (5.8.14) and (5.8.15) :

$$\frac{d\alpha_{mn}}{dt} - A_{mn}k_{mn}\sinh k_{mn}H = 0\ , \qquad m,n = 0,\ 1,\ 2,\ \cdots\ , \tag{5.8.16}$$

$$\frac{dA_{mn}}{dt}\cosh k_{mn}H - A_{mn}k_{mn}\frac{da}{dt}\sinh k_{mn}H + \left(g + \frac{d^2a}{dt^2}\right)\alpha_{mn}$$

$$+\frac{da}{dt}A_{mn}k_{mn}\sinh k_{mn}H = 0\ ,$$

or

$$\frac{dA_{mn}}{dt}\cosh k_{mn}H + \left(g + \frac{d^2a}{dt^2}\right)\alpha_{mn} = 0\ ,\ \ m,n = 0,\ 1. \tag{5.8.17}$$

From (5.8.16) and (5.8.17), we thus obtain

$$\frac{d^2\alpha_{mn}}{dt^2} + \left(\omega_{mn}^2 + \frac{d^2a}{dt^2}k_{mn}\tanh k_{mn}H\right)\alpha_{mn} = 0\ , \tag{5.8.18}$$

where

$$\omega_{mn}^2 = gk_{mn}\tanh k_{mn}H\ , \tag{5.8.19}$$

is the frequency relation which we have encountered when we discussed the water wave.

Equations (5.8.18) determine the linear stability of the primary state of the system. If *a(t)* is a periodic function of *t*, (5.8.18) is a Hill equation. If *a(t)* is a sinusoidal function of *t*, then (5.8.18) is a Mathieu equation.

We shall refer the property of Mathieu equation to standard texts on differential equations, which usually includes the well-known stability diagram. We shall mention only some important features which are relevant to this problem.

Take the Mathieu equation :

$$\frac{d^2y}{dt^2}+(a+b\cos 2t)y=0\ , \quad \text{(a, b constants)}\ . \quad (5.8.20)$$

If b=0, then $y(t)$ is stable for a>0, and unstable for a<0. Moreover $y(t)$ is periodic with period 2π for a=0, 1, 4, 9, 16, $\cdots$. For $b \neq 0$, then $y(t)$ becomes unstable for $|b|$ large enough even for a>0. The diagram showing the stability region in (a, b)-plane can usually be found in standard texts on differential equations. An important point to note is that $y(t)$ is unstable for even very small b when $a = n^2$, where n is an integer.

Return to equation (5.8.18), let us consider the case that

$$a=-F\cos\Omega t\ . \quad (5.8.21)$$

Then (5.8.18) becomes

$$\frac{d^2\alpha_{mn}}{dt^2}+\left(\omega_{mn}^2+F\Omega^2k_{mn}\tanh k_{mn}H\cos\Omega t\right)\alpha_{mn}=0\ . \quad (5.8.22)$$

Denote

$$\tau=\frac{\Omega}{2}t\ , \quad (5.8.23)$$

then we can transform (5.8.22) into the form of (5.8.20) :

$$\frac{d^2\alpha_{mn}}{d\tau^2}+(a+b\cos 2\tau)\alpha_{mn}=0\ , \quad (5.8.24)$$

where

$$a=\left(\frac{2\omega_{mn}}{\Omega}\right)^2, \quad (5.8.25)$$

$$b=4Fk_{mn}\tanh k_{mn}H\ . \quad (5.8.26)$$

Thus, given a forcing frequency Ω , the most likely mode to be excited is the mode with

$$\omega_{mn} \approx \frac{j}{2}\Omega\,, \quad j = 1,\ 2,\ 3,\ \cdots\,. \tag{5.8.27}$$

A few remarks may be made :

(1) Although the inviscid theory implies that instability occurs with arbitrary small b when $a = n^2$, the inclusion of viscous effect will displace the instability to a higher threshold b_c .

(2) The inclusion of viscous effect will also make the modes with higher (m, n) less likely to be excited. Therefore, the mode expected to be excited will have frequency close to half the driving frequency Ω .

(3) The spatial pattern of the excited mode is given by $J_m(k_{mn}r)\cos m\theta$.

(4) As the driving amplitude b increases, the unstable frequency band around $\dfrac{\Omega}{2}$ becomes wider, and more than one ω_{mn} may fall in the unstable band. Thus, more than one mode may be excited.

(5) A nonlinear theory is required to deal with the situation when the system is linearly unstable. When there is only one linearly unstable mode, the system may settle down to some frequency with finite amplitude somewhat resembling the linearly unstable mode. However, when there are more than one linearly unstable mode, then the nonlinear interactions between the modes need to be considered.

Before we discuss the nonlinear theory, it may be mentioned that if the container is rectangular with sides L_x and L_y, instead of circular in cross section, the same equation (5.8.18) is also obtained. Now the surface elevation η is given by

$$\eta(x,y,t) = \sum_{m=0}^{\infty}\sum_{n=0}^{\infty} \alpha_{mn}(t)\cos\left(\frac{m\pi x}{L_x}\right)\cos\left(\frac{n\pi y}{L_y}\right), \quad (5.8.28)$$

and

$$k_{mn}^2 = \left[\left(\frac{m\pi}{L_x}\right)^2 + \left(\frac{n\pi}{L_y}\right)^2\right]^{\frac{1}{2}}. \quad (5.8.29)$$

Nonlinear Theory

We shall describe the essence of the nonlinear thoery. The basic idea is that, since there is inevitably dissipation in the system, only those modes satisfy the relation (5.8.27) will be excited and maintained. Other modes, even if excited initially, will decay eventually. Usually, the case that $j=1$ will be considered in the relation (5.8.27), since dissipation is stronger for higher modes. Thus, given the driving frequency Ω , we look for the mode associated with ω_l such that

$$\omega_l \approx \frac{\Omega}{2}, \quad (5.8.30)$$

where we have used a single subscript l to designate the (m, n)-mode. It is possible that there is another mode also with frequency close to $\frac{\Omega}{2}$, i.e.,

$$\omega_p \approx \frac{\Omega}{2}. \quad (5.8.31)$$

Suppose only the mode l is close to resonance, i.e, $\omega_l \approx \frac{\Omega}{2}$, then we may carry out a perturbation analysis from the nonlinear formulation by singling out the mode l. We may use the multiple scale expansion or the variational method to achieve the purpose. For example, for rectangular basins as exemplified by (5.8.28), the surface elevation associated with the mode l, i.e., the (m, n)-mode, will be given by

$$Z_l = \left[A_l(t)\cos\frac{\Omega}{2}t + B_l(t)\sin\frac{\Omega}{2}t \right]\cos\left(\frac{m\pi x}{L_x}\right)\cos\left(\frac{n\pi y}{L_y}\right). \quad (5.8.32)$$

It is to be noted that at the marginally stable state of the linear theory, both A_l and B_l are constants. Here A_l and B_l are slowly varying functions of t. Denote the complex amplitude C_l by

$$C_l = A_l - iB_l\,, \quad (5.8.33)$$

then the amplitude function $C_l(t)$ associated with the model l will satisfy an equation of the form :

$$\frac{dC_l}{dt} = \alpha_l C_l + \beta_l C_l^* + \gamma_l |C_l|^2 C_l\,, \quad (5.8.34)$$

where α_l , β_l and γ_l are coefficients that can be determined from the perturbation analysis. The exponential growth of the linear instability will normally be arrested by the nonlinear effects as given in (5.8.34).

When two or more spatial modes are excited simultaneously, the amplitude equations will contain coupling terms due to mutual interactions besides the respective individual dynamics. For the case of rectangular basin, the (n, m)-mode will also be close to resonance, since $\omega_{mn} = \omega_{nm}$. Call this mode to be mode p, thus the corresponding amplitude function will be designated as C_p. Then, due to the mutual interaction, it can be shown that the amplitude functions satisfy the following type of equations :

$$\frac{dC_l}{dt} = \alpha_l C_l + \beta_l C_l^* + \gamma_l |C_l|^2 C_l + \mu_{lp} |C_p|^2 C_l + \upsilon_{lp} C_l^* C_p^2\,, \quad (5.8.35)$$

$$\frac{dC_p}{dt} = \alpha_p C_p + \beta_p C_p^* + \gamma_p |C_p|^2 C_p + \mu_{pl} |C_l|^2 C_p + \upsilon_{pl} C_p^* C_l^2 \,. \quad (5.8.36)$$

It is to be noted that not all the possible cubic terms are present in the above equations. For instance, there is no $C_p^2 C_l^*$ term in (5.8.36). The term is absent because of symmetry requirement dictated by the geometrical shape of the rectangular basin.

Based on the nonlinear evolution equations as exemplified by (5.8.35) and (5.8.36) detailed bifurcation analyses can be carried out. When the driving amplitude F is sufficiently large, chaos will appear.

Studies have also been carried out for coupling between modes whose natural frequencies are not equal or nearly equal but in the ratio of two to one. These are the cases that

$$\omega_l \approx \frac{\Omega}{2} \,, \quad (5.8.37)$$

and either

$$\omega_p \approx 2\omega_l \quad (5.8.38)$$

or

$$\omega_p \approx \frac{1}{2}\omega_l \,. \quad (5.8.39)$$

Stability of Oscillating Bubbles

In Section 5.4, we have discussed the stability of spherical bubble in motion. The equations governing the linear stability are given by (5.4.31) and (5.4.32) which we reproduce here :

$$\frac{d^2 c_{lm}}{dt^2} + C_l c_{lm} = 0 \,, \quad l \geq 1 \,, \quad -l \leq m \leq l \,, \quad (5.8.40)$$

$$C_l = (l-1)(l+1)(l+2)\left[\frac{\sigma}{\rho R^3}\right]$$

$$-\frac{1}{R}\left[l+\frac{1}{2}\right]\frac{d^2R}{dt^2}-\frac{3}{4}\left[\frac{1}{R}\frac{dR}{dt}\right]^2 \; 2, \cdots . \qquad (5.8.41)$$

If the spherical bubble is oscillating sinusoidally with frequency ω, then we may write

$$R = R_0[1+\delta\cos\omega t] . \qquad (5.8.42)$$

Thus C_l is also a periodic function with period $\left(\frac{2\pi}{\omega}\right)$, and (5.8.40) is a Hill's equation. If the amplitude of oscillation is small, i.e., $\delta << 1$, then we can express C_l as

$$C_l = a_l + b_l\cos\omega t + 0\left(\delta^2\right) , \qquad (5.8.43)$$

where

$$a_l = (l-1)(l+1)(l+2)\left[\frac{\sigma}{\rho R_0^3}\right] , \qquad (5.8.44)$$

$$b_l = \delta\left[\left(l+\frac{1}{2}\right)\omega^2-(l-1)(l+1)(l+2)\left(\frac{\sigma}{\rho R_0^3}\right)\right] . \qquad (5.8.45)$$

Therefore, when $\delta << 1$, equation (5.8.40) becomes a Mathieu equation. Just like the case we have discussed in section 5.8, the most likely mode to be excited is the mode with

$$\frac{2\sqrt{a_l}}{\omega} \approx n , \quad n = 1,\ 2,\ 3,\ \cdots . \qquad (5.8.46)$$

Take $n=1$, i.e., the lowest mode, we obtain

$$4(l-1)(l+1)(l+2)\left(\frac{\sigma}{\rho R_0^3}\right)=\omega^2 . \qquad (5.8.47)$$

The nonlinear oscillation of bubble is quite complicated. Even for sinusoidal oscillations, normally it is the ambient pressure that is prescribed to execute sinusoidal motion. For nonlinear oscillations, the radial motion will not be in the form of (5.8.42). In fact, even for purely spherical oscillations of the bubble, the nonlinear theory is not simple. Bifurcations and chaos will arise as the amplitude increases. For the nonlinear stability analysis of the nonspherical modes, a perturbation analysis similar to that for Faraday problem can be carried out. The variational method is appropriate for this nonlinear problem.

5.9 The Rayleigh-Benard Problem

Consider a layer of incompressible, viscous fluid heated below in the gravitational field. Now the density of the fluid will vary with the temperature. Since the hotter the fluid, the lighter will be the fluid, therefore we have a situation that the lighter fluid is underneath the heavier fluid. Thus the Rayleigh-Taylor mechanism discussed in section 5.2 will be operating and the equilibrium state can no longer be maintained. A convective flow will result from this instability. If the fluid is inviscid, any thermal gradient will cause instability. For finite viscosity, a threshold value of thermal gradient is required for the onset of the thermal instability. This is the essence of the Rayleigh-Benard problem.

The variation of fluid density ρ with the temperature T can be expressed as follows :

$$\rho=\rho_0\left[1-\alpha\left(T-T_0\right)\right], \qquad (5.9.1)$$

where α is the coefficient of thermal expansion, ρ_0 and T_0 are the reference density and reference temperature respectively. We shall consider the cases that the density variation is small, i.e.,

$$\frac{\rho - \rho_0}{\rho_0} << 1 . \tag{5.9.2}$$

Then the equation of continuity (1.1.1) can be approximated by

$$\nabla \cdot \mathbf{v} = 0 . \tag{5.9.3}$$

However, the variation of density cannot be completely ignored in the Navier-Stokes equation (1.1.6). The variation of density coupled with the gravitational field results in the buoyancy force, which must be taken into account. For other aspects, the density can be considered as constant. Making use of (5.9.3), the equation (1.1.6) become thus :

$$\rho_0 \left[\frac{\partial \mathbf{v}}{\partial t} + (\mathbf{v} \cdot \nabla)\mathbf{v} \right] = -\nabla p - \mu \nabla \times (\nabla \times \mathbf{v}) + \rho_0 \left[1 - \alpha (T - T_0) \right] \mathbf{b} . \tag{5.9.4}$$

We also have the heat equation (1.1.16) :

$$\frac{\partial T}{\partial t} + (\mathbf{v} \cdot \nabla) T = D_T \nabla^2 T , \tag{5.9.5}$$

where

$$D_T = \frac{\kappa}{\rho_0 c_p} . \tag{5.9.6}$$

The approximation involved in deriving (5.9.3) - (5.9.5) is known as Boussinesq approximation.

Let us now consider in details the problem of heat transfer for a layer of fluid of depth H heated below in the gravitation field. Let the fluid layer be of infinite extent in horizontal directions. The gravitational force is pointing in the negative z direction. Thus $\mathbf{b} = (0,0,-g)$. The fluid occupies the region $0 \le z \le H$.

At equilibrium, we have $\mathbf{v} = 0$, and the heat is transported purely by thermal conduction. Thus we have from (5.9.5) that

$$T = T_e \equiv T_0 - \beta z \ , \qquad (5.9.7)$$

where T_0 is the temperature at z=0. Using (5.9.7), we obtain from (5.9.1) and (5.9.4) that

$$\rho = \rho_e \equiv \rho_0 (1 + \alpha\beta z) \ , \qquad (5.9.8)$$

$$p = p_e \equiv p_0 - g\rho_0 \left(z + \frac{1}{2}\alpha\beta z^2 \right) , \qquad (5.9.9)$$

where ρ_0 and p_0 are density and pressure at z=0 respectively. The equilibrium state given by (5.9.7) - (5.9.9) satisfies exactly the general governing equations (5.9.3) - (5.9.5).

We shall consider only the linear stability analysis in this section. Let

$$T = T_e + \theta \ , \qquad p = p_e + p' \ , \qquad (5.9.10)$$

and substitute into (5.9.4) and (5.9.5). Retaining only the linear terms of $\mathbf{v}$, θ and p', we thus obtain :

$$\frac{\partial \mathbf{v}}{\partial t} = -\frac{1}{\rho_0}\nabla p' - \upsilon \nabla \times (\nabla \times \mathbf{v}) + g\alpha\theta \mathbf{e}_z \ , \qquad (5.9.11)$$

and

$$\frac{\partial \theta}{\partial t} = \beta w + D_T \nabla^2 \theta \ , \qquad (5.9.12)$$

where $\upsilon = \dfrac{\mu}{\rho_0}$, and $\mathbf{v} = (u, v, w)$. (5.9.3), (5.9.11) and (5.9.12) are the basic equations for the linear stability analysis.

We eliminate p' by taking curl of equation (5.9.11), and obtain

$$\frac{\partial}{\partial t}(\nabla \times \mathbf{v}) = g\alpha\left(\frac{\partial \theta}{\partial y}\mathbf{e}_x - \frac{\partial \theta}{\partial x}\mathbf{e}_y\right) - \upsilon \nabla \times \nabla \times (\nabla \times \mathbf{v}) . \quad (5.9.13)$$

Since $\nabla \cdot \mathbf{v} = 0$, thus $\nabla \times (\nabla \times \mathbf{v}) = -\nabla^2 \mathbf{v}$. Then we obtain, by taking curl again of (5.9.13),

$$\frac{\partial}{\partial t}(\nabla^2 \mathbf{v}) = g\alpha\left[-\frac{\partial^2 \theta}{\partial x \partial z}\mathbf{e}_x - \frac{\partial^2 \theta}{\partial y \partial z}\mathbf{e}_y + \left(\frac{\partial^2 \theta}{\partial x^2} + \frac{\partial^2 \theta}{\partial y^2}\right)\mathbf{e}_z\right]$$
$$+ \upsilon \nabla^2(\nabla^2 \mathbf{v}) . \quad (5.9.14)$$

Let ξ be the z-component of the vorticity, i.e.,

$$\xi = (\nabla \times \mathbf{v})_z = \frac{\partial v}{\partial x} - \frac{\partial u}{\partial y} . \quad (5.9.15)$$

Take the z-components of (5.9.13) and (5.9.14), and we have

$$\frac{\partial \xi}{\partial t} = \upsilon \nabla^2 \xi , \quad (5.9.16)$$

$$\frac{\partial}{\partial t}(\nabla^2 w) = g\alpha\left(\frac{\partial^2 \theta}{\partial x^2} + \frac{\partial^2 \theta}{\partial y^2}\right) + \upsilon \nabla^2(\nabla^2 w) . \quad (5.9.17)$$

Taking derivatives with respect to x and y of (5.9.3) respectively, we obtain

$$\frac{\partial^2 u}{\partial x^2}+\frac{\partial^2 u}{\partial y^2}=-\frac{\partial^2 w}{\partial x \partial z}-\frac{\partial \xi}{\partial y}, \tag{5.9.18}$$

$$\frac{\partial^2 v}{\partial x^2}+\frac{\partial^2 v}{\partial y^2}=-\frac{\partial^2 w}{\partial y \partial z}+\frac{\partial \xi}{\partial x}. \tag{5.9.19}$$

Thus we can solve (5.9.16) independently. We can solve for θ and w from (5.9.12) and (5.9.17). With ξ and w now known, then we can solve for u and v from (5.9.18) and (5.9.19). This is the program of solution to determine $\mathbf{v}$ and θ. However as far as stability is concerned, we need only to pay attention to (5.9.12) and (5.9.17).

For the boundary conditions on the planes z=0 and z=H, we should remark that the emphasis here is the problem of heat transfer. We shall not consider the surface waves even if there is free surface. Therefore the planes z=0 and z=H are fixed boundary for the fluid region. Hence we have

$$w=0, \quad \text{on} \quad z=0 \quad \text{and} \quad z=H. \tag{5.9.20}$$

We shall consider the case that the temperature will be kept unchanged from the equilibrium values on both planes z=0 and z=H. Then we have

$$\theta=0, \quad \text{on} \quad z=0 \quad \text{and} \quad z=H. \tag{5.9.21}$$

The other boundary conditions will depend on whether the boundary is rigid or free. Let us now investigate the cases separately.

Rigid Surface

For rigid surface, we have the non-slip boundary conditions for the fluid flow, i.e., u=v=0. Thus we also have

$$\frac{\partial u}{\partial x} = \frac{\partial u}{\partial y} = \frac{\partial v}{\partial x} = \frac{\partial v}{\partial y} = 0 \ .$$

We then obtain from (5.9.3) and (5.9.15) the following additional boundary conditions for w and ξ on rigid surface :

$$\frac{\partial w}{\partial z} = 0 \ , \tag{5.9.22}$$

$$\xi = 0 \ . \tag{5.9.23}$$

Free Surface

On free surface, the stresses vanish. Thus $\sigma_{xz}' = \sigma_{yz}' = 0$. Now

$$\sigma_{xz}' = \mu\left(\frac{\partial u}{\partial z} + \frac{\partial w}{\partial x}\right), \quad \sigma_{yz}' = \mu\left(\frac{\partial v}{\partial z} + \frac{\partial w}{\partial y}\right) \ .$$

Also since w=0 on these boundary surfaces, we thus have

$$\frac{\partial w}{\partial x} = \frac{\partial w}{\partial y} = 0 \ .$$

Therefore, we have on these plane surfaces that

$$\frac{\partial u}{\partial z} = \frac{\partial v}{\partial z} = 0 \ ,$$

and also

$$\frac{\partial^2 u}{\partial x \partial z} = \frac{\partial^2 u}{\partial y \partial z} = \frac{\partial^2 v}{\partial x \partial z} = \frac{\partial^2 v}{\partial y \partial z} = 0 \ .$$

Now we obtain from differentiation of (5.9.3) with respect to z:

$$\frac{\partial^2 u}{\partial x \partial z} + \frac{\partial^2 v}{\partial y \partial z} + \frac{\partial^2 w}{\partial z^2} = 0 .$$

Therefore we obtain the following additional boundary conditions for w and ξ on free surface :

$$\frac{\partial^2 w}{\partial z^2} = 0 , \tag{5.9.24}$$

$$\frac{\partial \xi}{\partial z} = 0 . \tag{5.9.25}$$

For stability analysis, we shall now concentrate on equations (5.9.12) and (5.9.17) with boundary conditions (5.9.21) and either (5.9.22) or (5.9.24) on z=0 and z=H. The normal mode analysis will be adopted for the linear stability study. Let

$$w = W(z)e^{i(k_x x + k_y y) + st} , \tag{5.9.26}$$

$$\theta = \Theta(z)e^{i(k_x x + k_y y) + st} , \tag{5.9.27}$$

and

$$k^2 = k_x^2 + k_y^2 . \tag{5.9.28}$$

Substitute (5.9.26) and (5.9.27) into (5.9.12) and (5.9.17), we obtain

$$s\Theta = \beta W + D_T\left(\frac{d^2}{dz^2} - k^2\right)\Theta , \tag{5.9.29}$$

$$s\left(\frac{d^2}{dz^2} - k^2\right)W = -g\alpha k^2\Theta + \upsilon\left(\frac{d^2}{dz^2} - k^2\right)^2 W . \tag{5.9.30}$$

Let us now non-dimensionalize the problem by taking H as unit of linear spatial dimension, and $\left(\dfrac{H^2}{\upsilon}\right)$ as unit of time. Also denote

$$a = kH \,, \quad q = \frac{sH^2}{\upsilon} \,, \quad \sigma = \frac{\upsilon}{D_T} \,, \quad D = H\frac{d}{dz} \,, \tag{5.9.31}$$

where σ is known as the Prandtl number, which is a measure of the relative importance of the viscousity versus the thermal conductivity. In terms of the dimensionless variables and parameters, equations (5.9.29) and (5.9.30) become

$$\left(D^2 - a^2 - \sigma q\right)\Theta = -\left(\frac{\beta H^2}{D_T}\right)W \,, \tag{5.9.32}$$

$$\left(D^2 - a^2\right)\left(D^2 - a^2 - q\right)W = \left(\frac{g\alpha H^2}{\upsilon}\right)a^2\Theta \,. \tag{5.9.33}$$

The boundary conditions at z=0 and z=1 are

$$\Theta = 0 \,. \quad W = 0 \,, \tag{5.9.34}$$

and

$$DW = 0 \,, \quad \text{if the surface is rigid,} \tag{5.9.35}$$

$$D^2W = 0 \,, \quad \text{if the surface is free.} \tag{5.9.36}$$

We may eliminate Θ from (5.9.32) and (5.9.33) to obtain

$$\left(D^2 - a^2\right)\left(D^2 - a^2 - q\right)\left(D^2 - a^2 - \sigma q\right)W = -R_a a^2 W \,, \tag{5.9.37}$$

where

$$R_a = \left(\frac{g\alpha\beta}{D_T \upsilon} \right) H^4 , \qquad (5.9.38)$$

is known as the Rayleigh number. Θ also satisfies exactly the same equation as (5.9.37).

Principle of Exchange of Stabilities

The stability problem now amounts to solving (5.9.32) and (5.9.33) with the boundary conditions (5.9.34), and (5.9.35) or (5.9.36). When the physical parameters as well as a are given, this is an eigenvalue problem for the eigenvalue q. If Rl $q > 0$, then the system is linearly unstable. Otherwise the system is stable. We shall now show that if $R_a > 0$, then q is always real. Therefore, the eigenvalue problem will become much simpler since only real eigenvalues need to be considered. Moreover, the critical state is defined by q=0. Let us now proceed to prove this statement.

Denote

$$G = \left(D^2 - a^2\right)W , \quad F = \left(D^2 - a^2 - q\right)G .$$

Then (5.9.37) can be written as

$$\left(D^2 - a^2 - \sigma q\right)F = -R_a a^2 W . \qquad (5.9.39)$$

Multiply (5.9.39) by F^* , and then integrate with respect to z from 0 to 1, we obtain

$$\int_0^1 F^* \left(D^2 - a^2 - \sigma q\right) F dz = -R_a a^2 \int_0^1 F^* W dz . \quad (5.9.40)$$

Using integration by parts, we obtain that

$$\int_0^1 F^* \left(D^2 F\right) dz = -\int_0^1 |DF|^2 dz ,$$

since $F = \left(\frac{g\alpha H^2}{\upsilon}\right)a^2\Theta = 0$ on boundary and thus $F^*(DF) = 0$ also on boundary. Therefore the left-hand side of (5.9.40) can be rewritten as :

$$\int_0^1 F^*\left(D^2 - a^2 - \sigma q\right)F dz = -\int_0^1 \left[|DF|^2 + \left(a^2 + \sigma q\right)|F|^2\right]dz \ .$$

Now

$$\int_0^1 F^* W dz = \int_0^1 W\left(D^2 - a^2 - q^*\right)G^* dz \ .$$

Using integration by parts, we obtain

$$\int_0^1 W\left(D^2 G^*\right)dz = -\int_0^1 (DW)\left(DG^*\right)dz$$

$$= \int_0^1 G^*\left(D^2 W\right)dz \ ,$$

since on the boundary, W=0 thus $W\left(DG^*\right) = 0$; and either DW=0 or $D^2 W = 0$ thus $G^* = 0$, and thus $G^*\left(D^2 W\right) = 0$. Therefore,

$$\int_0^1 F^* W dz = \int_0^1 G^*\left(D^2 - a^2 - q^*\right)W dz$$

$$= \int_0^1 G^*\left[G - q^* W\right]dz$$

$$= \int_0^1 |G|^2 dz - q^* \int_0^1 W\left(D^2 - a^2\right)W^* dz$$

$$= \int_0^1 |G|^2 dz + q^* \int_0^1 \left[|DW|^2 + a^2 |W|^2 \right] dz ,$$

where we have used integration by parts again for the last step.

Therefore the equation (5.9.40) can be rewritten as

$$\int_0^1 \left[|DF|^2 + \left(a^2 + \sigma q\right)|F|^2 \right] dz$$

$$- R_a a^2 \int_0^1 \left[|G|^2 + q^* \left(|DW|^2 + a^2 |W|^2 \right) \right] dz = 0 . \quad (5.9.41)$$

Let us now take the real part of (5.9.41), we obtain :

$$\int_0^1 \left[|DF|^2 + a^2 |F|^2 - R_a a^2 |G|^2 \right] dz$$

$$+ (\mathrm{Rl}\, q) \int_0^1 \left[\sigma |F|^2 - R_a a^2 \left(|DW|^2 + a^2 |W|^2 \right) \right] dz = 0 . \quad (5.9.42)$$

Thus if $R_a \le 0$ then $\mathrm{Rl}\, q \le 0$. Therefore, without solving the eigenvalue problem, we can conclude that the system is stable if $R_a \le 0$, i.e., if the fluid is heated from top.

Now let us take the imaginary part of (5.9.41), we then obtain :

$$(\mathrm{Im}\, q) \left[\sigma \int_0^1 |F|^2 dz + R_a a^2 \int_0^1 \left(|DW|^2 + a^2 |W|^2 \right) dz \right] = 0 . \quad (5.9.43)$$

Thus if $R_a > 0$, then $\mathrm{Im}\, q = 0$. That completes the proof.

We say the principle of exchange of stabilities is valid if the critical state is defined by $q=0$. Then the instability sets in as a steady secondary flow. The principle of exchange of stabilities is actually also valid for the Rayleigh-Taylor stability.

Critical Rayleigh Number

Since the principle of exchange of stabilities is valid for this problem. We may set $q=0$ for the study of the critical case. Thus (5.9.32), (5.9.33) and (5.9.37) become respectively :

$$\left(D^2 - a^2\right)\Theta = -\left(\frac{\beta H^2}{D_T}\right)W , \tag{5.9.44}$$

$$\left(D^2 - a^2\right)^2 W = -\left(\frac{g\alpha H^2}{\upsilon}\right)a^2\Theta , \tag{5.9.45}$$

and

$$\left(D^2 - a^2\right)^3 W = -R_a a^2 W . \tag{5.9.46}$$

For (5.9.46), the boundary conditions are

$$W = 0 , \quad \left(D^2 - a^2\right)^2 W = 0 , \quad \text{at } z\text{=0 and } z\text{=1}, \tag{5.9.47}$$

and

$$DW = 0 , \quad \text{if the surface is rigid,} \tag{5.9.48}$$

$$D^2 W = 0 , \quad \text{if the surface is free.} \tag{5.9.49}$$

Equation (5.9.46) is a six order homogeneous differential equation, and we have six homogeneous boundary conditions. Therefore it is an eigenvalue problem for R_a if a is prescribed. We shall describe the results when different boundary conditions are applied.

(1) Both upper and lower surfaces are free

For this case, the boundary conditions are

$$W = D^2W = D^4W = 0 \,, \quad \text{at } z=0 \text{ and } z=1 \,. \tag{5.9.50}$$

If we make use of (5.9.46) and its successive derivatives, we can obtain that

$$D^{2m}W = 0 \,, \quad m=1, 2, \ldots \,, \quad \text{at } z=0 \text{ and } z=1 \,. \tag{5.9.51}$$

Therefore the eigensolutions are

$$W_n = \sin n\pi z \,, \quad n = 1, 2, \cdots \,. \tag{5.9.52}$$

Substitute (5.9.52) into (5.9.46), we obtain the eigen-Rayleigh number :

$$R_n = \frac{1}{a^2}\left(n^2\pi^2 + a^2\right)^3 \,. \tag{5.9.53}$$

The lowest eigen-Rayleigh number for non-trivial eigensolution is:

$$R_1 = \frac{1}{a^2}\left(\pi^2 + a^2\right)^3 \,. \tag{5.9.54}$$

So far, we have not said anything about a. Since the horizontal dimension is infinite in extent, any a is permissible. To find the minimum of R_1 , we take

$$\frac{dR_1}{d\left(a^2\right)} = 0 \,,$$

and find that

$$a_c^2 = \frac{\pi^2}{2} , \quad \text{or} \quad a_c \approx 2.2214 . \tag{5.9.55}$$

Substitute (5.9.55) into (5.9.54), we obtain the critical Rayleigh number :

$$R_c = \frac{27}{4}\pi^4 \approx 657.511 . \tag{5.9.56}$$

Therefore, for $R_a > R_c$, the conductive equilibrium state becomes unstable, and convection cells with non-dimensional wave number a start to appear. Returning to the dimensional quantities, the wavelength of the convection cell is

$$\lambda_c = \frac{2\pi}{k_c} = \frac{2\pi H}{a_c} = 2^{\frac{3}{2}} H . \tag{5.9.57}$$

(2) *Both upper and lower surfaces are rigid*

The boundary conditions for this case are

$$W = DW = \left(D^2 - a^2\right)^2 W = 0 , \quad \text{at } z=0 \text{ and } z=1. \tag{5.9.58}$$

The solutions of (5.9.46) have the form :

$$W = e^{\pm \gamma z} , \tag{5.9.59}$$

where γ satisfies the relation :

$$\left(\gamma^2 - a^2\right)^3 = -R_a a^2 . \tag{5.9.60}$$

Introduce τ such that

$$R_a a^2 = \tau^3 a^6 . \tag{5.9.61}$$

Then the six roots of (5.9.60) and $\pm \gamma_0$, $\pm \gamma_1$ and $\pm \gamma_1^*$, where

$$\gamma_0 = ia(\tau - 1)^{\frac{1}{2}} , \tag{5.9.62}$$

$$\gamma_1^2 = a^2\left[1 + \frac{\tau}{2}\left(1 + i\sqrt{3}\right)\right] . \tag{5.9.63}$$

The general solution is a linear combination of the solutions given by (5.9.59) with γ taking these six values. Substitute the solution into the boundary conditions (5.9.58) we can solve for the eigen Rayleigh number when a is prescribed. Just like the case for free-free boundaries, we take the eigen-Rayleigh number $R_1(a)$, and then find the minimum $R_1(a)$ to be R_c , the critical Rayleigh number. Numerical computation is needed to determine $R_1(a)$ and R_c , and it is found that R_c and the corresponding wave number are

$$R_c = 1707.762 , \quad a_c = 3.117 . \tag{5.9.64}$$

(3) Upper surface free, lower surface rigid

Similar approach and numerical computation can also determine the critical Rayleigh number and the corresponding wave number, and it is found that

$$R_c = 1100.65 , \quad a_c = 2.682 . \tag{5.9.65}$$

The same result is valid for the case that the upper surface is rigid and the lower surface free, although the case is difficult to realize physically.

The Cell Patterns

When the Rayleigh number is just above R_c , only the mode associated with a_c is unstable and the flow pattern will be convection cells given by (5.9.26) with $k^2H^2 = a_c^2$. In other words, the horizontal spatial pattern will be specified by $f(x,y)$ which satisfies the equation :

$$\frac{\partial^2 f}{\partial x^2}+\frac{\partial^2 f}{\partial y^2}+a^2 f=0 . \qquad (5.9.66)$$

Even though $a = a_c$ or $k = k_c$, since $k^2 = k_x^2 + k_y^2$, various spatial pattern is possible for the satisfaction (5.9.66). The following are some important possible patterns :

(1) The rolls

$$f=\cos ax , \quad \text{or} \quad f=\cos ay . \qquad (5.9.67)$$

This is the simplest case, for which f depends only on x or y. It represents infinitely long rolls; or the three-dimensional flow now reduces to two-dimensional flow.

(2) Rectangular cells

$$f=\cos a_x x \cos a_y y , \quad \text{with } a^2=a_x^2+a_y^2 . \qquad (5.9.68)$$

When $a_x = a_y = \dfrac{a}{\sqrt{2}}$, the rectangular cell becomes the square cell.

(3) Hexagonal cells

$$f(x,y)=\cos\frac{a}{2}\left(\sqrt{3}x+y\right)+\cos\frac{a}{2}\left(\sqrt{3}x-y\right)+\cos ay . \qquad (5.9.69)$$

It may be readily verified that the above f satisfies (5.9.66).

To demonstrate that this represents hexagonal cells, we may note that

$$f(x,y) = f(\pm x, \pm y)$$

$$f\left(x+\frac{4m\pi}{a\sqrt{3}}, y+\frac{4n\pi}{a}\right) = f(x,y) \;, \quad \text{for} \quad m,n = 0,\pm 1,\pm 2,\cdots \,.$$

And also if we put $x = r\cos\theta$, $y = r\sin\theta$, it can be verified that

$$f(r,\theta) = f\left(r, \theta+\frac{\pi}{3}\right) ,$$

thus the pattern has symmetry with respect to rotation of angle 60° about the origin. With these symmetries, it can be readily shown that the f given by (5.9.69) represents hexagonal patterns.

The detailed flow patterns require the solutions of u and v from (5.9.18) and (5.9.19) with $w = f(x,y)W(z)$ and $\theta = f(x,y)\Theta(z)$. However, the periodicity of the patterns given above indicates that the general flow patterns are alike without going into detailed calculations. The linear theory cannot determine which of the possible patterns will be chosen in reality. Whether nonlinear theories can solve these problem of pattern selections is still debatable.

Non-uniform Heating

If the lower surface is heated non-uniformly, i.e., if the temperature distribution on the lower surface is not uniform, then the problem becomes quite complex. First of all, with non-uniform heating, the state of rest and pure conduction is no longer a solution of the governing equations. Cellular motion, or plume flow, appears even for very low Rayleigh number. No simple analytical solution for such cellular flows can be found in general. We need to resort to numerical solutions to begin with. Then, no corresponding linear stability analyses as presented above can be carried out. But in application to real problems, we often need to deal with problems with non-uniform heating. Some numerical studies of this type of problems have been carried out. It has been found that extra localized heating may delay the transition which corresponds to the onset of Rayleigh-Benard instability for uniform heating.

5.10 The Taylor-Couette Problem

The Taylor-Couette problem, i.e., the stability of the viscous fluid between two rotating cylinders, is analytically somewhat similar to the Rayleigh-Benard problem. When we discuss the Rayleigh-Benard problem, we claimed that its instability is due to a kind of Rayleigh-Taylor mechanism. The temperature difference causes the fluid on top to become heavier than the fluid underneath, and thus the instability. The Taylor-Couette problem is also a kind of generalization of Rayleigh-Taylor problem. Here the centrifugal force replaces the gravitational force. Although the fluid density is constant, the centrifugal force is variable. So is also the angular momentum density. The system is unstable, if it is not in the state with minimal energy consistent with the conservation of angular momentum.

We consider an incompressible, viscous fluid between two infinite, concentric, circular cylinders. Let the inner cylinder with radius r_1 rotate with angular velocity Ω_1 , and the outer cylinder with radius r_2 rotate with Ω_2 . Take the cylindrical coordinates (r,θ,z) , and let $\mathbf{v}=(v_r,v_\theta,v_z)$, then the continuity equation (1.1.1) and the Navier-Stokes equations (1.1.6) have the following forms :

$$\frac{\partial v_r}{\partial r}+\frac{v_r}{r}+\frac{1}{r}\frac{\partial v_\theta}{\partial \theta}+\frac{\partial v_z}{\partial z}=0\ , \tag{5.10.1}$$

$$\frac{\partial v_r}{\partial t}+v_r\frac{\partial v_r}{\partial r}+\frac{v_\theta}{r}\frac{\partial v_r}{\partial \theta}+v_z\frac{\partial v_r}{\partial z}-\frac{v_\theta^2}{r}$$

$$=-\frac{1}{\rho}\frac{\partial p}{\partial r}+\upsilon\left(\nabla^2 v_r-\frac{2}{r^2}\frac{\partial v_\theta}{\partial \theta}-\frac{v_r}{r^2}\right), \tag{5.10.2}$$

$$\frac{\partial v_\theta}{\partial t}+v_r\frac{\partial v_\theta}{\partial r}+\frac{v_\theta}{r}\frac{\partial v_\theta}{\partial \theta}+v_z\frac{\partial v_\theta}{\partial z}+\frac{v_r v_\theta}{r}$$

$$=-\frac{1}{\rho r}\frac{\partial p}{\partial \theta}+\upsilon\left(\nabla^2 v_\theta+\frac{2}{r^2}\frac{\partial v_r}{\partial \theta}-\frac{v_\theta}{r^2}\right), \tag{5.10.3}$$

$$\frac{\partial v_z}{\partial t}+v_r\frac{\partial v_z}{\partial r}+\frac{v_\theta}{r}\frac{\partial v_z}{\partial \theta}+v_z\frac{\partial v_z}{\partial z}$$

$$=-\frac{1}{\rho}\frac{\partial p}{\partial z}+\upsilon\nabla^2 v_z . \tag{5.10.4}$$

The boundary conditions are

$$\mathbf{v}=(0,\Omega_1 r_1,0), \quad \text{on } r=r_1 , \tag{5.10.5}$$

$$\mathbf{v}=(0,\Omega_2 r_2,0), \quad \text{on } r=r_2 . \tag{5.10.6}$$

The primary flow state is to be given by

$$\mathbf{v}=(0,V(r),0), \quad p=p_0(r) . \tag{5.10.7}$$

Then it can be readily verified that

$$V(r)=\left(\frac{\Omega_2 r_2^2-\Omega_1 r_1^2}{r_2^2-r_1^2}\right)r+\left[\frac{(\Omega_1-\Omega_2)r_1^2 r_2^2}{r_2^2-r_1^2}\right]\frac{1}{r} . \tag{5.10.8}$$

The flow as represented by (5.10.7) and (5.10.8) is called the Couette flow. If $r_1=0$, i.e., for the internal flow within a rotating cylinder, (5.10.8) reduces to

$$V(r)=\Omega_2 r . \tag{5.10.9}$$

Therefore the fluid is rotating like a rigid body. If $\Omega_2 = 0$ and $r_2 \to \infty$, then it is the flow outside a rotating cylinder, and (5.10.8) reduces to

$$V(r) = \frac{\Omega_1 r_1^2}{r} , \qquad (5.10.10)$$

which is actually irrotational.

Now we consider the linear stability of the Couette flow. We shall limit the scope of the study to axisymmetric flows, i.e., $\mathbf{v}$ and p are independent of θ . Then from the continuity equation, we see that v_r and v_z may be represented by the Stokes stream function $\psi(r,z,t)$:

$$v_r = -\frac{\partial \psi}{\partial z} , \quad v_z = D_* \psi \equiv \left(\frac{\partial}{\partial r} + \frac{1}{r} \right) \psi . \qquad (5.10.11)$$

The flow field with axisymmetry can then be expressed as

$$\mathbf{v} = \left(-\frac{\partial \psi}{\partial z} , V(r) + v(r,z,t) , D_* \psi \right) . \qquad (5.10.12)$$

Before we write the governing equations in terms of ψ and v, let us first non-dimensionalize the system by taking r_2 as the unit of linear spatial dimension, $\left(\frac{r_2^2}{\upsilon} \right)$ as the unit of time, $\left(\frac{\upsilon}{r_2} \right)$ as the unit of $\mathbf{v}$, υ as the unit ψ. Also denote the Reynolds number R:

$$R = \frac{\Omega_1^2 r_1^2}{\upsilon} , \qquad (5.10.13)$$

and

$$\mu = \frac{\Omega_2}{\Omega_1}\,, \quad \eta = \frac{r_1}{r_2}\,. \tag{5.10.14}$$

Then $V(r)$ in (5.10.8) can be rewritten as

$$V(r) = \frac{R}{1-\eta^2}\left[\frac{(1-\mu)}{r} - \left(1 - \frac{\mu}{\eta^2}\right)r\right]. \tag{5.10.15}$$

Substitute (5.10.12) into (5.10.2) - (5.10.4) and eliminate p, we obtain

$$\frac{\partial}{\partial t}\mathcal{D}^2\psi = \frac{\partial\left(\psi, \mathcal{D}^2\psi\right)}{\partial(z,r)} - \frac{1}{r}\frac{\partial}{\partial z}\left(\psi\mathcal{D}^2\psi\right) + \mathcal{D}^4\psi$$

$$-\frac{2V(r)}{r}\frac{\partial v}{\partial z} - \frac{1}{r}\frac{\partial}{\partial z}\left(v^2\right), \tag{5.10.16}$$

$$\frac{\partial v}{\partial t} = \frac{\partial \psi}{\partial z}D_* v - \frac{\partial v}{\partial z}D_*\psi + \mathcal{D}^2 v - \frac{2R}{1-\eta^2}\left(1-\frac{\mu}{\eta^2}\right)\frac{\partial\psi}{\partial z}\,, \tag{5.10.17}$$

where

$$\mathcal{D}^2 = \left(\nabla^2 - \frac{1}{r^2}\right) = \frac{\partial}{\partial r}D_* + \frac{\partial^2}{\partial z^2}\,.$$

The boundary conditions are

$$\psi = \frac{\partial\psi}{\partial r} = v = 0\,, \quad \text{on } r = \eta\,, \quad \text{and } r = 1\,. \tag{5.10.18}$$

Rayleigh's Criterion

Before we deal with the problem of the viscous fluid, let us take a look at the corresponding inviscid fluid problem. If we again restrict our study to the problem with axial symmetry, then we can rewrite (5.10.3) when $\upsilon = 0$ as :

$$\frac{D}{Dt}(rv_\theta) \equiv \left(\frac{\partial}{\partial t} + v_r \frac{\partial}{\partial r} + v_z \frac{\partial}{\partial z}\right)(rv_\theta) = 0 \ , \quad (5.10.19)$$

where $\frac{D}{Dt}$ is the time derivative following the fluid particle. The quantity (rv_θ) is the angular momentum per unit mass. (5.10.19) states that the angular momentum of the fluid particle remains constant if the fluid flow field is axisymmetric. Let us denote (rv_θ) by L. Then the kinetic energy per unit mass associated with the azimuthal motion is $\left(\frac{1}{2}v_\theta^2\right)$ or $\left(\frac{1}{2}\frac{L^2}{r^2}\right)$. Equivalently, $\left(\frac{1}{2}\frac{L^2}{r^2}\right)$ is also the potential energy per unit mass associated with the fluid particle at r, since a centrifugal force per unit mass $\left(\frac{v_\theta^2}{r}\right) = \frac{L^2}{r^3}$ is acting on the fluid element in the outward radial direction.

Suppose now there are two fluid elements each with mass Δm , one situated at $r = r_l$, the other at $r = r_s$, with $r_l > r_s$. The fluid element at r_l has angular momentum $L_l(\Delta m)$ and kinetic energy $\frac{1}{2}\frac{L_l^2}{r_l^2}(\Delta m)$, while the fluid element at r_s has the corresponding quantities $L_s(\Delta m)$ and $\frac{1}{2}\frac{L_s^2}{r_s^2}(\Delta m)$ respectively. Now we interchange the positions of these two elements. The fluid elements carry with them the original angular momentum because of (5.10.19). But the kinetic energy will change. The net change of kinetic energy is

$$\Delta K = \frac{\Delta m}{2}\left[\left(\frac{L_l^2}{r_s^2} + \frac{L_s^2}{r_l^2}\right) - \left(\frac{L_l^2}{r_l^2} + \frac{L_s^2}{r_s^2}\right)\right]$$

$$= \frac{\Delta m}{2}\left(L_l^2 - L_s^2\right)\left(\frac{1}{r_s^2} - \frac{1}{r_l^2}\right) . \qquad (5.10.20)$$

Since $r_l > r_s$, therefore

$$\Delta K \gtrless 0 \quad \Leftrightarrow \quad L_l^2 \gtrless L_s^2 . \qquad (5.10.21)$$

If $\Delta K > 0$, then external work is required to supply the extra energy required by this interchange. Hence the original state is stable. If $\Delta K < 0$, then this interchange will lead to release of energy. Hence the original state is unstable. To insure stability, the requirement is that, for any pair of r_l and r_s , $L_l^2 > L_s^2$. In other words, we require everywhere that :

$$\frac{d}{dr}\left(L^2\right) > 0 . \qquad (5.10.22)$$

This is the Rayleigh's criterion.

With $V(r)$ given by (5.10.15) and since $L=rV(r)$, the condition (5.10.22) becomes for this case

$$\frac{-4\Omega_1^2\eta^2(1-\mu)\left(1-\dfrac{\mu}{\eta^2}\right)\left[1-\dfrac{\left(1-\dfrac{\mu}{\eta^2}\right)r^2}{(1-\mu)}\right]r}{\left(1-\eta^2\right)^2} > 0 . \qquad (5.10.23)$$

As $\eta \le r < 1$, thus the condition (5.10.23) is satisfied if

$$\mu > \eta^2 , \quad \text{or} \quad \frac{\Omega_2}{\Omega_1} > \left(\frac{r_1}{r_2} \right)^2 . \tag{5.10.24}$$

In general, viscosity tends to enhance the stability. Therefore the Rayleigh criterion (5.10.22) or (5.10.24) should be a conservative stability criterion for axisymmetric disturbances.

Linear Stability Analysis

Let us now return to (5.10.16) and (5.10.17), and study the linear problem. The linearized equations are

$$\mathcal{D}^4 \psi - \frac{\partial}{\partial t} \mathcal{D}^2 \psi = \frac{2R(1-\mu)}{1-\eta^2} \left(\frac{1}{r^2} - K \right) \frac{\partial v}{\partial z} , \tag{5.10.25}$$

and

$$\mathcal{D}^2 v - \frac{\partial v}{\partial t} = \frac{2R}{1-\eta^2} \left(1 - \frac{\mu}{\eta^2} \right) \frac{\partial \psi}{\partial z} , \tag{5.10.26}$$

where

$$K = \frac{1 - \dfrac{\mu}{\eta^2}}{1 - \mu} . \tag{5.10.27}$$

Comparing (5.10.25) and (5.10.26) with (5.9.17) and (5.9.12) of the Rayleigh-Benard problem, we may see the mathematical resemblence of these two cases.

Let us again employ the normal mode analysis. For convenience, let

$$\psi = \left(1 - \eta^2\right) \left[2Ra \left(1 - \frac{\mu}{\eta^2} \right) \right]^{-1} \psi'(r) e^{qt} \sin az , \tag{5.10.28}$$

$$v = v'(r)e^{qt}\cos az \; . \tag{5.10.29}$$

Substitute (5.10.28) and (5.10.29) into (5.10.25) and (5.10.26), we obtain

$$\left(\frac{d}{dr}D_* - a^2 - q\right)\left(\frac{d}{dr}D_* - a^2\right)\psi' = -T'a^2\left(\frac{1}{r^2} - K\right)v' \; , \tag{5.10.30}$$

$$\left(\frac{d}{dr}D_* - a^2 - q\right)v' = \psi' \; , \tag{5.10.31}$$

where

$$T' = \frac{4R^2(1-\mu)\left(1 - \dfrac{\mu}{\eta^2}\right)}{\left(1-\eta^2\right)^2} \tag{5.10.32}$$

is also known as Taylor number, although another parameter defined later is more generally known as the Taylor number. It is generally accepted that the principle of exchange of stability is also valid for this case. It can be analytically shown that it is valid if $0 < \mu < \eta^2$, and it has been numerically demonstrated that it is true also for $\mu < 0$. Therefore we shall take $q = 0$ for the discussion of the critical state. Thus the problem we need to solve is the following :

$$\left(\frac{d}{dr}D_* - a^2\right)^2 \psi' = -T'a^2\left(\frac{1}{r^2} - K\right)v' \; , \tag{5.10.33}$$

$$\left(\frac{d}{dr}D^* - a^2\right)v' = \psi' \; , \tag{5.10.34}$$

with

$$\psi' = \frac{d\psi'}{dr} = v' = 0 \ , \quad \text{on } r = \eta \quad \text{and } r = 1 \ . \quad (5.10.35)$$

Given a, this is an eigenvalue problem for the eigenvalue T'. To deal with this eigenvalue problem, let us first study the following self-adjoint eigenvalue problems :

$$\frac{d}{dr} D_* \phi = -\beta^2 \phi \ , \quad \text{with } \phi(\eta) = \phi(1) = 0 \ , \quad (5.10.36)$$

and

$$\left(\frac{d}{dr} D_* \right)^2 \phi = \alpha^4 \phi \ , \quad \text{with}$$

$$\phi(\eta) = \phi(1) = \frac{d\phi}{dr}(\eta) = \frac{d\phi}{dr}(1) = 0 \ . \quad (5.10.37)$$

We shall denote the eigenvalues and eigenfunctions of (5.10.36) by β_j and $\mathcal{B}_{1,j}(\beta_j r)$; and the eigenvalues and eigenfunctions of (5.10.37) by α_j and $C_{1,j}(\alpha_j r)$. $\mathcal{B}_{1,j}$ and $C_{1,j}$ are in fact suitable combinations of Bessel functions of the first order. We can express $\mathcal{B}_{1,j}$ and $C_{1,j}$ as follows :

$$\mathcal{B}_{1,j}(\beta_j r) = J_1(\beta_j r) + A_j Y_1(\beta_j r) \ , \quad (5.10.38)$$

$$C_{1,j}(\alpha_j r) = J_1(\alpha_j r) + B_j Y_1(\alpha_j r) + D_j I_1(\alpha_j r) + E_j K_1(\alpha_j r) \ . \quad (5.10.39)$$

$\mathcal{B}_{1,j}$ and $C_{1,j}$ each has $(j-1)$ nodes for r between η and 1. $C_{1,j}$ is generally known as the Chandrasekhar functions.

The system (5.10.30) and (5.10.31) is not self-adjoint. It is not as easy to analyze as the self-adjoint problem. Our approach is to express ψ' in terms of Chandrasekhar functions, i.e., let

$$\psi' = \sum_{j=1}^{\infty} P_j C_{1,j}(\alpha_j r). \tag{5.10.40}$$

Then ψ' satisfies automatically the boundary conditions in (5.10.35). Substitute (5.10.40) into (5.10.34), we obtain

$$\left(\frac{d}{dr}D_* - a^2\right)v' = \sum_{j=1}^{\infty} P_j C_{1,j}(\alpha_j r). \tag{5.10.41}$$

Solve the equation (5.10.41) for v' with the boundary condition $v'(\eta) = v'(1) = 0$. Then substitute this solution v' and the ψ' from (5.10.40) into (5.10.30). Now we multiply the equation we thus obtained by $rC_{1,k}(\alpha_k r)$, and integrate with respect to r from η to 1. Thus we obtain an infinite set of algebraic equations for P_j. The eigenvalues are then determined from the characteristic determinant. The characteristic determinant has infinitely many rows and columns. In practice, we need to truncate to have $n \times n$ determinant, and to obtain the nth order approximate solution. Fortunately for this problem, the convergence is reasonably fast.

The Taylor-Couette problem involves many parameters. What we have done is to fix η and μ first, thus fixing the value K. Then set the value of a. Solve the eigenvalue problem to obtain the lowest T'. For various values of a, we obtain different lowest $T'(a)$. We then choose the minimum of these lowest $T'(a)$ and obtain the critical value T_c' and the corresponding a_c. Numerical computations are required to obtain T_c' and a_c. Table 5.10.1 lists a few such values for the case of $\eta = \dfrac{1}{2}$. Note that the case $\mu = \dfrac{1}{4}$ is the limiting case for which the Rayleigh criterion (5.10.24) holds.

TABLE 5.10.1 Critical Taylor numbers and corresponding wave number for various values of μ $\left(\eta = \frac{1}{2}\right)$.

μ	a_c	T_c'
$\frac{1}{4}$	6.2	1.533×10^4
$\frac{1}{6}$	6.2	1.954×10^4
0	6.4	3.310×10^4
$-\frac{1}{8}$	6.4	5.328×10^4
$-\frac{1}{4}$	6.6	9.883×10^4
$-\frac{1}{2}$	9.6	4.286×10^5

Narrow Gap

If the gap between the rotating cylinders are small, i.e., if $(1-\eta) << 1$; then equations (5.10.30) and (5.10.31) may be simplified. Let

$$\xi = \frac{r-\eta}{1-\eta}, \quad \alpha = (1-\eta)a . \tag{5.10.42}$$

As $\dfrac{d}{dr} = \dfrac{1}{1-\eta}\dfrac{d}{d\xi}$, we have $D_* \approx \dfrac{1}{1-\eta}\dfrac{d}{d\xi}$, since $\eta \approx 1$. Also we have

$$\frac{1}{r^2} \approx \frac{1}{\eta^2}\left[1 - \frac{2(1-\eta)}{\eta}\xi\right],$$

$$\left(\frac{1}{r^2} - K\right) \approx \frac{1-\eta^2}{\eta^2(1-\mu)} - \frac{2(1-\eta)}{\eta^3}\xi \approx \frac{2(1-\eta)}{1-\mu}\left[1-(1-\mu)\xi\right].$$

Therefore (5.10.30) and (5.10.31) can be rewritten as

$$\frac{1}{(1-\eta)^2}\left(\frac{d^2}{d\xi^2} - \alpha^2\right)^2 \psi' = -\frac{2T\alpha^2(1-\eta)}{1-\mu}\left[1-(1-\mu)\xi\right]v', \quad (5.10.43)$$

$$\frac{1}{(1-\eta)^2}\left(\frac{d^2}{d\xi^2} - \alpha^2\right)v' = \psi'. \quad (5.10.44)$$

Now denote

$$v'' = -\frac{2T(1-\eta)^3\alpha^2}{1-\mu}v'. \quad (5.10.45)$$

Then (5.10.43) and (5.10.44) can be rewritten as

$$\left(\frac{d^2}{d\xi^2} - \alpha^2\right)^2 \psi' = \left[1-(1-\mu)\xi\right]v'', \quad (5.10.46)$$

$$\left(\frac{d^2}{d\xi^2} - \alpha^2\right)v'' = -T\alpha^2\psi', \quad (5.10.47)$$

where

$$T = \frac{(1-\eta)^4(1-\eta^2)}{\eta^2(1-\mu)} T' \approx \frac{2(1-\eta)^5}{1-\mu} T' . \qquad (5.10.48)$$

T is generally known as the Taylor number. (5.10.46) and (5.10.47) is not self-adjoint again. Therefore, as before, we can consider the following eigenvalue problem for the self-adjoint problem :

$$\frac{d^4}{d\xi^4}\phi = \lambda^4 \phi , \qquad (5.10.49)$$

with

$$\phi\left(-\frac{1}{2}\right) = \phi\left(\frac{1}{2}\right) = \frac{d\varphi}{d\xi}\left(-\frac{1}{2}\right) = \frac{d\varphi}{d\xi}\left(\frac{1}{2}\right) = 0 . \qquad (5.10.50)$$

Let the eigenvalues be λ_m , then the corresponding eigenfunctions $C_m(\xi)$ are of the form

$$C_m(\xi) = \frac{\cosh \lambda_m \xi}{\cosh\left(\frac{\lambda_m}{2}\right)} - \frac{\cos \lambda_m \xi}{\cos\left(\frac{\lambda_m}{2}\right)} . \qquad (5.10.51)$$

$C_m(\xi)$ is another type of Chandrasekhar functions. As before we let

$$\psi' = \sum_{m=1}^{\infty} P_m C_m\left(\xi - \frac{1}{2}\right) , \qquad (5.10.52)$$

and substitute into (5.10.47) to solve for v''. Then eventually we again obtain an infinite set of algebraic equations for P_m , and in a similar way determines the critical Taylor number T_c , and the corresponding α_c . In Table 5.10.2, we list some of the computed T_c and α_c for various values of μ .

TABLE 5.10.2 Critical Taylor number and corresponding wave number for various values of μ (narrow gap).

μ	α_c	T_c
1.00	3.12	1.708×10^3
0.50	3.12	2.275×10^3
0.25	3.12	2.725×10^3
0.00	3.12	3.390×10^3
-0.25	3.13	4.462×10^3
-0.50	3.20	6.417×10^3
-1.00	4.00	1.868×10^4
-2.00	6.10	9.558×10^4
-3.00	8.14	3.025×10^5

Returning to the case of wide gap, the critical Taylor number depends on the parameters μ and η, i.e.,

$$T_c' = T_c'(\mu, \eta) = T_c'\left(\frac{\Omega_2}{\Omega_1}, \frac{r_1}{r_2}\right).$$

Therefore when r_1 and r_2 are fixed, the critical state defines a curve relating Ω_1 and Ω_2. Computations and experiments are carried out by G.I. Taylor (1923), and Fig. 5.10.1 shows such a curve for the case that $r_1 = 3.55$ cm and $r_2 = 4.035$ cm. The agreement between theory and experiment is remarkable.

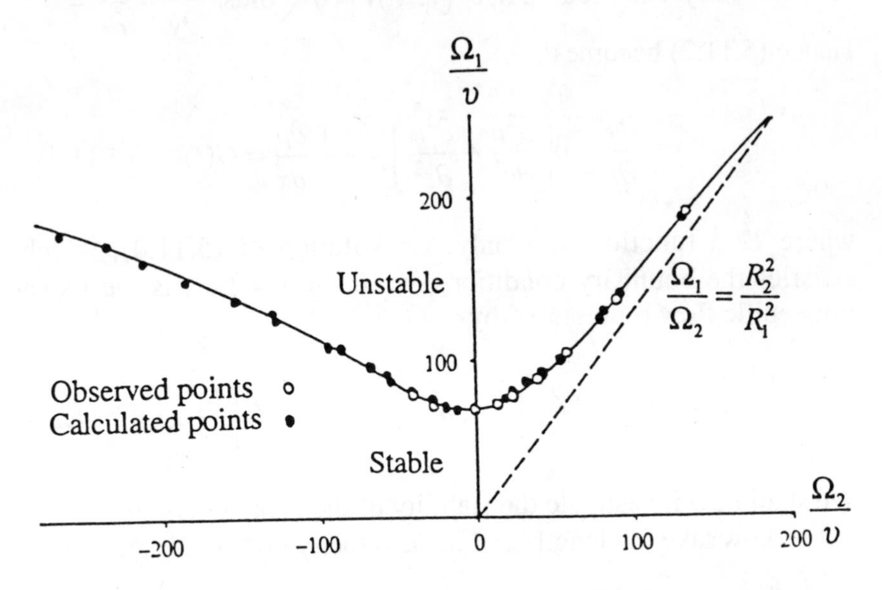

Fig. 5-7 Stability characteristics of Couette motion (after Taylor, 1923).

5.11 Stability of Parallel Flows

Consider the flow of an incompressible, viscous fluid between two parallel plates which we set to be $y = \pm h$. The governing equations are (1.1.1) and (1.1.6) :

$$\nabla \cdot \mathbf{v} = 0 \ , \tag{5.11.1}$$

$$\frac{\partial \mathbf{v}}{\partial t} + (\mathbf{v} \cdot \nabla)\mathbf{v} = -\frac{1}{\rho}\nabla p + \upsilon \nabla^2 \mathbf{v} \ . \tag{5.11.2}$$

Let

$$\mathbf{v} = \left(u(y,z,t),0,0\right) \ . \tag{5.11.3}$$

This represents the unidirectional flow in x-direction. (5.11.1) is automatically satisfied. Also $(\mathbf{v}\cdot\nabla)\mathbf{v}=0$, thus $\dfrac{\partial p}{\partial y}=\dfrac{\partial p}{\partial z}=0$. Hence (5.11.2) becomes

$$\frac{\partial u}{\partial t}-\upsilon\left(\frac{\partial^2 u}{\partial y^2}+\frac{\partial^2 u}{\partial z^2}\right)=-\frac{1}{\rho}\frac{\partial p}{\partial x}=G(t)\,, \quad (5.11.4)$$

where G a function of t only. A solution of (5.11.4), which satisfies the boundary conditions $\mathbf{v}=0$ on $y=\pm h$, is the plane Poisseuille flow represented by :

$$u=u_0(y)=U_c\left(1-\frac{y^2}{h^2}\right), \quad p=p_0(x)=-\frac{2\rho\upsilon U_c}{h^2}x\,. \quad (5.11.5)$$

We shall now investigate the stability of the plane Poisseuille flow. Let us now take the length unit be h, velocity unit be U_c , time unit be $\left(\dfrac{h}{U_c}\right)$, and the unit of pressure be $\left(\rho U_c^2\right)$, then the non-dimensional form of the equations (5.11.1), (5.11.2) and (5.11.3) become

$$\nabla\cdot\mathbf{v}=0\,, \quad (5.11.6)$$

$$\frac{\partial \mathbf{v}}{\partial t}+(\mathbf{v}\cdot\nabla)\mathbf{v}=-\nabla p+\frac{1}{R}\nabla^2\mathbf{v}\,, \quad (5.11.7)$$

and

$$u=u_0(y)=\left(1-y^2\right), \quad p=p_0(x)=-\frac{2x}{R}\,, \quad (5.11.8)$$

where

$$R=\frac{U_c h}{\upsilon}\,, \quad (5.11.9)$$

is the Reynolds number.

To investigate the linear stability, we let

$$\mathbf{v} = \left(u_0 + u' , v' , w'\right) , \quad p = p_0 + p' , \quad (5.11.10)$$

and substitute into (5.11.6) and (5.11.7), and retain only terms linear in perturbed quantities. We shall again employ the normal mode analysis, thus let

$$u' = \tilde{u}(y)\exp\left\{i\left(\alpha_x x + \alpha_z z\right) - i\alpha_x ct\right\} , \quad (5.11.11)$$

$$v' = \tilde{v}(y)\exp\left\{i\left(\alpha_x x + \alpha_z z\right) - i\alpha_x ct\right\} , \quad (5.11.12)$$

$$w' = \tilde{w}(y)\exp\left\{i\left(\alpha_x x + \alpha_z z\right) - i\alpha_x ct\right\} , \quad (5.11.13)$$

$$p' = \tilde{p}(y)\exp\left\{i\left(\alpha_x x + \alpha_z z\right) - i\alpha_x ct\right\} . \quad (5.11.14)$$

Then the perturbed quantities $\tilde{u}(y)$, $\tilde{v}(y)$, $\tilde{w}(y)$ and $\tilde{p}(y)$ satisfy the following equations

$$i\left(\alpha_x \tilde{u} + \alpha_z \tilde{w}\right) + D\tilde{v} = 0 , \quad (5.11.15)$$

$$\left[D^2 - \left(\alpha_x^2 + \alpha_z^2\right) - i\alpha_x R\left(u_0 - c\right)\right]\tilde{u} = R\left(Du_0\right)\tilde{v} + i\alpha_x R\tilde{p} , \quad (5.11.16)$$

$$\left[D^2 - \left(\alpha_x^2 + \alpha_z^2\right) - i\alpha_x R\left(u_0 - c\right)\right]\tilde{v} = R\left(D\tilde{p}\right) , \quad (5.11.17)$$

$$\left[D^2 - \left(\alpha_x^2 + \alpha_z^2\right) - i\alpha_x R\left(u_0 - c\right)\right]\tilde{w} = i\alpha_z R\tilde{p} , \quad (5.11.18)$$

where $D = \dfrac{d}{dy}$.

Squire Theorem

We have formulated the problem of linear stability in terms of three-dimensional disturbances. It can be shown that the problem of three-dimensional disturbances is equivalent to a two-dimensional problem at a lower Reynolds number. This is known as Squire Theorem.

Let us introduce the new variables $\overline{u}$ and $\overline{p}$, and new parameters $\overline{R}$ and α as follows :

$$\alpha^2 = \alpha_x^2 + \alpha_z^2 , \qquad \alpha\overline{R} = \alpha_x R , \tag{5.11.19}$$

$$\alpha\overline{u} = \alpha_x\tilde{u} + \alpha_z\tilde{w} , \qquad \overline{p}\overline{R} = \tilde{p}R . \tag{5.11.20}$$

Then (5.11.15) and (5.11.17) become

$$i\alpha\overline{u} + D\tilde{v} = 0 , \tag{5.11.21}$$

$$\left[D^2 - \alpha^2 - i\alpha\overline{R}(u_0 - c)\right]\tilde{v} = \overline{R}(D\overline{p}) , \tag{5.11.22}$$

and from (5.11.16) and (5.11.18) we obtain

$$\left[D^2 - \alpha^2 - i\alpha\overline{R}(u_0 - c)\right]\overline{u} = \overline{R}(Du_0)\tilde{v} + i\alpha\overline{R}\overline{p} . \tag{5.11.23}$$

The system (5.11.21) - (5.11.23) has the same structure as the system (5.11.15) - (5.11.18) with $\tilde{w} = 0$ and $\alpha_z = 0$. Therefore the system (5.11.21) - (5.11.23) corresponds to the problem with a two-dimensional disturbance, but with the Reynolds number $\overline{R} = \dfrac{\alpha_x}{\alpha} R \le R$. For stability analysis, the most critical case is the one that associates with minimum Reynolds number. Therefore it is sufficient for that purpose to consider just the two-dimensional problem.

Orr-Sommerfeld Equation

We shall now consider the two-dimensional problem. For two-dimensional flow of incompressible fluids, we can introduce the streamfunction $\psi(x,y,t)$, such that

$$u = \frac{\partial \psi}{\partial y} , \quad v = -\frac{\partial \psi}{\partial x} . \tag{5.11.24}$$

The elimination of p from the two equations in (5.11.7) then leads to :

$$\frac{\partial}{\partial t}\left(\nabla^2 \psi\right) + \frac{\partial\left(\nabla^2 \psi, \psi\right)}{\partial(x,y)} = \frac{1}{R}\nabla^2\left(\nabla^2 \psi\right) . \tag{5.11.25}$$

The boundary conditions that $\mathbf{v} = 0$ on $y = \pm 1$ now becomes

$$\frac{\partial \psi}{\partial x} = \frac{\partial \psi}{\partial y} = 0 , \quad \text{on } y = \pm 1 . \tag{5.11.26}$$

The parallel flow solution $\mathbf{v} = \left(u_0(y), 0\right)$ is now equivalent to

$$\psi = \psi_0(y) .$$

For the linear stability analysis, we write

$$\psi = \psi_0(y) + \psi'(x,y,t) , \tag{5.11.27}$$

and substitute into (5.11.25). Retaining only terms linear in ψ' , we obtain :

$$\frac{\partial}{\partial t}\left(\nabla^2 \psi'\right) + \left(\frac{d\psi_0}{dy}\right)\left(\nabla^2 \psi'\right) - \left(\frac{d^3 \psi_0}{dy^3}\right)\frac{\partial \psi'}{\partial x}$$

$$= \frac{1}{R}\nabla^2\left(\nabla^2 \psi'\right) . \tag{5.11.28}$$

Now consider a normal mode :

$$\psi'(x,y,t) = \phi(y)e^{i\alpha(x-ct)} . \tag{5.11.29}$$

Then the equation (5.11.28) becomes

$$\left(D^2-\alpha^2\right)^2\varphi = i\alpha R\left[\left(u_0-c\right)\left(D^2-\alpha^2\right)\varphi - \left(D^2u_0\right)\varphi\right] , \tag{5.11.30}$$

and the boundary conditions (5.11.26) become now

$$\varphi = D\varphi = 0 , \quad \text{on } y = \pm 1 . \tag{5.11.31}$$

Equation (5.11.30) is known as the Orr-Sommerfeld equation. It is in fact valid for any admissible parallel flows and not just Poiseuille flow given by (5.11.8).

Sufficient Conditions for Stability

If we multiply (5.11.30) by φ^* and integrate the equation with respect to y from $y = -1$ to $y = 1$, we obtain, after applying the boundary conditions (5.11.31),

$$I_2^2 + 2\alpha^2 I_1^2 + \alpha^4 I_0^2 = -i\alpha RQ + i\alpha Rc\left(I_1^2 + \alpha^2 I_0^2\right) , \tag{5.11.32}$$

where

$$I_0^2 = \int_{-1}^{1} |\varphi|^2 dy , \quad I_1^2 = \int_{-1}^{1} \left|\frac{d\varphi}{dy}\right|^2 dy , \quad I_2^2 = \int_{-1}^{1} \left|\frac{d^2\varphi}{dy^2}\right|^2 dy , \tag{5.11.33}$$

and

$$Q = \int_{-1}^{1} \left[u_0 \left|\frac{d\varphi}{dy}\right|^2 + \left(\alpha^2 u_0 + \frac{d^2u_0}{dy^2}\right)|\varphi|^2 \right] dy + \int_{-1}^{1} \frac{du_0}{dy}\frac{d\varphi}{dy}\varphi^* dy . \tag{5.11.34}$$

Now let us add (5.11.32) to its complex conjugate, then we obtain

$$2\left(I_2^2 + 2\alpha^2 I_1^2 + \alpha^4 I_0^2\right)$$

$$= -i\alpha R\left(Q - Q^*\right) - 2\alpha R c_i\left(I_1^2 + \alpha^2 I_0^2\right), \qquad (5.11.35)$$

where c_i is the imaginary part of c. Since u_0 is real, thus

$$\left|Q - Q^*\right| = \left|\int_{-1}^{1} \frac{du_0}{dy}\left(\frac{d\varphi}{dy}\varphi^* - \frac{d\varphi^*}{dy}\varphi\right)dy\right|$$

$$\leq 2\int_{-1}^{1}\left|\frac{du_0}{dy}\right| \cdot \left|\frac{d\varphi}{dy}\right| \cdot |\varphi| dy$$

$$\leq 2qI_0 I_1 ,$$

where q is the maximum of $\left|\frac{du_0}{dy}\right|$ and the last inequality results from the Schwarz's inequality. Therefore, (5.11.35) leads to

$$\left(I_1^2 + \alpha^2 I_0^2\right)\alpha R c_i \leq \alpha R q I_0 I_1 - \left(I_2^2 + 2\alpha^2 I_1^2 + \alpha^4 I_0^2\right). \quad (5.11.36)$$

Thus if αR is small enough, c_i will be negative, and the flow is stable. A sufficient condition for stability can be stated as follows :

$$q\alpha R < \mathrm{Min}\left[\frac{I_2^2 + 2\alpha^2 I_1 + \alpha^4 I_0^2}{I_0 I_1}\right]. \qquad (5.11.37)$$

Other sufficient conditions for stability may also be derived. One essential consequence is that the flow is stable for given α if the Reynolds number is small enough.

Parallel Flow of Inviscid Fluids

Before we discuss the case for general R, let us investigate the case with $R \to \infty$, i.e., the case of inviscid fluids. Then equation (5.11.30) becomes

$$(u_0 - c)(D^2 - \alpha^2)\varphi - (D^2 u_0)\varphi = 0 , \qquad (5.11.38)$$

with boundary conditions reduced to

$$\varphi = 0 , \quad \text{on} \quad y = \pm 1 . \qquad (5.11.39)$$

Equation (5.11.38) is second order, much simpler than the Orr-Sommerfeld equation which is fourth order. Let us denote

$$L\varphi = (D^2 - \alpha^2)\varphi - \frac{(D^2 u_0)\varphi}{u_0 - c} , \qquad (5.11.40)$$

and

$$c = c_r + ic_i . \qquad (5.11.41)$$

Since u_0 and α are real, we obtain from (5.11.38) that

$$\varphi^* L\varphi - \varphi L^* \varphi^* = \varphi^*(D^2\varphi) - \varphi(D^2\varphi^*) - \frac{2ic_i (D^2 u_0)\varphi\varphi^*}{|u_0 - c|^2} . \quad (5.11.42)$$

Using the boundary conditions (5.11.39), we obtain

$$\int_{-1}^{1} (\varphi^* L\varphi - \varphi L^* \varphi^*) dy = -2ic_i \int_{-1}^{1} \frac{(D^2 u_0)|\varphi|^2}{|u_0 - c|^2} dy . \quad (5.11.43)$$

As $L\varphi = L^*\varphi^* = 0$ by (5.11.38), we obtain

$$-2ic_i\int_{-1}^{1}\frac{(D^2u_0)|\varphi|^2}{|u_0-c|^2}dy=0\ . \tag{5.11.44}$$

Therefore $c_i \neq 0$ only if $D^2u_0 = 0$ for some y between $y=-1$ and y=1. In other words, the parallel flow of the inviscid fluids is unstable only if the velocity profile $u_0(y)$ has point of inflection in the interval $(-1,\ 1)$. In a sense, this result is a generalization of the Kelvin-Helmholtz instability, because a step function can be considered as having a point of inflection. We also note now that the parabolic velocity profile of the Poiseuille flow has no point of inflection in the interval $(-1,\ 1)$. Thus for Poiseuille flow, if the fluid is inviscid, it is stable.

Now let us return to the Orr-Sommerfeld equation (5.11.30). For plane Poiseuille flow, since $u_0(y)=1-y^2$ is symmetric in y, thus equation (5.11.30) is also symmetric in y. Therefore we can decompose the solution $\varphi(y)$ into two parts :

$$\varphi(y)=\varphi_o(y)+\varphi_e(y)\ , \tag{5.11.45}$$

where $\varphi_o(y)$ is odd, and φ_e is even in y. Both φ_o and φ_e satisfy the boundary conditions (5.11.31). If we discuss φ_o and φ_e separately, we need only to deal with the half-region from $y=-1$ to y=0 for the problem. On y=0, φ_o and φ_e satisfy the following conditions according to their respective odd and even characters :

$$\varphi_o(0)=\varphi_o''(0)=0\ , \tag{5.11.46}$$

$$\varphi_e'(0)=\varphi_e'''(0)=0\ , \tag{5.11.47}$$

where we have used the superscript prime to denote $\dfrac{d}{dy}$ or D. We shall also use this notation in what follows.

Let us now concentrate on ϕ_e. So the problem is to solve the equation (5.11.30) subject to the boundary conditions :

$$\varphi(-1) = \varphi'(-1) = \varphi'(0) = \varphi'''(0) = 0\ . \qquad (5.11.48)$$

The equation (5.11.30) is a 4th order linear differential equation. Therefore there are four linearly independent solutions: $\{\varphi_1, \varphi_2, \varphi_3, \varphi_4\}$. The general solution φ can be expressed as a linear combination of these linearly independent solutions , i.e.,

$$\varphi = C_1\varphi_1 + C_2\varphi_2 + C_3\varphi_3 + C_4\varphi_4\ . \qquad (5.11.49)$$

We now substitute (5.11.49) into (5.11.48) and obtain four homogeneous linear equations for C_1, C_2, C_3, and C_4. In order that we have nontrivial solution, the characteristic determinant has to vanish, and we obtain

$$\begin{vmatrix} \varphi_1(-1) & \varphi_2(-1) & \varphi_3(-1) & \varphi_4(-1) \\ \varphi_1'(-1) & \varphi_2'(-1) & \varphi_3'(-1) & \varphi_4'(-1) \\ \varphi_1'(0) & \varphi_2'(0) & \varphi_3'(0) & \varphi_4'(0) \\ \varphi_1'''(0) & \varphi_2'''(0) & \varphi_3'''(0) & \varphi_4'''(0) \end{vmatrix} = 0\ . \qquad (5.11.50)$$

The equation (5.11.50) can be put in the form :

$$F(\alpha, R, c) = 0\ . \qquad (5.11.51)$$

Given real R and α , (5.11.51) can be solved to obtain a characteristic value c. If the imaginary part of c is positive, i.e., $c_i > 0$, then the flow state is unstable. The marginal state between stability and instability is given by $c_i = 0$. The marginal state defines a relation betwen α and R. In other words, it can be represented by a curve in the (α, R) plane. This curve is the neutral curve, i.e., the curve for neutral stability.

Asymptotic Analysis and Comparison Equations

In order to obtain the relation (5.11.51), we need to find $\{\varphi_1, \varphi_2, \varphi_3, \varphi_4\}$ first. Preliminary analysis and experimental observation indicate that (αR) is large on the neutral curve. Since the solutions for general R is difficult to find, we shall try to find the asymptotic solutions for large αR.

Let us take a close look at (5.11.30). When (αR) is very large, the equation may be approximated by the inviscid equation (5.11.38) :

$$(u_0 - c)(D^2 - \alpha^2)\varphi - (D^2 u_0)\varphi = 0 . \qquad (5.11.52)$$

This is a second order equation. Hence it can only furnish two linearly independent solutions which we shall denote as $\varphi_1^{(0)}$ and $\varphi_2^{(0)}$. To find the other two linearly independent solutions, we have to take into account of the fourth order terms. In this sense, the operator D is considered to be large together with (αR). Therefore, we should consider the following approximate equation:

$$\left[D^4 - i\alpha R(u_0 - c)D^2\right]\varphi = 0 . \qquad (5.11.53)$$

This equation will furnish the other two linearly independent solutions $\varphi_3^{(0)}$ and $\varphi_4^{(0)}$.

There are two singular points for the equation (5.11.52) given by

$$u_0(y) - c = 1 - y^2 - c = 0 . \qquad (5.11.54)$$

The singularities are located at $y = \pm(1-c)^{\frac{1}{2}}$. When $c_i = 0$, y_c is real, and only one of the two can be in the interval (–1, 0). If $c_i \neq 0$, then y_c is complex. Now this singularity is a regular singularity. Thus we can use the Frobenius method to find the solutions, and obtain

$$\varphi_1^{(0)}(y) = (y - y_c) + \frac{u_0''(y_c)}{2u_0'(y_c)}(y - y_c)^2 + \cdots, \quad (5.11.55)$$

$$\varphi_2^{(0)}(y) = 1 + b_1(y - y_c) + \cdots + \frac{u_0''(y_c)}{u_0'(y_c)}\varphi_1^{(0)}(y)\ln(y - y_c). \quad (5.11.56)$$

For equation (5.11.53), we use the WKB method to obtain the approximate solutions for large (αR) :

$$\varphi_3^{(0)}(y) \approx (u_0 - c)^{\frac{5}{4}} \exp\left\{-\int_{y_c}^{y} [i\alpha R(u_0 - c)]^{\frac{1}{2}} dy\right\}, \quad (5.11.57)$$

and

$$\varphi_4^{(0)}(y) \approx (u_0 - c)^{-\frac{5}{4}} \exp\left\{+\int_{y_c}^{y} [i\alpha R(u_0 - c)]^{\frac{1}{2}} dy\right\}. \quad (5.11.58)$$

Substitute the above $\{\varphi_1^{(0)}, \varphi_2^{(0)}, \varphi_3^{(0)}, \varphi_4^{(0)}\}$ into (5.11.50), then we obtain the relation (5.11.51) in approximation when (αR) is large. However, in carrying out the substitution, we need values of $\varphi_1^{(0)}$ evaluated at both $y = -1$ and $y=0$. As we can see from (5.11.55) - (5.11.58), $\varphi_2^{(0)}, \varphi_3^{(0)}$ and $\varphi_4^{(0)}$ all have branch point at $y = y_c$. Therefore it is not clear which branch to take when we connect the solutions between $y = -1$ and $y=0$. This ambiguity arises from the asymptotic nature of the solutions. For $\varphi_2^{(0)}$, we have approximated the original equation by its asymptotic version (5.11.52), while $\varphi_3^{(0)}$ and $\varphi_4^{(0)}$ are asymptotic because they are WKB solutions. In other words, the singularity y_c is introduced by the asymptotic approximation. Indeed, if we go back to original equation (5.11.30), $y = y_c$ is not a singular point of the differential equation. The full original equation (5.11.30) is too difficult to handle. But we shall seek help from the original equation to resolve the difficulties faced by the asymptotic solutions. The method we shall employ is the method of camparison equations.

Let us introduce new variables and parameter :

$$z = \left[\frac{3}{2}\int_{y_c}^{y}\left\{-i(u_0 - c)\right\}^{\frac{1}{2}} dy\right]^{\frac{2}{3}} , \qquad (5.11.59)$$

$$w = \left[\frac{-i(u_0 - c)}{z}\right]^{\frac{3}{4}} \varphi , \qquad (5.11.60)$$

$$\lambda = (\alpha R)^{\frac{1}{3}} . \qquad (5.11.61)$$

It may be remarked that in the neighborhood of $y = y_c$, $z \propto (y - y_c)$, $w \propto \varphi$. Therefore the transformation of variable is quite regular near $y = y_c$. The Orr-Sommerfeld equation (5.11.30) can then be transformed into the following form :

$$\frac{d^4 w}{dz^2} + \lambda^2\left[z\frac{d^2 w}{dz^2} + q_0\frac{dw}{dz} + r_0 w\right]$$

$$= \lambda\left[p_1\frac{d^2 w}{dz^2} + q_1\frac{dw}{dz} + r_1 w\right] , \qquad (5.11.62)$$

where q_0, r_0, p_1, q_1, r_1 are all functions of z. Now λ is a large parameter. Therefore the solution of the following equation

$$\frac{d^4 w}{dz^4} + \lambda^2\left[z\frac{d^2 w}{dz^2} + q_0\frac{dw}{dz} + r_0 w\right] = 0 , \qquad (5.11.63)$$

is the first approximation of solution of (5.11.62). A regular perturbation scheme can be developed to find the successive approximations of the solution. Equation (5.11.63) is a comparison equation. Its solution is the asymptotic solution of the original equation for large λ , uniformly valid in a region that contains z=0 or $y = y_c$. Further transformation can make q_0 and r_0 constants. Then (5.11.63) is a Laplace's linear differential equation. Hence Laplace's method can be employed to find the solutions in terms of contour integrals.

Having established the validity of the method of comparison equation to resolve the difficulties of our problem, it is not necessary for us to make use of such elaborate comparison equation as (5.11.63) for our purpose. We need only to use much simpler comparison equations like :

$$\frac{d^4w}{dz^4} + \lambda^2\left[z\frac{d^2w}{dz^2} + r_0 w\right] = 0 , \qquad (5.11.64)$$

or even

$$\frac{d^4w}{dz^4} + \lambda^2 z\frac{d^2w}{dz^2} = 0 . \qquad (5.11.65)$$

Our discussion will now base on the comparison equation (5.11.65). Then it is even more convenient to go back to (5.11.30). Let us introduce new variable η and parameter ε :

$$y - y_c = \varepsilon\eta , \quad \varepsilon = (\alpha R)^{-\frac{1}{3}} . \qquad (5.11.66)$$

Expand $(u_0 - c)$ in terms of powers of $(y - y_c)$. Also expand φ as follows :

$$\varphi = \chi^{(0)} + \varepsilon\chi^{(1)} + \varepsilon^{(2)}\chi^{(2)} + \cdots , \qquad (5.11.67)$$

and substitute into (5.11.30). Equating terms of the same order in ε , we obtain

$$i\frac{d^4\chi^{(0)}}{d\eta^4}+\eta\frac{d^2\chi^{(0)}}{d\eta^2}=0\ , \tag{5.11.68}$$

$$i\frac{d^4\chi^{(1)}}{d\eta^4}+\eta\frac{d^2\chi^{(1)}}{d\eta^2}=\frac{u_0''(y_c)}{u_0'(y_c)}\chi^{(0)}-\frac{1}{2}\left(\frac{u_0''(y_c)}{u_0'(y_c)}\right)\eta^2\frac{d^2\chi^{(0)}}{d\eta^2}\ , \tag{5.11.69}$$

$$\cdots .$$

Equation (5.11.68) is essentially the same as the comparison equation (5.11.65). It has the following 4 linearly independent solutions :

$$\chi_1^{(0)}=\eta\ , \tag{5.11.70}$$

$$\chi_2^{(0)}=1\ , \tag{5.11.71}$$

$$\chi_3^{(0)}=\int_\infty^\eta d\eta\int_\infty^\eta \eta^{\frac{1}{2}}H_{\frac{1}{3}}^{(1)}\left[\frac{2}{3}(i\eta)^{\frac{3}{2}}\right]d\eta\ , \tag{5.11.72}$$

$$\chi_4^{(0)}=\int_{-\infty}^\eta d\eta\int_{-\infty}^\eta \eta^{\frac{1}{2}}H_{\frac{1}{3}}^{(2)}\left[\frac{2}{3}(i\eta)^{\frac{3}{2}}\right]d\eta\ . \tag{5.11.73}$$

These four linearly independent solutions are chosen so that they correspond to the four solutions (5.11.55) - (5.11.58). $H_{\frac{1}{3}}^{(1)}$ and $H_{\frac{1}{3}}^{(2)}$ are Hankel functions of order $\frac{1}{3}$. They are directly related to the Airy functions. In fact, we can immediately see that $\frac{d^2\chi^{(0)}}{d\eta^2}$ satisfies the Airy equation. (5.11.72) and (5.11.73) are repeated integrals over the Airy functions, and $\eta=0$ is actually not a branch point.

We can now compare the solutions $\left\{\varphi_1^{(0)},\varphi_2^{(0)},\varphi_3^{(0)},\varphi_4^{(0)}\right\}$ with $\left\{\chi_1^{(0)},\chi_2^{(0)},\chi_3^{(0)},\chi_4^{(0)}\right\}$. For $y > y_c$, $\varphi_i^{(0)}$ should correspond to $\chi_i^{(0)}(\eta\to\infty)$, and for $y < y_c$, $\varphi_1^{(0)}$ should correspond to $\chi_i^{(0)}(\eta\to\infty)$ since ε is a small parameter. The solution $\varphi_1^{(0)}$ has no problem for connection. The comparison of $\varphi_3^{(0)}$ and $\varphi_4^{(0)}$ with $\chi_3^{(0)}$ and $\chi_4^{(0)}$ then determines which branch in $\varphi_3^{(0)}$ and $\varphi_4^{(0)}$ should be chosen. The result of detailed analysis is as follows. If we let

$$Q=\int_{y_c}^{y}\left[i\left(u_0-c\right)\right]^{\frac{1}{2}}dy\ . \tag{5.11.74}$$

Then the integration path is the one that would make the real part of Q monotonously increasing from $y=-1$ to $y=0$. Let $i=e^{\frac{i\pi}{2}}$, then the integration path in the complex y plane passes underneath the point $y=y_c$. If c is real, thus y_c is also real. Then we have

$$u_0-c=\left|u_0-c\right|\,,\qquad \arg Q=\frac{\pi}{4}\,,\qquad \text{for } y>y_c\,, \tag{5.11.75}$$

$$u_0-c=\left|u_0-c\right|e^{-i\pi}\,,\qquad \arg Q=-\frac{5\pi}{4}\,,\qquad \text{for } y<y_c\,. \tag{5.11.76}$$

To establish the correspondence between $\varphi_2^{(0)}$ and $\chi_2^{(0)}$, we need to solve (5.11.69) and find $\chi_2^{(1)}$. Now $\chi_2^{(1)}$ satisfies the equation

$$i\frac{d^4\chi_2^{(1)}}{d\eta^4}+\eta\frac{d^2\chi_2^{(1)}}{d\eta^2}=\frac{u_0''\left(y_c\right)}{u_0'\left(y_c\right)}\ . \tag{5.11.77}$$

We can use the method of variation of parameter to find the particular integral $N(\eta)$ such that $N(\eta)\to\left(\dfrac{u_0''\left(y_c\right)}{u_0'\left(y_c\right)}\right)\eta\ln\eta$ as

$\eta \to \infty$, because then the solution will correspond to the logarithm term in $\varphi_2^{(0)}$. Such a particular integral can indeed be found, and it is also found that $N(\eta) \to \eta \ln|\eta| - \pi i \eta$ as $\eta \to -\infty$. The result of this comparison is then in the expression $\varphi_2^{(0)}$:

$$\ln(y - y_c) = \ln|y - y_c| , \qquad \text{for } y > y_c , \qquad (5.11.78)$$

$$\ln(y - y_c) = \ln|y - y_c| - \pi i , \quad \text{for } y < y_c . \qquad (5.11.79)$$

The path of connection in the complex y plane is again passing underneath y_c .

After settling the question of how to take the correct branches, the relation (5.11.51) can then be computed. The result of numerical computation is shown in Fig. 5-8. The curve $c_i = 0$ is the neutral curve. The region inside the neutral curve is unstable. The growth rate of instability increases with c_i . There is a range in α for given Reynolds number R that the flow is unstable. For R<5300, the flow is stable for any α . This Reynolds number is the critical Reynolds number R_c .

Experiments show that the parallel laminar flow becomes unstable at Reynolds number much smaller than R_c . Perhaps it is due to finite disturbances. Then nonlinear theories need to be developed for the study of this stability problem.

It may be remarked that viscous effects generally inhibit the development of flow instability. However for this parallel flow problem, the introduction of viscosity apparently is the cause of instability. Because the inviscid flow with parabolic velocity profile, which has no point of inflation, is stable; whereas the viscous flow is unstable for large enough R. It is indeed quite surprising.

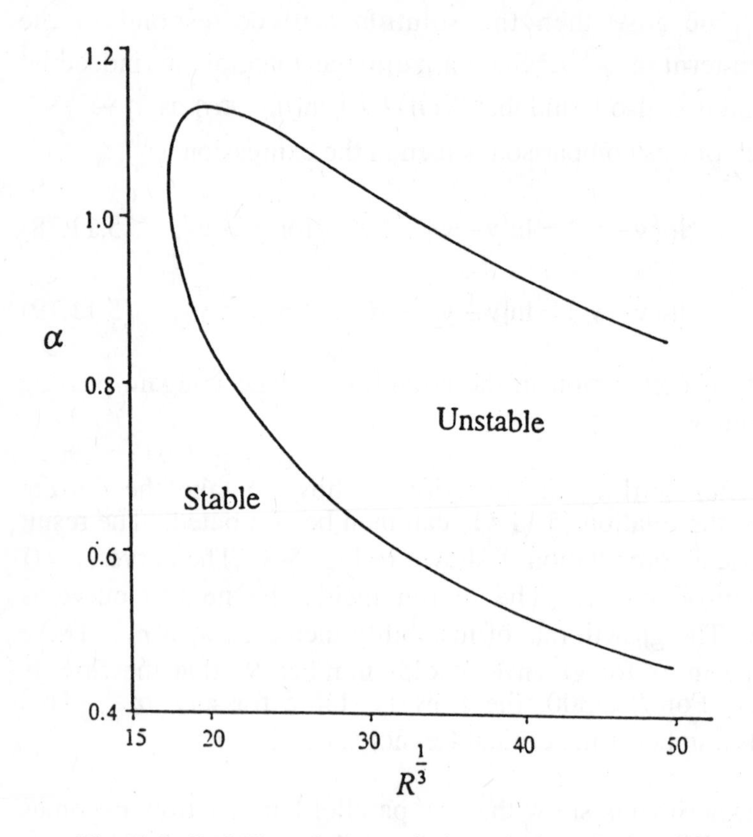

Fig. 5-8 Stability characteristics of plane Poiseuille motion (after Shen, 1954).

5.12 Stability of Flow Down an Inclined Plane

Consider an incompressible, viscous fluid flowing down an inclined plane that makes an angle θ with the horizontal in the gravitational field. The primary flow configuration is shown in Fig. 5-9.

We shall consider only the two-dimensional flow problem, thus $\mathbf{v} = (u, v)$. The continuity equation and the Navier-Stokes equations are as follows :

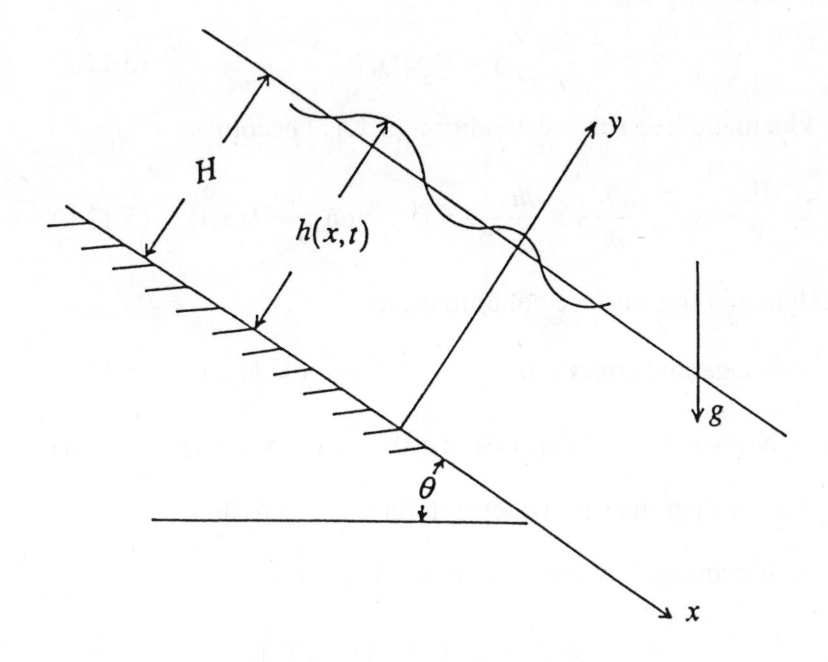

Fig. 5-9 The schematic flow configuration of a fluid flowing down an inclined plane.

$$\frac{\partial u}{\partial x}+\frac{\partial v}{\partial y}=0\ , \tag{5.12.1}$$

$$\frac{\partial u}{\partial t}+u\frac{\partial u}{\partial x}+v\frac{\partial u}{\partial y}=-\frac{1}{\rho}\frac{\partial p}{\partial x}+g\sin\theta+\upsilon\left(\frac{\partial^2 u}{\partial x^2}+\frac{\partial^2 u}{\partial y^2}\right), \tag{5.12.2}$$

$$\frac{\partial v}{\partial t}+u\frac{\partial v}{\partial x}+v\frac{\partial v}{\partial y}=-\frac{1}{\rho}\frac{\partial p}{\partial y}-g\cos\theta+\upsilon\left(\frac{\partial^2 v}{\partial x^2}+\frac{\partial^2 v}{\partial y^2}\right). \tag{5.12.3}$$

For this problem, we have the nonslip boundary condition at the bottom :

$$u=0\ , \quad v=0\ , \quad \text{on } y=0\ . \tag{5.12.4}$$

The free surface is given by

$$y = h(x,t). \tag{5.12.5}$$

The kinematic free surface condition (5.1.17) becomes :

$$\frac{\partial h}{\partial t} + u\frac{\partial h}{\partial x} - v = 0\ , \quad \text{on } y = h(x,t)\ . \tag{5.12.6}$$

The dynamic free surface conditions are

$$\text{Tangential stress} = 0, \qquad \text{on } y = h(x,t)\ , \tag{5.12.7}$$

$$\text{Normal stress and pressure} = 0, \quad \text{on } y = h(x,t)\ . \tag{5.12.8}$$

We shall neglect the surface tension for this discussion.

A primary flow state exists and is given by

$$u = U(y) = U_0\left[2\left(\frac{y}{h}\right) - \left(\frac{y}{h}\right)^2\right], \tag{5.12.9}$$

$$v = 0, \tag{5.12.10}$$

$$p = p_0(y) = \rho g(h-y)\cos\theta\ , \tag{5.12.11}$$

$$h = H\ , \tag{5.12.12}$$

and

$$U_0 = \frac{gh^2\sin\theta}{2\upsilon}\ . \tag{5.12.13}$$

It can be readily verified that the primary flow state satisfies exactly the governing equations and conditions (5.12.1) - (5.12.8). The velocity profile given by (5.12.9) is a parabolic profile, the same as that of the plane Poiseuille flow. But there is an important difference : there is a free surface for this problem whereas for the Poiseuille flow the fluid is confined between rigid boundaries. We have also purposedly written h in the expressions of (5.12.9), (5.12.11) and (5.12.13). In the primary flow state, we could have written simply H.

For linear stability analysis, we write

$$u = U(y) + u'(x,y,t) \ , \tag{5.12.14}$$

$$p = p_0(y) + p'(x,y,t) \ , \tag{5.12.15}$$

and

$$h = H + \xi(x,t) \ . \tag{5.12.16}$$

Keeping only the terms linear in u', v, p' and ξ , the equations and conditions (5.12.1) - (5.12.8) become

$$\frac{\partial u'}{\partial x} + \frac{\partial v}{\partial y} = 0 \ , \tag{5.12.17}$$

$$\frac{\partial u'}{\partial t} + U\frac{\partial u'}{\partial x} + \frac{dU}{dy}v = -\frac{1}{\rho}\frac{\partial p'}{\partial x} + \upsilon\left(\frac{\partial^2 u'}{\partial x^2} + \frac{\partial^2 u'}{\partial y^2}\right) , \tag{5.12.18}$$

$$\frac{\partial v}{\partial t} + U\frac{\partial v}{\partial x} = -\frac{1}{\rho}\frac{\partial p'}{\partial y} + \upsilon\left(\frac{\partial^2 v}{\partial x^2} + \frac{\partial^2 v}{\partial y^2}\right) , \tag{5.12.19}$$

with

$$u'(x,0,t) = 0 \ , \tag{5.12.20}$$

$$v(x,0,t) = 0 \ , \tag{5.12.21}$$

$$\frac{\partial \xi}{\partial t} + U_0\frac{\partial \xi}{\partial x} - v(x,H,t) = 0 \ , \tag{5.12.22}$$

$$\frac{\partial v}{\partial x}(x,H,t)+\frac{\partial u'}{\partial y}(x,H,t)-\frac{2U_0\xi(x,t)}{H^2}=0\ , \quad (5.12.23)$$

$$\frac{1}{\rho}p'(x,H,t)=2\upsilon\frac{\partial v}{\partial y}(x,H,t)+(g\cos\theta)\xi(x,t)\ . \quad (5.12.24)$$

Proceed with a normal mode analysis, i.e., let the *(x, t)* dependence be proportional to $e^{ik(x-ct)}$, we shall again obtain an Orr-Sommerfeld equation. As we may recall for the study of the Plane Poiseuille flow, there we have used a large (αR) asymptotic analysis. Indeed, it is found that the critical Reynolds number is about 5300 and $\alpha = O(1)$. For this problem with free surface, it turns out that the most critical case is the one for which k is very small, i.e., the disturbance wavelength is very long. Therefore a perturbation approach based on small k will serve the purpose. This perturbation expansion is a regular perturbation, and the analysis is relatively straightforward. Let us introduce the stream function $\psi(x,y,t)=\Psi(y)e^{ik(x-ct)}$ so that

$$u'=\frac{\partial\psi}{\partial y}=\frac{d\Psi}{dy}e^{ik(x-ct)}\ , \qquad v=-\frac{\partial\psi}{\partial x}=-ik\Psi e^{ik(x-ct)}\ . \quad (5.12.25)$$

Non-dimensionalize the system by taking H as the length unit, and U_0 as the velocity unit. Then, by eliminate p', (5.12.18) and (5.12.19) reduce to

$$\frac{d^4\Psi}{dy^4}-\left[ikR(U-c)+2k^2\right]\frac{d^2\Psi}{dy^2}$$

$$+\left\{ikR\left[\frac{d^2U}{dy^2}-(U-c)k^2\right]+k^4\right\}\Psi=0\ , \quad (5.12.26)$$

where

$$R = \frac{U_0 H}{\upsilon} = \frac{gH^3 \sin\theta}{2\upsilon^2} = \frac{2U_0^2}{gH\sin\theta}\,, \qquad (5.12.27)$$

is the Reynolds number and now in the new units,

$$U(y) = 2y - y^2\,. \qquad (5.12.28)$$

Equation (5.12.26) is the same Orr-Sommerfeld equation which was derived in the last section as (5.11.30). Write $\xi(x,t) = \bar{\xi} e^{ik(x-ct)}$. The boundary and free surface conditions (5.12.20) - (5.12.24) now become

$$\frac{d\Psi}{dy}(0) = 0\,, \qquad (5.12.29)$$

$$\Psi(0) = 0\,, \qquad (5.12.30)$$

$$(1-c)\bar{\xi} - \Psi(1) = 0\,, \qquad (5.12.31)$$

$$\frac{d^2\Psi}{dy^2}(1) + k^2\Psi(1) + 2\bar{\xi} = 0\,, \qquad (5.12.32)$$

$$\frac{d^3\Psi}{dy^3}(1) - \left[ikR(1-c) + 3k^3\right]\frac{d\Psi}{dy}(1) + 2ik\cot\theta\,\bar{\xi} = 0\,. \qquad (5.12.33)$$

In deriving (5.12.33), we have made use of (5.12.18) to express p' in terms of ψ .

Small k Expansion

Now let us adopt a small k expansion, and write

$$\Psi = \Psi_0 + k\Psi_1 + k^2\Psi_2 + \cdots\,, \qquad (5.12.34)$$

$$\bar{\xi} = \xi_0 + k\xi_1 + k^2\xi_2 + \cdots\,, \qquad (5.12.35)$$

$$c = c_0 + kc_1 + k^2 c_2 + \cdots . \tag{5.12.36}$$

Substitute (5.12.34) - (5.12.36) into (5.12.26) and the conditions (5.12.29) - (5.12.33), we obtain, for successive orders of k, the following equations :

$O(1)$:

$$\frac{d^4\Psi_0}{dy^4} = 0 , \tag{5.12.37}$$

with

$$\frac{d\Psi_0}{dy}(0) = 0 , \tag{5.12.38}$$

$$\Psi_0(0) = 0 , \tag{5.12.39}$$

$$(1 - c_0)\xi_0 - \Psi_0(1) = 0 , \tag{5.12.40}$$

$$\frac{d^2\Psi_0}{dy^2}(1) + 2\xi_0 = 0 , \tag{5.12.41}$$

$$\frac{d^3\Psi}{dy^3}(1) = 0 . \tag{5.12.42}$$

$O(k)$

$$\frac{d^4\Psi_1}{dy^4} = iR\left[(U - c_0)\frac{d^2\Psi_0}{dy^2} - \frac{d^2U}{dy^2}\Psi_0\right], \tag{5.12.43}$$

with

$$\frac{d\Psi_1}{dy}(0) = 0 , \tag{5.12.44}$$

$$\Psi_1(0) = 0 \ , \tag{5.12.45}$$

$$(1 - c_0)\xi_1 - \Psi_1(1) - c_1\xi_0 = 0 \tag{5.12.46}$$

$$\frac{d^2\Psi_1}{dy^2}(1) + 2\xi_1 = 0 \ , \tag{5.12.47}$$

$$\frac{d^3\Psi_1}{dy^3}(1) = iR(1 - c_0)\frac{d\Psi_0}{dy}(1) - 2i\cot\theta\xi_0 \ . \tag{5.12.48}$$

$$\cdots .$$

From (5.12.37) we obtain

$$\Psi_0 = A_0 + B_0 y + C_0 y^2 + D_0 y^3 \ . \tag{5.12.49}$$

From (5.12.38), (5.12.39) and (5.12.42), we find $A_0 = B_0 = D_0 = 0$. Then (5.12.41) leads to

$$\Psi_0 = -\xi_0 y^2 \ , \tag{5.12.50}$$

and (5.12.40) leads to

$$c_0 = 2 \ . \tag{5.12.51}$$

Carry on to the order $O(k)$, the equation (5.12.43) is now :

$$\frac{d^4\Psi_1}{dy^4} = iR\left[(2y - y^2 - 2)(-2) - 2y^2\right]\xi_0 = -4iR\xi_0(y - 1) \ . \tag{5.12.52}$$

Therefore we obtain

$$\Psi_1 = A_1 + B_1 y + C_1 y^2 + D_1 y^3 - \left(\frac{iR\xi_0}{30}\right)(y - 1)^5 \ . \tag{5.12.53}$$

From (5.12.44) and (5.12.45), we obtain

$$A_1 = -\left(\frac{iR\xi_0}{30}\right), \tag{5.12.54}$$

$$B_1 = \frac{iR\xi_0}{6}. \tag{5.12.55}$$

From (5.12.48) we obtain

$$6D_1 = 2i(R - \cot\theta)\xi_0. \tag{5.12.56}$$

From (5.12.46) and (5.12.47) we obtain, since $c_0 = 2$, that

$$\begin{aligned} c_1 &= \frac{1}{2\xi_0}\left[\frac{d^2\Psi_1}{dy^2}(1) - 2\Psi_1(1)\right] \\ &= \frac{1}{2\xi_0}\left[6D_1 - 2(A_1 + B_1 + D_1)\right] \\ &= i\left[\frac{2}{3}(R - \cot\theta) + \frac{R}{30} - \frac{R}{6}\right] \\ &= \frac{8i}{15}\left(R - \frac{5}{4}\cot\theta\right). \end{aligned} \tag{5.12.57}$$

Since $c = c_0 + kc_1$ up to first order of k, thus c has a positive imaginary part if

$$R > \frac{5}{4}\cot\theta. \tag{5.12.58}$$

In other words, the flow is unstable if $R > R_c$, where

$$R_c = \frac{5}{4}\cot\theta \ . \tag{5.12.59}$$

In the normal range of θ , $R_c = O(1)$, which is much smaller than the critical Reynolds number for the Poiseuille flow. It is rather puzzling from the first sight. But if we take a closer look at the definition of Reynolds number R in (5.12.27), we see that we can write

$$R = \frac{2}{\sin\theta} F \ , \tag{5.12.60}$$

where

$$F = \frac{U_0^2}{gH} \ , \tag{5.12.61}$$

is the Froude number. The instability condition (5.12.58) can be rewritten as

$$F > \frac{5}{8}\cos\theta \ . \tag{5.12.62}$$

With the presence of free surface, the critical Froude number of $O(1)$ is indeed what we expect from roll waves in hydraulics. We can have a better understanding of this instability mechanism from the following illuminating analysis based on the hydraulic approximation.

Hydraulic Approximation

Let us go back to the nonlinear governing equations (5.12.1) - (5.12.8). Integrate (5.12.1) with respect to y from $y = 0$ to $y = h(x,t)$, we obtain, after using (5.12.6),

$$\frac{\partial h}{\partial t} + \frac{\partial}{\partial x}\int_0^h u(x,y,t)dy = 0 \ . \tag{5.12.63}$$

Integrating (5.12.2) with respect to y from $y = 0$ to $y = h$, we obtain, after using (5.12.1), (5.12.4) and (5.12.6),

$$\frac{\partial}{\partial t}\int_0^h u dy + \frac{\partial}{\partial x}\int_0^h u^2 dy = -\frac{1}{\rho}\frac{\partial}{\partial x}\int_0^h p dy + gh\sin\theta - f \ , \quad (5.12.64)$$

where

$$f = \upsilon\left[\frac{\partial u}{\partial y}(x,0,t) - \frac{\partial u}{\partial y}(x,h,t) + \frac{\partial v}{\partial x}(x,h,t)\right]$$

$$-\frac{1}{\rho}p(x,h,t)\frac{\partial h}{\partial x} \ . \quad (5.12.65)$$

The term f represents the frictional force. Its major contribution comes from $\upsilon\frac{\partial u}{\partial y}(x,0,t)$, i.e., the friction at the bottom.

In the hydraulic approximation, the velocity component v is assumed to be small. Therefore equation (5.12.3) can be approximated by

$$0 = -\frac{1}{\rho}\frac{\partial p}{\partial y} - g\cos\theta \ , \quad (5.12.66)$$

which leads to

$$p = -\rho g(y-h)\cos\theta \ , \quad (5.12.67)$$

if we take $p(x, h, t) = 0$ as an approximation of the free surface condition (5.12.8). Then (5.12.64) can be rewritten as

$$\frac{\partial}{\partial t}\int_0^h u dy + \frac{\partial}{\partial x}\int_0^h u^2 dy = -\frac{1}{2}g\cos\theta\frac{\partial}{\partial x}\left(h^2\right) + gh\sin\theta - f \ . \quad (5.12.68)$$

From the exact theory, we know the primary flow as given by (5.12.9) - (5.12.13). For the primary flow state, we obtain from (5.12.9) the rate of discharge Q :

$$Q = \int_0^h u dy = \frac{1}{3\upsilon} g h^3 \sin\theta \ . \tag{5.12.69}$$

Introducing $\overline{U}$ as the average of u, i.e.,

$$\int_0^h u dy = \overline{U} h \ , \tag{5.12.70}$$

then

$$\overline{U} = \left(\frac{g \sin\theta}{3\upsilon}\right) h^2 \ . \tag{5.12.71}$$

Now we shall write

$$\int_0^h u^2 dy = \alpha \overline{U}^2 h \ , \tag{5.12.72}$$

where α is a parameter. If we use (5.12.9) for the computation of (5.12.72), then

$$\alpha = \frac{6}{5} \ , \tag{5.12.73}$$

or sometimes we may just take

$$\alpha = 1 \ . \tag{5.12.74}$$

The primary flow state is a steady state, and it is achieved by balancing of $(gh \sin\theta)$ and f in (5.12.68). In order that $f = gh \sin\theta$ will be equivalent to (5.12.71), we set

$$f = \frac{3\upsilon\overline{U}}{h} \, . \tag{5.12.75}$$

Now substitute (5.12.70), (5.12.72) and (5.12.75) into (5.12.63) and (5.12.68), we obtain,

$$\frac{\partial h}{\partial t} + \frac{\partial}{\partial x}\left(\overline{U}h\right) = 0 \, , \tag{5.12.76}$$

$$\frac{\partial}{\partial t}\left(\overline{U}h\right) + \alpha\frac{\partial}{\partial x}\left(\overline{U}^2 h\right) + \frac{g}{2}\cos\theta\frac{\partial}{\partial x}\left(h^2\right) = gh\sin\theta - \frac{3\upsilon\overline{U}}{h} \, . \tag{5.12.77}$$

Equations (5.12.76) and (5.12.77) are the roll wave equations in hydraulics. The friction term f can be interpreted as a viscous stress of the form $3\upsilon\left(\frac{\Delta u}{\Delta l}\right)$, where we may identify (Δu) as $\overline{U}$ and Δl as h.

If we neglect the left-hand side of (5.12.77), then (5.12.76) becomes

$$\frac{\partial h}{\partial t} + \frac{gh^2 \sin\theta}{\upsilon}\frac{\partial h}{\partial x} = 0 \, . \tag{5.12.78}$$

Equation (5.12.78) represents the propagation of the kinematic wave with the speed of propagation $\frac{gh^2\sin\theta}{\upsilon}$, which is exactly the same as c_0 in (5.12.51) in dimensional form.

To analyze the stability of roll waves, we write

$$h = H + \overline{\xi} e^{ik(x-ct)} \, , \tag{5.12.79}$$

$$\overline{U} = \overline{U}_0 + \overline{u} e^{ik(x-ct)} \, , \tag{5.12.80}$$

where

$$\overline{U}_0 = \frac{gH^2 \sin\theta}{3\upsilon} = \frac{2}{3}U_0 \,. \tag{5.12.81}$$

Substitute (5.12.79) and (5.12.80) into (5.12.76) and (5.12.77), and keeping only terms linear in $\overline{\xi}$ and $\overline{u}$, we obtain

$$\left(\overline{U}_0 - c\right)\overline{\xi} + H\overline{u} = 0 \ , \tag{5.12.82}$$

$$-ikc\left(\overline{U}_0\overline{\xi} + H\overline{u}\right) + ik\alpha\left(\overline{U}_0^{\,2}\overline{\xi} + 2\overline{U}_0 H\overline{u}\right) + ikgH\overline{\xi}\cos\theta$$

$$= g\overline{\xi}\sin\theta - \frac{3\upsilon}{H}\overline{u} + \frac{3\upsilon\overline{U}_0}{H^2}\overline{\xi} \ . \tag{5.12.83}$$

Expressing $\overline{u}$ in terms of $\overline{\xi}$ from (5.12.82), then (5.12.83) becomes

$$ik\left\{-c\left[\overline{U}_0 - \left(\overline{U}_0 - c\right)\right] + \alpha\left[\overline{U}_0^{\,2} - 2\overline{U}_0\left(\overline{U}_0 - c\right)\right] + gH\cos\theta\right\}$$

$$= g\sin\theta + \frac{3\upsilon}{H^2}\left[\left(\overline{U}_0 - c\right) + \overline{U}_0\right] ,$$

or

$$ik\left[-c^2 + \alpha\overline{U}_0\left(2c - \overline{U}_0\right) + gH\cos\theta\right] = \frac{3\upsilon}{H^2}\left(3\overline{U}_0 - c\right) . \tag{5.12.84}$$

The critical state is the one for which c is real. Thus

$$c = 3\overline{U}_0 = 2U_0 \ , \tag{5.12.85}$$

and

$$c^2 - \alpha\overline{U}_0\left(2c - \overline{U}_0\right) = gH\cos\theta \ ,$$

or

$$\left(4-\frac{20}{9}\alpha\right)U_0^2 = gH\cos\theta\ . \tag{5.12.86}$$

(5.12.85) is exactly the same as (5.12.51), while (5.12.86) yields a critical Froude number F_c :

$$F_c = \frac{9\cos\theta}{4(9-5\alpha)}\ . \tag{5.12.87}$$

If we take $\alpha = 1$, then $F_c = \frac{9}{16}\cos\theta$; while for $\alpha = \frac{6}{5}$, then $F_c = \frac{3}{4}\cos\theta$. On the other hand, from (5.12.62), we have $F_c = \frac{5}{8}\cos\theta$.

In fact we can solve for c directly from (5.12.84). For the case kH<<1, i.e., when the wavelength is long, it is found that one mode is heavily damped, while the leading terms of the other mode are given by

$$c = c_0 + (kH)c_1\ , \tag{5.12.88}$$

where

$$\frac{c_0}{U_0} = 2\ , \tag{5.12.89}$$

and

$$\frac{c_1}{U_0} = \frac{i}{3}(kH)\left[\frac{4}{9}(9-5\alpha)R - 2\cot\theta\right]. \tag{5.12.90}$$

Comparing (5.12.90) with (5.12.37), we see that the growth rates of the instability are also the same except for the difference between $\frac{4}{9}(9-5\alpha)$ and $\frac{8}{5}$.

It is clear that the hydraulic approximation captures the essential feature of the long wavelength instability of the fluid flowing down an inclined plane. If we set $\alpha = \frac{27}{25}$, we can even recover the accurate stability criterion. The role of the viscosity, or the frictional force, in determining the stability of the system is also revealed clearly, and it is threefold :

(1) It determines the velocity profile of the primary flow, or the flow discharge rate.

(2) It determines the basic propagation speed c_0 of the critical perturbation wave. This wave is the so-called kinematic wave; and it is well known that $c_0 = \frac{dQ}{dh}$, once the discharge rate Q as a function of h is known.

(3) It yields explicitly the growth rate of the instability when $F > F_c$.

CHAPTER 6

Chaos

We have studied various stability problems of fluids in motion in Chapter 5. The discussions are mainly concerned with the linear stability analysis. If the fluid system is linearly unstable, the disturbances will grow indefinitely according to the linear theory. The linear analyses are of course in general no longer valid when the disturbances become finite. What actually happens then? Experiments and observations indicate there may be the following outcomes :

1. The system settles down to a regular organized flow state. For example, when the Rayleigh number exceeds the critical Rayleigh number in Rayleigh-Benard problem, convection cells will appear. This regular flow pattern can be maintained if the Rayleigh number is kept to be the same order of magnitude of R_c .

2. The system becomes turbulent. For example, the plane Poiseuille flow becomes turbulent when the Reynolds number exceeds a critical Reynolds number. The Rayleigh-Benard problem will also lead to turbulence for large Rayleigh numbers.

3. The system eventually settles down to a stable new state. For instance, for the Rayleigh-Taylor instability, the eventual state will be that the heavier fluid is underneath the lighter fluid.

4. The eventual state is out of the range of the original framework of mathematical formulation. For example, there are speculations that the solution of the Navier-Stokes equations is not unique for large Reynolds number.

5. The eventual state is out of the range of the original framework of physical model. For certain instability

problems, the Newtonian, continuum assumption of the fluid may no longer be valid.

Nonlinear studies of stability problems are difficult. Even with the rapid advances of computer technology, numerical studies are still limited to cases of relatively mild instabilities, i.e., for cases whose instability parameters are not much beyond the critical values. It is found that as the instability parameters increase, the system often first develops regular organized patterns, then the patterns become more and more complex, and then irregular and seemingly random but bounded patterns appear, or the system is in the chaotic state.

Since Lorenz presented the equations bearing his name in 1963, and Feigenbaum discovered universal numbers relating to nonlinear maps in 1978, the subject of chaos has been one of the most active fields of study in mathematical sciences. Numerous books have been written on this subject. We shall only select a few topics to introduce this exciting subject matter, mainly from the historical as well as the fluid dynamical perspectives.

6.1 The Lorenz Equations

In dealing with the Rayleigh-Benard problem, it is found that the thermally conductive equilibrium state is not stable when the Rayleigh number exceeds R_c. Convection cells start to appear. With further increase of the Rayleigh number, the flow will eventually become turbulent. Even when the convection cells start to appear, the flow state is not just a small deviation from the equilibrium state. Therefore the linear theory is no longer adequate to deal with the situation. A nonlinear study needs to be developed. In the following, we shall describe such an attempt and its important consequential developments.

Let us go back to equations (5.9.3) - (5.9.5), and use the notation (5.9.10). Then the nonlinear equations within the context of Boussinesq approximation are

$$\frac{\partial \mathbf{v}}{\partial t}+(\mathbf{v}\cdot\nabla)\mathbf{v}=-\frac{1}{\rho_0}\nabla p'-\upsilon\nabla\times(\nabla\times\mathbf{v})+g\alpha\theta\mathbf{e}_z \qquad (6.1.1)$$

$$\frac{\partial\theta}{\partial t}+(\mathbf{v}\cdot\nabla)\theta=\beta w+D_T\nabla^2\theta\,. \qquad (6.1.2)$$

Let us restrict our study to two-dimensional flow. Therefore, the convection cells will simply be rolls. Thus $\mathbf{v}=(u,0,w)$ and the variables depend only on (x, z, t). The continuity equation (5.9.3) implies that we may introduce the stream function $\psi(x,z,t)$ such that

$$w=\frac{\partial\psi}{\partial x}\,,\qquad u=-\frac{\partial\psi}{\partial z}\,. \qquad (6.1.3)$$

Substitute (6.1.3) into (6.1.1) and eliminate p', we obtain

$$\frac{\partial}{\partial t}\left(\nabla^2\psi\right)=-\frac{\partial\left(\psi,\nabla^2\psi\right)}{\partial(x,z)}+\upsilon\nabla^4\psi+g\alpha\frac{\partial\theta}{\partial x}\,, \qquad (6.1.4)$$

and (6.1.2) can be rewritten as

$$\frac{\partial\theta}{\partial t}=-\frac{\partial(\psi,\theta)}{\partial(x,z)}+\beta\frac{\partial\psi}{\partial x}+D_T\nabla^2\theta\,. \qquad (6.1.5)$$

Non-dimensionalize the equations in the same manner as (5.9.31), and let υ be the unit of ψ, (βH) be unit of θ, then equations (6.1.4) and (6.1.5) can be rewritten as

$$\frac{\partial}{\partial t}\left(\nabla^2\psi\right)=-\frac{\partial\left(\psi,\nabla^2\psi\right)}{\partial(x,z)}+\nabla^4\psi+\frac{R_a}{\sigma}\frac{\partial\theta}{\partial x}\,, \qquad (6.1.6)$$

$$\frac{\partial\theta}{\partial t}=-\frac{\partial(\psi,\theta)}{\partial(x,z)}+\frac{\partial\psi}{\partial x}+\frac{1}{\sigma}\nabla^2\theta\,, \qquad (6.1.7)$$

where R_a is the Rayleigh number defined in (5.9.38) and σ is the Prandtl number defined in (5.9.31). If we neglect the nonlinear terms, these are the equations governing the linear stability, now in terms of the stream function ψ .

We shall now consider the case that both upper and lower boundary surfaces are free. Then at the state of marginal linear stability, the flow state is defined by (5.9.52) and (5.9.47). We shall take $n = 1$, i.e., the most critical case, then the state is given by

$$\psi = \psi_0 \sin ax \sin \pi z \ , \tag{6.1.8}$$

$$\theta = \theta_0 \cos ax \sin \pi z \ . \tag{6.1.9}$$

For this critical state, we know from (5.9.54) - (5.9.56), that

$$a^2 = \frac{\pi^2}{2} \ , \qquad R_c = \frac{27}{4} \pi^4 \ . \tag{6.1.10}$$

To solve the nonlinear equations (6.1.6) and (6.1.7), we seek the solutions, as suggested by the solution of the critical state, in the following expansion :

$$\psi = \sum_m \sum_n \psi_{mn}(t) \sin(\max) \sin(n\pi z) \ , \tag{6.1.11}$$

$$\theta = \sum_m \sum_n \theta_{mn}(t) \cos(\max) \sin(n\pi z) \ , \tag{6.1.12}$$

and substitute into (6.1.6) and (6.1.7). Thus we are led to a set of coupled ordinary differential equations for $\psi_{mn}(t)$ and $\theta_{mn}(t)$. If the number of terms in the summation is infinite, then it is hopeless to solve the problem. We hope that a reasonable approximate solution can be found, when a suitably chosen finite number of terms in the summation is taken. Numerical experiments taking finite number of terms in the expansions (6.1.11) and (6.1.12) yields quite interesting results for R_a somewhat larger than R_c . It is found that as t increases, for certain cases, only three of the

$\psi_{mn}(t)$ and $\theta_{mn}(t)$ remain substantial, whereas all the other terms approach zero for large t. These three terms are ψ_{11}, θ_{11} and θ_{02}. In view of this result from the numerical experiment, Lorenz (1963) take the bold step and let

$$\psi = \left(\frac{\pi^2 + a^2}{\pi a \sigma} \right) \sqrt{2} X(\tau) \sin ax \sin \pi z \ , \qquad (6.1.13)$$

$$\theta = \frac{R_c}{\pi R_a} \left[\sqrt{2} Y(\tau) \cos ax \sin \pi z - Z(\tau) \sin 2\pi z \right] , \qquad (6.1.14)$$

where

$$\tau = \frac{1}{\sigma} \left(\pi^2 + a^2 \right) t \ . \qquad (6.1.15)$$

Substituting (6.1.13) and (6.1.14) into (6.1.6) and (6.1.7), and retaining only terms with $(\sin ax \sin \pi z)$, $(\cos ax \sin \pi z)$ and $(\sin 2\pi z)$, discarding all other terms, we obtain from this truncation exercise the following equations :

$$\frac{dX}{d\tau} = -\sigma X + \sigma Y \ , \qquad (6.1.16)$$

$$\frac{dY}{d\tau} = -XZ + rX - Y \ , \qquad (6.1.17)$$

$$\frac{dZ}{d\tau} = XY - bZ \ , \qquad (6.1.18)$$

where

$$r = \frac{R_a}{R_c} \ , \qquad b = \frac{4\pi^2}{\pi^2 + a^2} \ . \qquad (6.1.19)$$

The set of equations (6.1.16) - (6.1.18) are known as Lorenz equations.

The Lorenz equations is an autonomous system of three nonlinear ordinary differential equations. It is amazing that although autonomous systems of two nonlinear differential equations have been studied extensively, few works had devoted to the study of autonomous systems of three equations up to 1960s. The appearance of Lorenz equations literally opened up a new scientific field.

We begin the study of Lorenz equations by analyzing the equilibrium points of the system. It is readily found that the following equilibrium points exist :

$$(0): \qquad X = Y = Z = 0, \tag{6.1.20}$$

$$(C_1, C_2): \quad X = Y = \pm[b(r-1)]^{\frac{1}{2}}, \quad Z = r-1. \tag{6.1.21}$$

Going back to the Rayleigh-Benard problem, (0) represents the thermally conductive equilibrium state, while (C_1, C_2) represent the steady flows of convection cells. The convection cells appear only for $R_a > R_c$, and indeed it is only for $r > 1$, (C_1, C_2) are real and non-zero.

Next, we shall discuss the stability of these equilibrium points. Let (X_0, Y_0, Z_0) be an equilibrium point. Let

$$X = X_0 + x, \quad Y = Y_0 + y, \quad Z = Z_0 + z, \tag{6.1.22}$$

and substitute into (6.1.16) - (6.1.18), and retain only terms linear in *x, y,* and *z*. We thus obtain the stability equations for the equilibrium point :

$$\frac{dx}{d\tau} = -\sigma x + \sigma y, \tag{6.1.23}$$

$$\frac{dy}{d\tau} = (r - Z_0)x - y - X_0 z \ , \qquad (6.1.24)$$

$$\frac{dz}{d\tau} = Y_0 x + X_0 y - bz \ . \qquad (6.1.25)$$

The solution *(x, y, z)* will behave like $e^{\lambda\tau}$, and the eigenvalue λ satisfies the following characteristic equation :

$$(\lambda + \sigma)(\lambda + 1)(\lambda + b) - \sigma(\lambda + b)(r - Z_0)$$

$$+ X_0^2(\lambda + \sigma) + \sigma X_0 Y_0 = 0. \qquad (6.1.26)$$

Let us now discuss the stability of the equilibrium point (0). Thus $X_0 = Y_0 = Z_0 = 0$. Hence (6.1.26) becomes

$$(\lambda + b)\left[\lambda^2 + (\sigma + 1)\lambda + \sigma(1 - r)\right] = 0 \ . \qquad (6.1.27)$$

Since σ and b are both positive, we arrive at the following conclusions :

For $0 < r \leq 1$: all three eigenvalues are real and non-positive. Hence the equilibrium point (0) is stable.

For $r > 1$: Two eigenvalues are negative and real, but one eignevalue is real and positive. Therefore (0) is unstable in general.

Thus $r = 1$ is a transition point, or a bifurcation point. The equilibrium point (0) transits from stability to instability at this transition point; and at the same transition point, new equilibrium points (C_1, C_2) start to appear. These results are both qualitatively and quantitatively in agreement with the linear stability analysis of the Rayleigh-Benard problem.

Now, let us discuss the stability of equilibrium points (C_1, C_2). Let $X_0 = Y_0 = \pm[b(r-1)]^{\frac{1}{2}}$, $Z_0 = r-1$, and substitute into (6.1.26). We obtain

$$\lambda^3 + (\sigma + b + 1)\lambda^2 + (r + \sigma)b\lambda + 2\sigma b(r-1) = 0 \,. \quad (6.1.28)$$

It is clear that one of roots of (6.1.28) is real and negative. By inspecting the derivative with respect to λ of (6.1.28), i.e.,

$$3\lambda^2 + 2(\sigma + b + 1)\lambda + b(r + \sigma) = 0 \,, \quad (6.1.29)$$

we may deduce that the other two roots are also negative and real if $r > 1$, and r is close to 1. As r increases, these two real roots become complex conjugates but still with negative real parts. In this range of r, (C_1, C_2) is stable. Further increase of r will eventually cause the real parts of these two complex conjugate roots to become positive. We shall denote this transitional value of r, r_H. The value of r_H can be readily determined. At $r = r_H$, equation (6.1.28) has two purely imaginary roots $\lambda = \pm ih$, and h is real. The vanishing of both real and imaginary parts of (6.1.28) leads to the following two equations :

$$-h^3 + (r_H + \sigma)bh = 0 \,,$$

and

$$-(\sigma + b + 1)h^2 + 2\sigma b(r_H - 1) = 0 \,.$$

The consistency requirement thus leads to

$$(r_H + \sigma)b = \frac{2\sigma b(r_H - 1)}{(\sigma + b + 1)} \,,$$

or

$$r_H = \frac{\sigma(\sigma+b+3)}{\sigma-(b+1)} . \tag{6.1.30}$$

It can also be shown that indeed the real part of the complex conjugate roots is negative for $r < r_H$. Therefore the equilibrium points (C_1, C_2) are stable for $1 < r < r_H$, and unstable for $r > r_H$. The transition at $r = r_H$ is a Hopf bifurcation.

Stable equilibrium point is one of the well known attractors. All the neighboring trajectories $[X(\tau), Y(\tau), Z(\tau)]$ will eventually be attracted towards the attractor. There is another type of classical attractors, i.e., the limit cycle. Periodic orbit is a limit cycle. For the Lorenz equations, (0) is an attractor for $r < 1$, while (C_1, C_2) are attractors for $1 < r < r_H$. What about the situation for $r > r_H$?

For the Lorenz equations, by making use of the Lyapunov function

$$V = rX^2 + \sigma Y^2 + \sigma(Z-2r)^2 ,$$

it may be shown that all the trajectories $[X(\tau), Y(\tau), Z(\tau)]$ will eventually enter into a bounded ellipsoid. One estimate gives $V = c$ as a satisfactory ellipsoid, where

$$c = \begin{cases} \dfrac{b^2 r^2}{(b-\sigma)} , & \text{if } b \geq 2\sigma , \quad \sigma \leq 1 , \\[2ex] \dfrac{b^2 r^2}{(b-1)} , & \text{if } b \geq 2 , \quad \sigma \geq 1 , \\[4ex] 4\sigma r^2 , & \text{otherwise} . \end{cases}$$

Thus, even for $r > r_H$, there is a bounded attracting set. Moreover, from (6.1.16) - (6.1.18), we see that

$$\frac{\partial}{\partial X}\left[\frac{dX}{d\tau}\right]+\frac{\partial}{\partial Y}\left[\frac{dY}{d\tau}\right]+\frac{\partial}{\partial Z}\left[\frac{dZ}{d\tau}\right]=-(\sigma+b+1)\,, \quad (6.1.31)$$

i.e., the divergence of the flow is negative. Now all the trajectories outside the bounded ellipsoid $V = c$, which we shall call E, will eventually reach the surface of E. The volume of this ellipsoid, because of (6.1.31), will be contracted by the flow by a factor of $e^{-(\sigma+b+1)}$ at a time t. Eventually as $t \to \infty$, the volume of the set which attracts all the trajectories will approach zero. In this way, we have established the existence of a bounded globally attracting set of zero volume. Furthermore, since the surface of E is closed and connected, this attracting set is also closed and connected. It includes the equilibrium points (0) and (C_1, C_2).

In order to understand better the nature of this attracting set, numerical studies of the Lorenz equations have been carried out extensively. The parameters chosen by Lorenz are $\sigma = 10$, $b = \frac{8}{3}$. $b = \frac{8}{3}$ corresponds to $a^2 = \frac{\pi^2}{2}$ which is the linear critical value. For water at $20°\mathrm{C}$, $\sigma = 7.1$ which is not too far from the value of 10. All the subsequent discussions are based on these values of the parameters. From (6.1.30), we then obtain $r_H = 24.74$.

Numerical solutions for the trajectories at r = 28 was carried out by Lorenz. If we project such a trajectory onto the X-Z plane, it looks like the one shown in Fig. 6-1(b). It shows a trajectory moving sometimes around C_1, and sometimes around C_2, going on indefinitely in an irregular manner within a bounded region. The trajectory is attracted to some attracting set, which we shall call S, a strange attractor, since it is not like equilibrium points, or limit cycles which are the classical attractors we know well. It should be pointed out that although the trajectory looks very irregular, it is the unique solution of a set of deterministic equations. It is not random. However, the solution is very sensitive to initial conditions. A small change of initial conditions will result in a completely different trajectory, although it looks qualitatively similar.

Detailed numerical studies show that the strange attractor S starts to appear before r reaches r_H , the Hopf bifurcation. It starts to appear at $r = r_A = 24.06$, whereas $r_H = 24.74$. We shall now describe briefly the developments as r increases from $r = 1$ to $r = r_A$.

We have noted that the equilibrium point (0) becomes unstable at $r = 1$. For $r > 1$, as may be seen from (6.1.27), two of the eigenvalues are still real and negative. Therefore in the neighborhood of (0), there is still a stable two-dimensional manifold on which the trajectories are attracted to (0). Trajectories not on this stable two-dimensional manifold will spiral towards C_1 or C_2 . This stable two-dimensional manifold of (0) extends into the whole three-dimensional space for r moderately larger than one. It divides the whole space into two halves. Trajectories started in one half-space tend towards C_1 and trajectories started in the other half-space tend towards C_2 . The trajectories do not cross over the stable manifold of (0). But as r increases, something drastic happens at $r = r' = 13.926$. At $r = r'$, trajectories started on the unstable manifold of (0), i.e., in a direction transverse to the stable manifold, may also lie in the stable manifold of (0) and will therefore tend, in both forwards and backwards time, towards (0). These trajectories are called homoclinic orbits associated with (0). For $r > r'$, the trajectories will cross over from the region associated with C_1 to that associated with C_2, and vice versa. We say that there is a homoclinic explosion at $r = r'$. For $r > r'$, trajectories start to look like that shown in Fig. 6-1. However, the trajectories will eventually move towards C_1 or C_2 . But it takes longer and longer time for the trajectories to approach C_1 or C_2 as r increases, and as r approaches r_A , the time it takes to approach C_1 or C_2 approaches infinity. In other words, for $r \geq r_A$, there are trajectories that never reach C_1 or C_2 . These trajectories are then attracted by the strange attractor S.

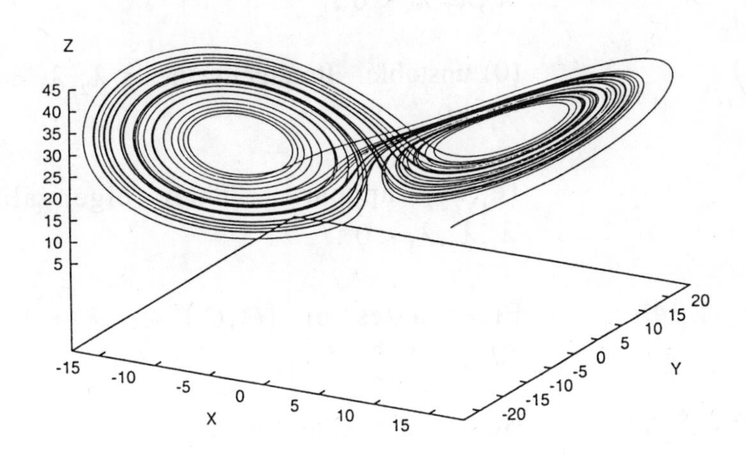

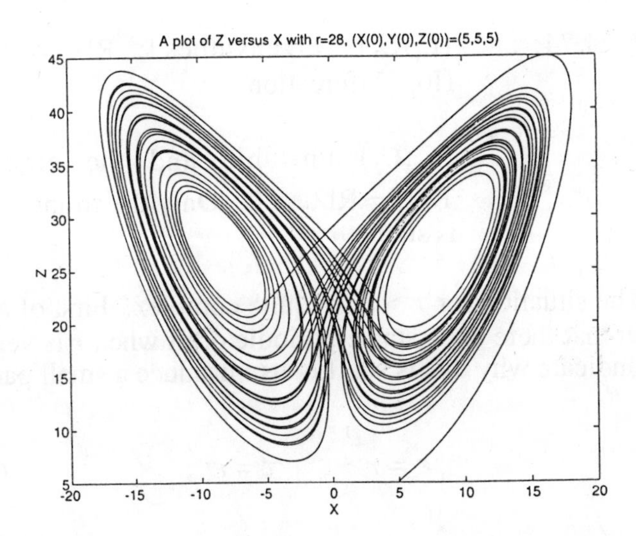

Fig. 6-1 A numerically computed solution to the Lorenz equations ($\sigma = 10$, $b = \frac{8}{3}$, $r = 28.0$).

We can now summarize the situation as follows :

$r < 1$:	(0) globally stable. Its eigenvalues : $\lambda_1, \lambda_2, \lambda_3 < 0$.
$r > 1$:	(0) unstable. Its eigenvalues : $\lambda_2, \lambda_3 < 0$, $\lambda_1 > 0$.
	(C_1, C_2) appear. Their eigenvalues $\lambda_1, \lambda_2, \lambda_3 < 0$.
$r > 1.346$:	Eigenvalues of (C_1, C_2) : $\lambda_1 = \lambda_2^*$, $\text{Rl}\lambda_1 < 0$, $\lambda_3 < 0$.
$r = r' = 13.926$:	Homoclinic explosion.
$r = r_A = 24.06$:	The strange attractor S appears.
$r = r_H = 24.74$:	(C_1, C_2) eigenvalues : $\text{Rl}\lambda_1 = \text{Rl}\lambda_2 = 0$. Hopf bifurcation.
$r > r_H$:	(C_1, C_2) unstable and the eigenvalues $\text{Rl}\lambda_1 = \text{Rl}\lambda_2 > 0$. Only the strange attractor is stable.

The situation for $r > r_H$ is quite complex. First of all it can be shown that there exits stable periodic orbit when r is very large. We can indicate why this is so. Let us introduce a small parameter

$$\varepsilon = r^{-\frac{1}{2}}, \qquad \tau = \varepsilon t , \tag{6.1.32}$$

and let

$$X = \varepsilon^{-1} u , \quad Y = \varepsilon^{-2} \sigma^{-1} v , \quad Z = \varepsilon^{-2}\left(\sigma^{-1} w + 1\right) . \tag{6.1.33}$$

Then the Lorenz equations (6.1.16) - (6.1.18) become :

$$\frac{du}{dt} = v - \varepsilon\sigma u , \tag{6.1.34}$$

$$\frac{dv}{dt} = -uw - \varepsilon v , \tag{6.1.35}$$

$$\frac{dw}{dt} = uv - \varepsilon b(w + \sigma) . \tag{6.1.36}$$

As $r \to \infty$, then $\varepsilon \to 0$. In the limit of $\varepsilon = 0$, the equations (6.1.34) - (6.1.36) become

$$\frac{du}{dt} = v , \tag{6.1.37}$$

$$\frac{dv}{dt} = -uw , \tag{6.1.38}$$

$$\frac{dw}{dt} = uv . \tag{6.1.39}$$

Two integrals of (6.1.37) - (6.1.39) can be readily found :

$$u^2 - 2w = 2A , \tag{6.1.40}$$

$$v^2 + w^2 = B^2 , \tag{6.1.41}$$

where A and B are constants. Thus the trajectories lie on the cylinder given by (6.1.41), and in general it is a closed orbit. Perturbation expansion based on small ε can be carried out. Then it may be shown that if $\left(\frac{\sigma+1}{b+1}\right) > 1$, there is a stable periodic orbit which winds once around the z-axis for large r, and this is the only orbit which persists for all r. For the case $\sigma = 10$, $b = \frac{8}{3}$, the requirement $\left(\frac{\sigma+1}{b+1}\right) > 1$ is satisfied.

Numerical experiments show that as r decreases, period doubling bifurcation will start to happen at $r \approx 313$. Successive stages of period doubling bifurcation will appear as r decreases and culminate at $r \approx 214.364$. At each period doubling bifurcation, the original periodic orbit becomes unstable, and new stable orbits which wind twice as much as the original appear. Numerical computations indicate that there are two more windows of period doubling at $145 < r < 166$ and $99.524 < r < 100.795$. For these phenomena of period doubling, if we denote r_n the value of r at n-th bifurcation, then it is found the constant

$$\delta = \lim_{n\to\infty} \frac{\left(r_{n-1} - r_n\right)}{\left(r_n - r_{n+1}\right)} , \tag{6.1.42}$$

agrees with the Feigenbaum constant which we shall discuss in the next section. Numerical experiments also reveal that there are other phenomena such as intermittent chaos and noisy periodicity in other ranges of the parameter r.

Chaos is the revelation of a strange attractor. Chaotic flow is irregular and appears to be random, but it is deterministic. Its details depend sensitively on the initial condition. It is sometimes termed as order in randomness. Chaos may represent a stage leading to turbulence. As far as Lorenz equations are concerned, they can at best be a model for Rayleigh-Benard problem when R is only moderately large in comparison of R_c , in view of the existence of stable periodic orbits as $R \to \infty$. Because we expect the flow to be turbulent when R becomes very large. However, the significance of the Lorenz equations lies not so much in the modeling of Rayleigh-Benard problem as the opening up of a new scientific field. Even within the context of fluid dynamics, experiments have since verified the existence of such phenomena as period doubling, intermittency and chaos, which were never known before 1960s.

6.2 The Logistic Map

The recursion relation

$$x_{n+1} = 4\lambda x_n(1-x_n)\ , \quad x_n \in [0,\ 1]\ , \quad \lambda \in [0,\ 1]\ , \tag{6.2.1}$$

which maps [0, 1] onto itself is known as the logistic map. It is the prototype of a more general class of nonlinear map of [0, 1] onto itself :

$$x_{n+1} = \lambda f(x_n)\ , \quad x_n \in [0,\ 1]\ , \quad \lambda \in [0,\ 1]\ , \tag{6.2.2}$$

which Feigenbaum (1978) discussed and generated so much excitement ever since. We shall limit our discussion to the logistic map for definiteness and also for illustration of general features of those nonlinear maps.

The logistic map takes its name from the corresponding differential equation

$$\frac{dx}{dt} = \mu x(1-x)\ , \quad \mu > 0\ , \tag{6.2.3}$$

originally proposed by Verhulst (1804 - 1849) for population studies. It is worth noting that for (6.2.3), there are two equilibrium points, or fixed points, one at $x = 0$ which is unstable, and another at $x = 1$ which is stable. But the behavior of (6.2.1) which is the difference equation version of (6.2.3) is quite different.

Returning to (6.2.1), or (6.2.2) now with $f(x) = 4x(1-x)$, the fixed points of the mapping are given by

$$x = \lambda f(x) \tag{6.2.4}$$

or for the logistic map,

$$x = 4\lambda x(1-x) , \tag{6.2.5}$$

which yields :

$$x = 0 , \tag{6.2.6}$$

and

$$x = x^* = \frac{4\lambda - 1}{4\lambda} . \tag{6.2.7}$$

x^* is in [0, 1] only for $\frac{1}{4} \leq \lambda \leq 1$.

To investigate the stability of fixed points, let us write $\lambda f(x) = g(x)$, and let x_f be the fixed point. Now write

$$x_n = x_f + \xi_n , \tag{6.2.8}$$

and substitute into (6.2.2), i.e.

$$x_{n+1} = g(x_n) , \tag{6.2.9}$$

we obtain

$$x_f + \xi_n = g(x_f + \xi_n) = g(x_f) + \xi_n g'(x_f) + \cdots .$$

Since $x_f = g(x_f)$, we thus obtain

$$\xi_n = g'(x_f)\xi_n + 0(\xi_n^2) . \tag{6.2.10}$$

Clearly, $\xi_n \to 0$ if $|g'(x_f)| < 1$, which is the criterion for local stability. For the logistic map, we have

$$g'(x) = 4\lambda(1-2x) . \tag{6.2.11}$$

Thus the fixed point 0 is stable for $0 \le \lambda < \frac{1}{4}$; while the fixed point x^* is stable for $|2-4\lambda| < 1$, or for $\frac{1}{4} < \lambda < \frac{3}{4}$. Figure 6-2 shows the case for $\lambda = \frac{1}{2}$, with fixed point at $x^* = \frac{1}{2}$, which is the intersection point of the straight line $x_{n+1} = x_n$ and the curve $x_{n+1} = g(x_n) = 2x_n(1-x_n)$. Notice that the magnitude of the slope of the curve $g'(x_n)$ at $x_n = \frac{1}{2}$ is less than 1.

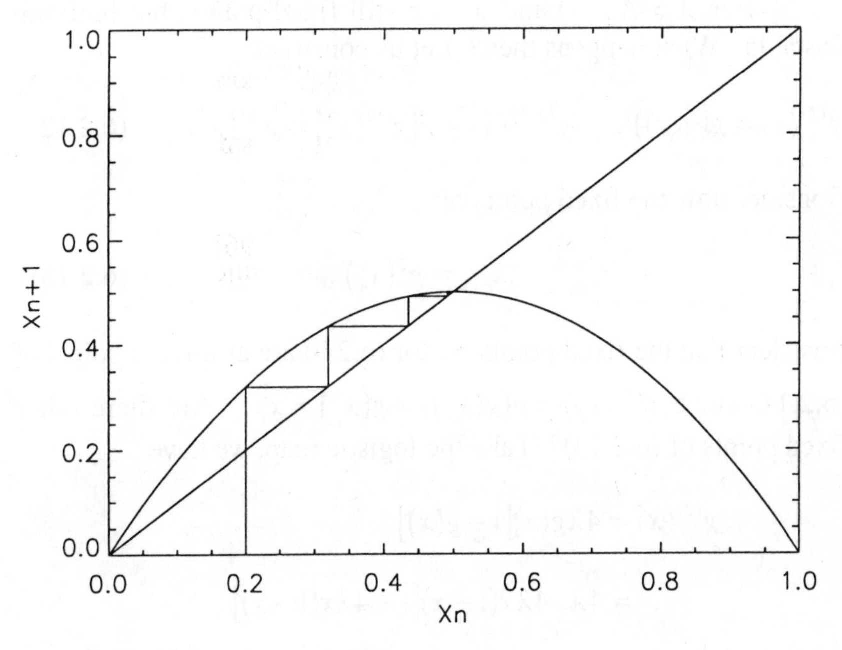

Fig. 6-2 Evolution of the logistic map for $\lambda = \frac{1}{2}$.

The fixed point is $x = \frac{1}{2}$.

Thus for the logistic map, as λ increases from $\lambda = 0$, we have for $0 \le \lambda < \frac{1}{4}$, the stable fixed point is $x = 0$. At $\lambda = \frac{1}{4}$, $x = 0$ is marginally stable. For $\lambda > \frac{1}{4}$, $x = 0$ is no longer stable, but another fixed point at $x = x^* = \dfrac{4\lambda - 1}{4\lambda}$ appears and is stable. x^* remains stable until λ reaches the value of $\frac{3}{4}$, which we shall denote as Λ_0. At $\lambda = \Lambda_0$, we have $x^* = \frac{2}{3}$, and $g'(x^*) = -1$.

For $\lambda > \Lambda_0$, 0 and x^* are still fixed points, but both are unstable. What happens then? Let us construct

$$g^{(2)}(x) = g(g(x)), \quad g^{(n+1)}(x) = g\left(g^{(n)}(x)\right) = g^{(n)}(g(x)). \quad (6.2.12)$$

Consider now the fixed points of

$$x_{n+1} = g^2(x_n). \quad (6.2.13)$$

It is clear that the fixed points x_f for (6.2.9) are also fixed points of (6.2.13) since $g^{(2)}(x_f) = g(g(x_f)) = g(x_f) = x_f$. Are there other fixed points of (6.2.13)? Take the logistic map, we have

$$\begin{aligned} g^{(2)}(x) &= 4\lambda g(x)[1 - g(x)] \\ &= 4\lambda \cdot 4\lambda x(1-x)[1 - 4\lambda x(1-x)] \\ &= 16\lambda^2\left[x(1-x) - 4\lambda x^2(1-x)^2\right]. \end{aligned} \quad (6.2.14)$$

It is readily verified that the equation $x = g^{(2)}(x)$ can be rewritten as

$$x[4\lambda x-(4\lambda-1)][16\lambda^2x^2-4\lambda(4\lambda+1)x+(4\lambda+1)]=0\ .$$

Therefore the four roots are $x = 0,\ x = \dfrac{4\lambda-1}{4\lambda} = x^*$, and

$$x = x_{1,2}^* = \frac{1}{8\lambda}\left\{(4\lambda+1)\pm[(4\lambda+1)(4\lambda-3)]^{\frac{1}{2}}\right\}\ . \quad (6.2.15)$$

Thus there are only two real fixed points for $\lambda < \Lambda_0 = \dfrac{3}{4}$, i.e., the two fixed points for (6.2.9). However, two more real fixed points appear for $\lambda > \Lambda_0$. At $\lambda = \Lambda_0 = \dfrac{3}{4}$, $x_{1,2}^* = \dfrac{4\lambda+1}{8\lambda} = \dfrac{2}{3} = x^*$. It is clear that the fixed points bifurcate from the fixed point x^* , when $\lambda = \Lambda_0$. It is a pitchfork bifurcation for $g^{(2)}$.

To investigate the stability of fixed points of (6.2.13), we need to compute $g^{(2)'}(x)$ at the fixed points. Now

$$g^{(2)'}(x) = \frac{d}{dx}g(g(x)) = g'(g(x))\cdot g'(x)\ . \quad (6.2.16)$$

At $x = x^*$, since $x^* = g(x^*)$, therefore

$$g^{(2)'}(x^*) = [g'(x^*)]^2 = 1\ , \quad (6.2.17)$$

which shows that the fixed point x^* is also marginally stable for $g^{(2)}$. Now, for $g^{(2)}(x)$ given by (6.2.14) we have

$$g^{(2)'}(x) = 16\lambda^2(1-2x)[1-8\lambda x(1-x)]\ . \quad (6.2.18)$$

Take $\mu = \lambda - \Lambda_0 = \lambda - \dfrac{3}{4}$, then from (6.2.15), we have

$$x_1^* = \frac{2}{3}\left[1+\mu^{\frac{1}{2}} - \frac{\mu}{3} + 0\left(\mu^{\frac{3}{2}}\right)\right]. \qquad (6.2.19)$$

Then from (6.12.18), we found

$$g^{(2)'}(x_1^*) = 1 - 16\mu + 0\left(\mu^{\frac{3}{2}}\right). \qquad (6.2.20)$$

Thus for $\mu << 1$, we have

$$\left|g^{(2)'}\left(x_1^*\right)\right| < 1 . \qquad (6.2.21)$$

Hence the fixed point $\left(x_1^*\right)$ of (6.2.13) is stable.

Now since $x_1^* = g^{(2)}\left(x_1^*\right) = g\left(g\left(x_1^*\right)\right)$, then there exists an x_a, such that $x_a = g\left(x_1^*\right)$ and $x_1^* = g\left(x_a\right)$. Thus $x_a = g\left(g\left(x_1^*\right)\right) = g^{(2)}\left(x_a\right)$. Therefore x_a is also a fixed point of $g^{(2)}$. But we only have four fixed points. Evidently 0 and x^* are not x_a , since they are fixed points of g . Therefore $x_a = x_2^*$. Hence we have for $\lambda \geq \Lambda_0$ that

$$x_1^* = g\left(x_2^*\right), \quad x_2^* = g\left(x_1^*\right). \qquad (6.2.22)$$

x_1^* and x_2^* form a two-point cycle as far as the original map *g(x)* is concerned. Then from (6.2.16), we have

$$g^{(2)'}\left(x_1^*\right) = g'\left(g\left(x_1^*\right)\right)\cdot g'\left(x_1^*\right) = g'\left(x_2^*\right)\cdot g'\left(x_1^*\right) = g^{(2)'}\left(x_2^*\right). \quad (6.2.23)$$

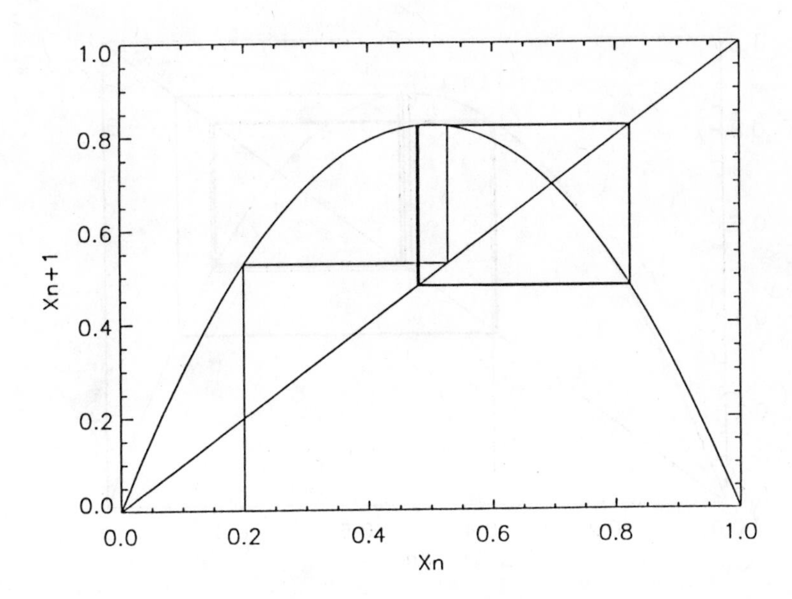

Fig. 6-3 The logistic map for $\lambda = 0.825$ showing an oscillation between $x = 0.479$ and $x = 0.824$.

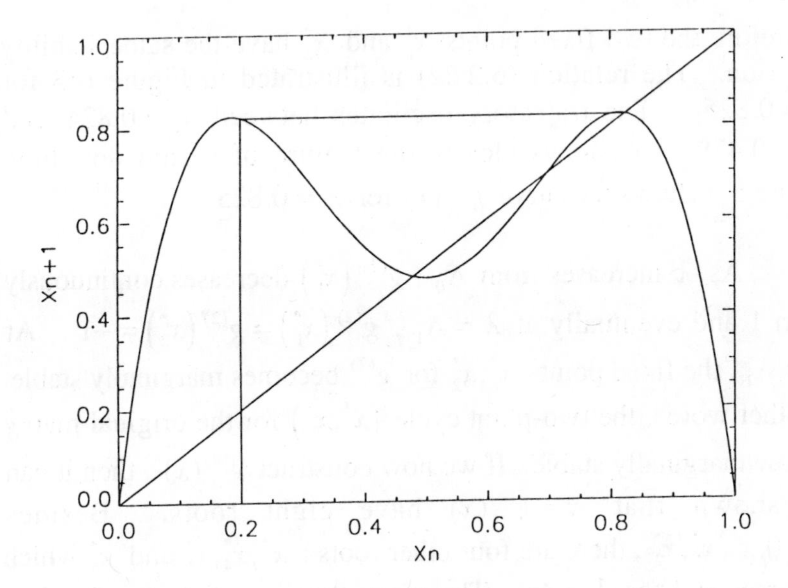

Fig. 6-4 The curve $g^{(2)}(x)$ for $\lambda = 0.825$.

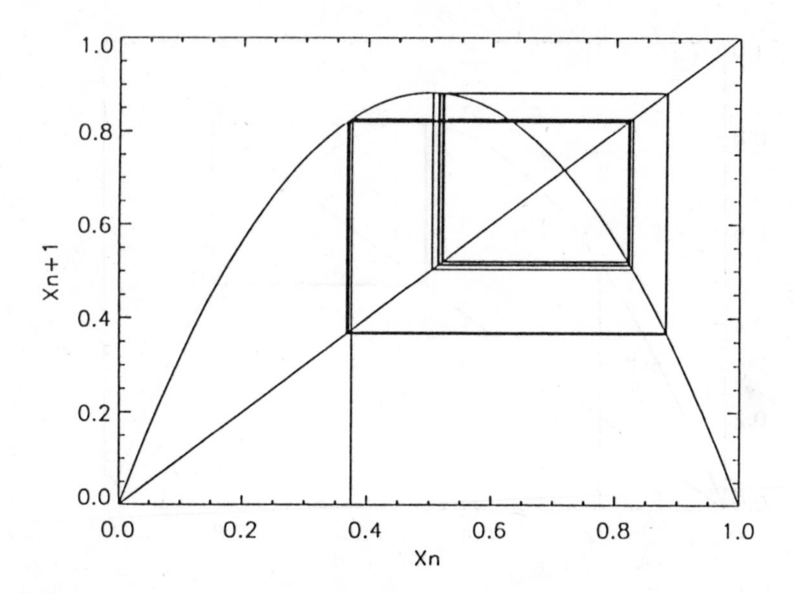

Fig. 6-5 A period-4 logistic map at $\lambda = 0.8825$ cycling between four values : $x = 0.37, 0.52, 0.83, 0.88$.

Therefore the two fixed points x_1^* and x_2^* have the same stability criterion. The relation (6.2.22) is illustrated in Figure 6-3 for $\lambda = 0.825$. The trajectory oscillates between $x_1 = 0.824$ and $x_2 = 0.479$. It shows clearly the feature of period doubling. Figure 6-4 shows the curve $g^{(2)}(x)$ for $\lambda = 0.825$.

As λ increases from Λ_0 , $g^{(2)\prime}\left(x_1^*\right)$ decreases continuously from 1 and eventually at $\lambda = \Lambda_1$, $g^{(2)\prime}\left(x_1^*\right) = g^{(2)\prime}\left(x_2^*\right) = -1$. At $\lambda = \Lambda_1$, the fixed points x_1^*, x_2^* for $g^{(2)}$ becomes marginally stable. In other words, the two-point cycle $\left(x_1^*, x_2^*\right)$ for the original map g is now marginally stable. If we now construct $g^{(4)}(x)$, then it can be shown that $x = g^{(4)}(x)$ have eight roots. Besides $x = 0, x^*, x_1^*, x_2^*$, there are four other roots : x_3^*, x_4^*, x_5^* and x_6^* which becomes real for $\lambda \geq \Lambda_1$. Therefore, there are eight fixed points with respect to $g^{(4)}$, and for $\lambda > \Lambda_1$, $0, x^*, x_1^*, x_2^*$ are unstable; whereas x_3^*, x_4^*, x_5^* and x_6^* can be shown to be stable for $\Lambda_1 < \lambda < \Lambda_2$, where Λ_2 is the value of λ at which x_3^*, x_4^*, x_5^* and x_6^* becomes marginally stable. In terms of $g^{(4)}$, x_3^* and x_4^* bifurcate from x_1^*, while x_5^* and x_6^* bifurcate from x_2^*. Again they are pitchfork bifurcations for $g^{(4)}$. These four fixed points x_3^*, x_4^*, x_5^* and x_6^* then form a 4-point cycle for the original map g. It represents the second period doubling. Figure 6-5 shows the 4-point cycle at $\lambda = 0.8825$.

This process of successive period-doubling bifurcations continue at $\lambda = \Lambda_2, \Lambda_3, \cdots$. It can be shown numerically that $\Lambda_\infty = 0.892486\cdots$. One remarkable finding is that there exists a universal constant δ which is defined as

$$\delta = \lim_{k\to\infty} \frac{\Lambda_k - \Lambda_{k-1}}{\Lambda_{k+1} - \Lambda_k} = 4.6692016091029909\cdots. \quad (6.2.24)$$

The constant is called "universal", because it is valid for a general class of maps. The existence of such universal transcendental number is indeed fascinating and leads people to ponder the deeper meaning of the phenomena.

For $\lambda > \Lambda_\infty$, the map is chaotic. Then at $\lambda \cong 0.957$, another period-doubling window starts to appear. It begins with a 3-point cycle, i.e., the fixed points for $g^{(3)}(x)$. Then as λ increases, a stable 6-point cycle takes the place of the 3-point cycle. This period-doubling process continues until λ reaches the value around 0.963. Afterwards it is chaotic.

It may be remarked that, from the numerical studies of the Lorenz equations, a geometric model can be constructed, which can be defined by a nonlinear map on interval. Follow a trajectory given by the Lorenz equations and consider the plane $z = r - 1$. The trajectory will cross from the side $z > r - 1$ to the other side at various points on $z = r - 1$. The successive points on the plane section $z = r - 1$, defines a return map which is known as Poincaré map. The plane $z = r - 1$ is a Poincaré section.

The development of the Lorenz system, in terms of the Poincaré map, for $1<r<28$ apparently can be understood by looking at a map, on interval $[-1, 1]$, of the following type :

$$g(x) = \begin{cases} 1 - \beta|x|^{\alpha}; & x \in [-1, 0], \\ -1 + \beta|x|^{\alpha}; & x \in [0, 1]. \end{cases} \tag{6.2.25}$$

We shall not go into details of discussion on this map. We mention this development to show the interrelationship between the Lorenz equations which is a set differential equations to the nonlinear map on intervals as exemplified by the logistic map.

6.3 Characterization of Chaos

Since the advent of the chaotic studies, many phenomena hitherto considered irregular and incomprehensible can be studied from the perspective of chaos. Not every phenomenon can be modeled in terms of differential equations or simple mappings. However there are ways devised to characterize the nature of the chaos in a general manner. We now introduce some of these approaches.

Lyapunov Exponents

Consider a map such as the logistic map

$$x_{n+1} = f(x_n) = f^{(n)}(x_0), \tag{6.3.1}$$

which has initial value x_0 . If the initial value is a neighboring value $(x_0 + \varepsilon)$, then corresponding to (6.3.1), we have

$$x_{n+1}' = f(x_n') = f^{(n)}(x_0 + \varepsilon). \tag{6.3.2}$$

Let us denote

$$\delta x_n = x_n' - x_n \; . \tag{6.3.3}$$

Thus we have

$$\delta x_0 = \varepsilon \; , \tag{6.3.4}$$

$$\delta x_1 = f(x_0 + \varepsilon) - f(x_0) \approx f'(x_0)\varepsilon \; , \tag{6.3.5}$$

and

$$\delta x_n \approx f'(x_{n-1})\delta x_{n-1} \approx \left(\prod_{m=0}^{n-1} f'(x_m) \right) \varepsilon \; . \tag{6.3.6}$$

The quantity

$$\lambda = \lim_{n \to \infty} \frac{1}{n} \sum_{m=0}^{n-1} \ln\left(\left| f'(x_m) \right| \right) \; , \tag{6.3.7}$$

is known as the Lyapunov exponent. In other words, comparing (6.3.6) and (6.3.7), we have, if n is not too large,

$$\delta x_n \approx \varepsilon e^{n\lambda} \; , \tag{6.3.8}$$

thus e^{λ} is the average rate of growth of deviation per iteration.

We can generalize the concept of Lyapunov exponent to p-dimensional maps :

$$\mathbf{x}_{n+1} = \mathbf{f}(\mathbf{x}_n) \; , \tag{6.3.9}$$

with

$$\delta \mathbf{x}_0 = \boldsymbol{\varepsilon} \; , \tag{6.3.10}$$

where $\mathbf{x}_n$, $\mathbf{f}$, $\delta\mathbf{x}_0$ and ε are now p-dimensional vectors. Corresponding to (6.3.5) and (6.3.6) we then have

$$\delta\mathbf{x}_1 \approx \mathbf{F}_0 \cdot \varepsilon \,, \tag{6.3.11}$$

and

$$\delta\mathbf{x}_n \approx \left(\prod_{m=0}^{n-1} \mathbf{F}_m \right) \varepsilon \,, \tag{6.3.12}$$

where $\mathbf{F}_m$ is a matrix, whose components are

$$\left(\mathbf{F}_m\right)_{ij} = \partial_j f_i(\mathbf{x}_m) \,, \quad i, j = 1, \cdots, p \,. \tag{6.3.13}$$

Formally, let us write

$$\mathbf{F}_{(n)} = \prod_{m=0}^{n-1} \mathbf{F}_m \,. \tag{6.3.14}$$

Then

$$\begin{aligned} |\delta\mathbf{x}_n|^2 &= \left(\delta\mathbf{x}_n^T\right)\left(\delta\mathbf{x}_n\right) \\ &= \varepsilon^T \mathbf{F}_{(n)}^T \mathbf{F}_{(n)} \varepsilon \,. \end{aligned} \tag{6.3.15}$$

Define the Lyapunov exponent λ , in a corresponding spirit as (6.3.8) :

$$|\delta\mathbf{x}_n|^2 = e^{2n\lambda} |\varepsilon|^2 \,, \tag{6.3.16}$$

we thus have

$$\lambda = \lim_{n\to\infty} \frac{1}{2n} \ln\left[\frac{\varepsilon^T \mathbf{F}_{(n)}^T \mathbf{F}_{(n)} \varepsilon}{|\varepsilon|^2} \right] . \tag{6.3.17}$$

If each $\mathbf{F}_{(m)}$ is the same and constant, and k is the largest eigenvalue in magnitude, then $\lambda = \ln|k|$. In fact $\delta\mathbf{x}_n$ will be aligned with the eigenvector of $\left(\mathbf{F}_{(m)}\right)^n$, corresponding to the eigenvalue k^n . But now each $\mathbf{F}_{(m)}$ is different and changes with each iteration. We need to follow the eigenvector corresponding to the largest eigenvalue in each step.

Dimensions

Dimension refers to the number of parameters or conditions required to specify the state of the system. Take the Lorenz system. The general state is given by $(x(t), y(t), z(t))$. If there is no condition to specify *x, y* and *z*, we need 3 parameters to describe the state. Thus the dimension is 3. However, if the state is the equilibrium point, say, $x = 0$, $y = 0$, $z = 0$. Then, we need 0 parameter to characterize the state, and the dimension is 0. Or equivalently, we say it has codimension 3, since there are 3 conditions specified. Similarly, a periodic orbit, say, $x^2 - 2z = A$, and $y^2 + z^2 = B^2$, has dimension 1 and codimension 2, since it needs 1 parameter to characterize the state and there are 2 conditions specified. The dimensions of these attractors are integral. However, for chaotic attractors, the dimensions are usually non-integral, or fractal.

To illustrate the notion of fractal dimension, we may refer to the example of Cantor set. The Cantor set is generated by iteration of an operation on the unit interval [0, 1]. The operation consists of removing the middle third of the interval, which is $\left(\frac{1}{3}, \frac{2}{3}\right)$ for [0, 1]. Repeating again of this process on the remaining intervals $\left[0, \frac{1}{3}\right]$ and $\left[\frac{2}{3}, 1\right]$ would yield 4 intervals : $\left[0, \frac{1}{9}\right]$, $\left[\frac{2}{9}, \frac{1}{3}\right]$, $\left[\frac{2}{3}, \frac{7}{9}\right]$ and $\left[\frac{8}{9}, 1\right]$. Repeating this process indefinitely, then we obtain the Cantor set. The Cantor set has infinite number of points but has zero length. The structure of the set is self-similar under a change of scale. A question to be asked is : what is the dimension of the set? It is presumably between 0 and 1.

One way to characterize such a set is the capacity dimension d_c. Let the set be covered by $N(\varepsilon)$ balls of diameter ε, the capacity dimension is defined by

$$d_c = \lim_{\varepsilon \to 0} \frac{\ln N(\varepsilon)}{\ln\left(\frac{1}{\varepsilon}\right)} . \tag{6.3.18}$$

Using the Cantor set as an example, we start with $\varepsilon = 1$ and $N = 1$. Next, we have $\varepsilon = \frac{1}{3}$, $N = 2$; and next $\varepsilon = \left(\frac{1}{3}\right)^2$, $N = 2^2$, etc . Thus when $N = 2^n$, we have $\varepsilon = \left(\frac{1}{3}\right)^N$. Therefore

$$\frac{\ln N(\varepsilon)}{\ln\left(\frac{1}{\varepsilon}\right)} = \frac{\ln(2^n)}{\ln(3^n)} = \frac{\ln 2}{\ln 3} = 0.631 . \tag{6.3.19}$$

Another dimension, the information dimension d_I, is related to the concept of entropy. In statistical mechanics, the entropy is defined as

$$\sigma = -\sum_i p_i \ln p_i , \tag{6.3.20}$$

where p_i is the probability that the system is in i^{th} state. Here we define the p_i be the occupation probability of the ith cell with size ε. Let the set be covered by N such cells, then the information dimension is defined as

$$d_I = \lim_{\varepsilon \to 0} \left(\frac{\sum_{i=1}^{N} p_i \ln p_i}{\ln \varepsilon} \right). \tag{6.3.21}$$

Take the example of Cantor set. At the N-th step, the cell size is $\left(\frac{1}{3}\right)^N$, while the occupational probability p_i is either $\left(\frac{1}{2}\right)^N$ or 0. There are 2^n cells with $p_i = \left(\frac{1}{2}\right)^N$. Therefore $\sum_{i=1}^{N} p_i \ln p_i = 2^N \cdot \left(\frac{1}{2}\right)^N \ln\left(\frac{1}{2}\right)^N = -N \ln 2.$ Now $\varepsilon = \left(\frac{1}{3}\right)^N$. Therefore

$$d_I = \frac{\ln 2}{\ln 3}, \tag{6.3.22}$$

the same as (6.3.19).

We may also define another dimension, i.e., the Lyapunov dimension, if we know the Lyapunov exponents. The Lyapunov exponent defined by (6.3.17) corresponds to the largest eigenvalue. Once the largest eigenvalue and the corresponding eigenvector have been extracted, the next-largest eigenvalue and its eigenvector can be found in principle, using the similar approach for vectors in the orthogonal subspace. Thus we may obtain the whole set of the Lyapunov exponents : $\lambda_1, \lambda_2, \cdots, \lambda_p$. Let us order the λ_i's by $\lambda_1 \geq \lambda_2 \geq \cdots \geq \lambda_p$, and let $\lambda_n \geq 0$, $\lambda_{n+1} < 0$. Then the Lyapunov dimension, d_L, is defined by

$$d_L = n + \frac{1}{|\lambda_{n+1}|} \left(\sum_{i=1}^{n} \lambda_i \right). \tag{6.3.23}$$

For experimental data or high dimensional systems, we may consider the correlation dimension d_r . The two-point correlation function of an attractor is

$$C(R) = \lim_{N\to\infty} \frac{1}{N^2} \sum_{i,j=1}^{N} H\left(R - \left|\mathbf{x}_i - \mathbf{x}_j\right|\right) , \qquad (6.3.24)$$

where $\mathbf{x}_i$ and $\mathbf{x}_j$ are points on the attractor and H is the Heaviside function, i.e., $H(y) = 1$, for $y > 0$, and $H(y) = 0$ for $y < 0$. Thus $C(R)$ is a measure of average fraction of points in a ball of radius R. Then d_r is defined by

$$d_r = \lim_{R\to 0} \frac{\ln C(R)}{\ln R} . \qquad (6.3.25)$$

Dynamical Information in Experimental Data

Given an attractor represented by $\{\mathbf{x}_n\}$, various dimensions defined above can be computed. Indeed, for the case of logistic map for instance, the variation of dimension versus the parameter λ can be obtained. As far as the technical determination of dimension is concerned, we are just dealing with a large set of points in p-dimensional space. For Lorenz attractor represented by $\mathbf{x}(t)$, we need to write $\mathbf{x}_n = \mathbf{x}(t_n)$ for successive t_n's, which should be appropriately chosen. For system governed by partial differential equation, the underlying degrees of freedom is infinite. How can we characterize the attractor, if a set of data for the system is given? Even more generally, if an experimental data set is given for a dynamical system, how can we characterize the attractor?

A scheme to extract hidden dynamical information from experimental data has been devised for this purpose. It is remarkable that only the data for a single degree of freedom, say only $X(t)$ for the Lorenz system, is needed for the construction of the attractor, even though the underlying degrees of freedom of the system is large or infinite. The rationale is that because the system is nonlinear, all the dynamical information is contained in any particular degree of freedom.

Let $u(t)$ be the experimental data for a property of the system. We now construct a p-dimensional vector $\mathbf{x}(t)$ from $u(t)$ by time-delayed copies of u :

$$\mathbf{x}(t) = \big(u(t),\ u(t+\tau),\ \cdots,\ u\big(t+[p-1]\tau\big)\big)\ , \quad (6.3.26)$$

where τ is the delay time. In this way, the time series $u(t)$ is now embedded in a p-dimensional vector space. Now we shall write

$$\mathbf{x}_n = \mathbf{x}\big(t + nt_s\big)\ , \quad (6.3.27)$$

where t_s is the sampling time of the data. The delay time τ is usually chosen to be a multiple of t_s in practice. Although there are some rules of thumb for the choice of delay time τ and the embedding dimension p, ultimately they are determined by various trials in practice. For instance, given p, we can obtain a correlation dimension $d_r^{(p)}$. If by successively increasing p , $d_r^{(p)}$ stabilizes to a definite value, then we have found the appropriate p. Similarly, by trying different τ, we may be able to settle for an appropriate value which will separate nearby trajectories in phase and yet not result in fuzzy appearance.

6.4 Almost Ill-posed Problems and Chaos

For instabilities of dynamical systems, in terms of partial differential equations, the simplest three instability mechanisms are negative damping, negative diffusion, and ellipticity.

For negative damping, the elemental representative equation is

$$\frac{\partial u}{\partial t} - \alpha u = 0\ , \quad \alpha > 0\ , \quad -\infty < x < \infty\ , \quad t > 0. \quad (6.4.1)$$

Thus, with initial condition, say,

$$u(x,0)=\frac{1}{k}\sin kx\ , \tag{6.4.2}$$

the solution is

$$u(x,t)=\frac{1}{k}\sin kxe^{\alpha t}\ . \tag{6.4.3}$$

Let us note that the growth rate of the instability, α , is independent of k.

For negative diffusion, the elemental representative equation is

$$\frac{\partial u}{\partial t}+\alpha\frac{\partial^2 u}{\partial x^2}=0\ ,\quad \alpha>0\ ,\ -\infty<x<\infty\ ,\quad t>0. \tag{6.4.4}$$

Take the initial condition again

$$u(x,0)=\frac{1}{k}\sin kx\ , \tag{6.4.5}$$

then the solution is

$$u(x,t)=\frac{1}{k}\sin kxe^{\alpha k^2 t}\ . \tag{6.4.6}$$

Note that the growth rate, αk^2, is proportional to k^2 . Also as $k\to\infty$, although $|u(x,0)|\to 0$, yet $|u(x,t)|\to\infty$ for any t.

For ellipticity, we may take Hadamard's example :

$$\frac{\partial^2 u}{\partial t^2}+\frac{\partial^2 u}{\partial x^2}=0\ ,\quad -\infty<x<\infty\ ,\quad t>0, \tag{6.4.7}$$

with initial conditions :

$$u(x,0)=\frac{1}{k^2}\sin kx\ , \qquad \frac{\partial u}{\partial t}(x,0)=\frac{1}{k}\sin kx\ . \tag{6.4.8}$$

Then the solution is

$$u(x,t)=\frac{1}{k^2}e^{kt}\sin kx\ . \tag{6.4.9}$$

The growth rate is k , and again as $k\to\infty$, even though $|u(x,0)|\to 0$, and $|u_t(x,0)|\to 0$, yet $|u(x,t)|\to\infty$ for any t .

For negative diffusion and ellipticity, the solutions are sensitively depending on the initial conditions, since infinitesimal initial conditions could result in unbounded and wild behavior a finite time later. Such problems are ill-posed. Systems that are ill-posed will collapse, and hence do not exist practically.

The ill-posed problems can be made well-posed if we add higher order diffusive terms. For instance, the equation

$$\frac{\partial u}{\partial t}+\alpha\frac{\partial^2 u}{\partial x^2}+\beta\frac{\partial^4 u}{\partial x^4}=0\ , \qquad \alpha,\ \beta>0 \tag{6.4.10}$$

with the initial condition (6.4.5) will lead to the solution

$$u(x,t)=\frac{1}{k}\sin kx e^{(\alpha k^2-\beta k^4)t}\ . \tag{6.4.11}$$

Thus the solution is very well behaved for large k and also the solution is bounded for all k at any finite t . Still, the system is unstable for small k , and the underlying elemental instability mechanism is negative diffusion. For small β, we may call the problem an almost ill-posed problem. Many hydrodynamic instabilities belong to this category, and their underlying elemental instability mechanisms can be identified as negative damping, negative diffusion or ellipticity.

Model partial differential equations have been constructed or derived for the study of the essential features of instability problems. We may mention the Ginzburg-Landau equation and the Kuramoto-Sivashinsky equation. The Ginzburg-Landau equation may be put in the following form :

$$\frac{\partial u}{\partial t} - \alpha u = (1 + i\beta)\frac{\partial^2 u}{\partial x^2} + (a + ib)|u|^2 u \,, \qquad (6.4.12)$$

where α , β , a and b are all real and $\alpha > 0$, $a < 0$. If we seek the normal mode solution $u(x, t) = v(t) \sin kx$ for the linearized equation, we obtain

$$v(t) = e^{\left[\alpha - (1+i\beta)k^2\right]t} \,. \qquad (6.4.13)$$

Therefore we may see that the elemental mechanism of the instability is the negative damping associated with the term α. The diffusion and nonlinear terms will make the problem well-behaved.

The Kuramoto-Sivashinsky equation can be put in the following form :

$$\frac{\partial u}{\partial t} + u\frac{\partial u}{\partial x} + \alpha\frac{\partial^2 u}{\partial x^2} + \beta\frac{\partial^4 u}{\partial x^4} = 0 \,, \qquad \alpha > 0 \,, \qquad \beta > 0 \,, \quad (6.4.14)$$

whose linear version is (6.4.10). Thus we see that its elemental mechanism of instability is negative diffusion.

Both Ginzburg-Landau equation and Kuramoto-Sivashinsky equation have been studied extensively. However, not much work has been carried out on model equations which exhibits ellipticity as the underlying elemental instability mechanism. In hydrodynamics, many stability problems have indeed the ellipticity as their elemental instability mechanism.

Examples of Underlying Ellipticity

Take the case of Rayleigh-Taylor instability. If we neglect the effect of surface tension, and consider the long wavelength disturbances, i.e., for $kh^{(1)} << 1$ and $kh^{(2)} << 1$, then the dispersion relation (5.2.29) becomes

$$n = \left[\frac{gh^{(1)}h^{(2)}\left(\rho^{(2)} - \rho^{(1)}\right)}{\rho^{(1)}h^{(2)} + g^{(2)}h^{(1)}} \right]^{\frac{1}{2}} k \ . \tag{6.4.15}$$

Thus the growth rate is proportional to k, indicating that the underlying elemental instability mechanism is ellipticity.

We may actually demonstrate this connection more explicitly from another perspective. Consider the Kelvin-Helmholtz problem for two incompressible fluid layers with thickness $h_1(x,t)$ and $h_2(x,t)$ moving with $u_1(x,t)$ and $u_2(x,t)$ respectively in the gravitational field. We shall again consider only the long wavelength disturbances. Thus we shall employ the approximate procedure which was employed in Section 3.5 and Section 5.12. The governing equations are then

$$\frac{\partial h_1}{\partial t} + \frac{\partial}{\partial x}(u_1 h_1) = 0 \ , \tag{6.4.16}$$

$$\frac{\partial h_2}{\partial t} + \frac{\partial}{\partial x}(u_2 h_2) = 0 \ , \tag{6.4.17}$$

$$\frac{\partial u_1}{\partial t} + u_1 \frac{\partial u_1}{\partial x} = -\frac{1}{\rho_1}\frac{\partial p}{\partial x} - g\frac{\partial h_1}{\partial x} \ , \tag{6.4.18}$$

$$\frac{\partial u_2}{\partial t} + u_2 \frac{\partial u_2}{\partial x} = -\frac{1}{\rho_2}\frac{\partial p}{\partial x} + g\frac{\partial h_2}{\partial x} \ , \tag{6.4.19}$$

and

$$h_1(x,t) + h_2(x,t) = H \ , \tag{6.4.20}$$

where H is the constant total thickness of the fluid layers. We can discuss the characteristics of the nonlinear system (6.4.16) -

(6.4.20) without any difficulty. But, for our purpose, it is more illuminating to discuss the linear case. Let

$$h_1 = H_1 + h\ , \quad h_2 = H_2 - h\ , \tag{6.4.21}$$

$$u_1 = U_1 + u_1'\ , \quad u_2 = U_2 + u_2'\ , \tag{6.4.22}$$

where H_1 , H_2 , U_1 and U_2 are the steady state quantities. Then the linearized equations are

$$\frac{\partial h}{\partial t} + U_1 \frac{\partial h}{\partial x} + H_1 \frac{\partial u_1'}{\partial x} = 0\ , \tag{6.4.23}$$

$$\frac{\partial h}{\partial t} + U_2 \frac{\partial h}{\partial x} - H_2 \frac{\partial u_2'}{\partial x} = 0\ , \tag{6.4.24}$$

$$\rho_1\left[\frac{\partial u_1'}{\partial t} + U_1 \frac{\partial u_1'}{\partial x} + g\frac{\partial h}{\partial x}\right] = \rho_2\left[\frac{\partial u_2'}{\partial t} + U_2 \frac{\partial u_2'}{\partial x} + g\frac{\partial h}{\partial x}\right]. \tag{6.4.25}$$

Eliminating u_1' and u_2' from (6.4.23) - (6.4.25), we obtain

$$A\frac{\partial^2 h}{\partial t^2} + 2B\frac{\partial^2 h}{\partial x \partial t} + C\frac{\partial^2 h}{\partial x^2} = 0\ , \tag{6.4.26}$$

where

$$A = -\left(\frac{\rho_1}{H_1} + \frac{\rho_2}{H_2}\right), \tag{6.4.27}$$

$$B = -\left(\frac{\rho_1 U_1}{H_1} + \frac{\rho_2 U_2}{H_2}\right), \tag{6.4.28}$$

$$C = g(\rho_1 - \rho_2) - \left(\frac{\rho_1 U_1^2}{H_1} + \frac{\rho_2 U_2^2}{H_2}\right). \tag{6.4.29}$$

Equation (6.4.26) is clearly elliptic if $B^2 - AC < 0$. It can be readily verified that

$$B^2 - AC$$

$$= \frac{1}{H_1 H_2}\left[g(\rho_1 - \rho_2)(\rho_1 H_1 + \rho_2 H_2) - \rho_1\rho_2(U_1 - U_2)^2\right]. \quad (6.4.30)$$

Thus the condition that $B^2 - AC < 0$ is equivalent to $(U_1 - U_2)^2 > U_c^2$, where U_c^2 is given in (5.5.39).

Another example is the Rayleigh-Benard instability. We shall start with (6.1.6) and (6.1.7) :

$$\frac{\partial}{\partial t}(\nabla^2\psi) = -\frac{\partial(\psi, \nabla^2\psi)}{\partial(x,z)} + \nabla^4\psi + \frac{R_a}{\sigma}\frac{\partial\theta}{\partial x}, \quad (6.4.31)$$

$$\frac{\partial\theta}{\partial t} = -\frac{\partial(\psi,\theta)}{\partial(x,z)} + \frac{\partial\psi}{\partial x} + \frac{1}{\sigma}\nabla^2\theta. \quad (6.4.32)$$

In view of (6.1.8) and (6.1.9), it may be a reasonable approximation to replace $\nabla^2\psi$ by $(-b^2\psi)$ and $\nabla^2\theta$ by $(-b^2\theta)$ for the purpose to extract information about the elemental mechanism for instability. Thus, (6.4.31) and (6.4.32) become

$$\frac{\partial\psi}{\partial t} = -b^2\psi - \frac{R_a}{\sigma b^2}\frac{\partial\theta}{\partial x}, \quad (6.4.33)$$

$$\frac{\partial\theta}{\partial t} = -\frac{\partial\psi}{\partial x}\frac{\partial\theta}{\partial z} + \frac{\partial\psi}{\partial z}\frac{\partial\theta}{\partial x} + \frac{\partial\psi}{\partial x} - \frac{b^2}{\sigma}\theta. \quad (6.4.34)$$

Let us further replace $\frac{\partial\theta}{\partial z}$ by $(\alpha\theta)$ and $\frac{\partial\psi}{\partial z}$ by $(\alpha\psi)$. We do not have legitimate justification to make such approximations. We could argue that we would be interested in the equations averaged over z . The terms $\frac{\partial\theta}{\partial z}$ and $\frac{\partial\psi}{\partial z}$ would in some sense leave an influence which is proportional to θ and ψ respectively. If we accept this simplification, then (6.4.34) becomes

$$\frac{\partial\theta}{\partial t}=\alpha\left(\psi\frac{\partial\theta}{\partial x}-\theta\frac{\partial\psi}{\partial x}\right)+\frac{\partial\psi}{\partial x}-\frac{b^2}{\sigma}\theta . \quad (6.4.35)$$

The characteristics equation for the system (6.4.33) and (6.4.35) is given by

$$\begin{vmatrix} -\lambda & \frac{R_a}{\sigma b^2} \\ \alpha\theta-1 & -\lambda-\alpha\psi \end{vmatrix}=0 , \quad (6.4.36)$$

or

$$\lambda^2+\alpha\psi\lambda+\frac{R_a}{\sigma b^2}(1-\alpha\theta)=0 . \quad (6.4.37)$$

Therefore

$$\lambda=\frac{1}{2}\left\{-\alpha^2\psi^2\pm\left[\alpha\psi-4(1-\alpha\theta)\frac{R_a}{\sigma b^2}\right]^{\frac{1}{2}}\right\} . \quad (6.4.38)$$

It is clear that no matter what is the value of α , when $|\psi|$ and $|\theta|$ are small, λ is not real. Therefore the system is elliptic for small disturbances. In other words the elemental mechanism of the instability is the ellipticity.

Model Partial Differential Equations with Underlying Ellipticity

A system of partial equation with underlying ellipticity has been proposed [Hsieh, 1987], which has the following form :

$$\frac{\partial \psi}{\partial t} = -\sigma\psi - \sigma\frac{\partial \theta}{\partial x}, \qquad (6.4.39)$$

$$\frac{\partial \theta}{\partial t} = -(1-\beta)\theta + \beta\frac{\partial^2 \theta}{\partial x^2} + r\frac{\partial \psi}{\partial x} + 2\psi\frac{\partial \theta}{\partial x}. \qquad (6.4.40)$$

This system resembles the system (6.4.33) and (6.4.35), except that a diffusive term $\beta\frac{\partial^2 \theta}{\partial x^2}$ is included in (6.4.40) to make the system well-posed. The system, like the Rayleigh-Benard problem, also contains the Lorenz equation. To be more specific, let us take

$$\psi = 2^{-\frac{1}{2}} X(t)\sin x\ , \qquad (6.4.41)$$

$$\theta = 2^{-\frac{1}{2}} Y(t)\cos x + \frac{1}{2}Z(t)\cos 2x\ , \qquad (6.4.42)$$

and substitute (6.4.41) and (6.4.42) into (6.4.39) and (6.4.40). If we retain only the coefficients of sin x, cos x and cos $2x$, then we obtain the following Lorenz equations :

$$\frac{dX}{dt} = -\sigma X + \sigma Y\ , \qquad (6.4.43)$$

$$\frac{dY}{dt} = -Y + rX - XZ\ , \qquad (6.4.44)$$

$$\frac{dZ}{dt} = -(1+3\beta)Z + XY\ . \qquad (6.4.45)$$

Thus it is reasonable to expect chaotic solutions from the system (6.4.39) and (6.4.40). We may also remark that if we replace (6.4.39) with the following equation,

$$\frac{\partial \psi}{\partial t} = -(\sigma-\alpha)\psi - \sigma\frac{\partial \theta}{\partial x} + \alpha\frac{\partial^2 \psi}{\partial x^2}, \qquad (6.4.46)$$

the system will be even more well behaved, yet still contains the Lorenz equations.

The variable θ can be eliminated from (6.4.39) and (6.4.40) by differentiating (6.4.40) with respect to x, and we obtain

$$\frac{\partial^2\psi}{\partial t^2}+\sigma(r-\beta)\frac{\partial^2\psi}{\partial x^2}-2\psi\frac{\partial^2\psi}{\partial x\partial t}-\beta\frac{\partial^3\psi}{\partial x^2\partial t}+(\sigma+1-\beta)\frac{\partial\psi}{\partial t}$$

$$-2\frac{\partial\psi}{\partial x}\frac{\partial\psi}{\partial t}-4\sigma\psi\frac{\partial\psi}{\partial x}+\sigma(1-\beta)\psi=0\,. \tag{6.4.47}$$

To investigate the underlying instability mechanism of the system, we shall set $\beta=0$, i.e., remove the "good" diffusive term. Then (6.4.47) becomes a second order partial differential equation of the mixed type, whose characteristics are given by

$$\frac{dx}{dt}=-\psi\pm\left[\psi^2-\sigma r\right]^{\frac{1}{2}}\,. \tag{6.4.48}$$

Thus the equation is elliptic when $\psi^2<\sigma r$ and hyperbolic when $\psi^2>\sigma r$. In other words, when r is large or when ψ is small, the equation is elliptic. Suppose initial values with small ψ is prescribed. The instability due to ellipticity will cause ψ^2 to grow with t. But when ψ^2 exceeds the value of σr , the equation becomes hyperbolic, hence the initial unstable growth will be arrested. However the system is actually dissipative since it has also lower order damping terms. These dissipative mechanisms will tend to diminish the magnitude of ψ^2, and drive the system back to the elliptic regime. When the equation is elliptic, the instability mechanism will operate again and raise the magnitude of ψ^2 over the value of σr , thus push the system to the hyperbolic regime. This type of switching back and forth from ellipticity to hyperbolicity may be what corresponds to the chaotic behavior of the Lorenz system.

The underlying ellipticity of the system (6.4.39) and (6.4.40) makes it difficult to carry out reliable numerical computation of solutions of the system. The system (6.4.40) and (6.4.46) is more manageable both from analytical and computational perspectives. When reliable solutions are available, we may treat $\psi(x,t)$ for some definite x as a time series from some "experiment", and use the scheme described in Section 6.3 to extract dynamical information of the system. For instance, it may then be possible to obtain the dimension of the system as a function of the parameters r , σ and β of the system. A dynamical system characterization of the partial differential equation can thus be established.

APPENDIX 1

Some Properties of Spherical Bessel Functions

Spherical Bessel functions $j_l(x)$ and $n_l(x)$ are defined as follows :

$$j_l(x) = \left(\frac{\pi}{2x}\right)^{1/2} J_{l+\frac{1}{2}}(x) = (-x)^l \left(\frac{1}{x}\frac{d}{dx}\right)^l \left(\frac{\sin x}{x}\right), \quad \text{(A1.1)}$$

$$n_l(x) = \left(\frac{\pi}{2x}\right)^{1/2} Y_{l+\frac{1}{2}}(x) = -(-x)^l \left(\frac{1}{x}\frac{d}{dx}\right)^l \left(\frac{\cos x}{x}\right), \quad \text{(A1.2)}$$

where $J_l(x)$ and $Y_l(x)$ are Bessel functions of first and second kind respectively.

The spherical Hankel functions are given by :

$$h_l^{(1)} = j_l + in_l, \quad h_l^{(2)} = j_l - in_l \, . \quad \text{(A1.3)}$$

As examples :

$$j_0(x) = \frac{\sin x}{x}, \quad j_1(x) = \frac{\sin x}{x^2} - \frac{\cos x}{x}, \quad \text{(A1.4)}$$

$$n_0(x) = -\frac{\cos x}{x}, \quad n_1(x) = -\frac{\cos x}{x^2} - \frac{\sin x}{x}, \quad \text{(A1.5)}$$

$$h_0^{(1)}(x) = \frac{e^{ix}}{ix}, \quad h_1^{(1)}(x) = -\frac{e^{ix}}{x}(1 + \frac{i}{x}) \, . \quad \text{(A1.6)}$$

For $x << l$:

$$j_l(x) \approx \frac{x^l}{(2l+1)!!}, \quad n_l(x) \approx -\frac{(2l-1)!!}{x^{l+1}}, \quad \text{(A1.7)}$$

where

$$(2l+1)!! = (2l+1)(2l-1)\cdot\cdots\cdot 5\cdot 3\cdot 1 .$$

For $x >> l$:

$$j_l(x) \approx \frac{1}{x}\sin\left(x - \frac{l\pi}{2}\right), \quad n_l(x) \approx -\frac{1}{x}\cos\left(x - \frac{l\pi}{2}\right), \quad \text{(A1.8)}$$

$$h_l^{(1)}(x) \approx (-i)^{l+1}\frac{e^{ix}}{x}, \qquad h_l^{(2)}(x) \approx i^{l+1}\frac{e^{-ix}}{x} . \qquad \text{(A1.9)}$$

APPENDIX 2

On Surface Tension

Figure A-1 shows schematically an interface. Let A be any region on the

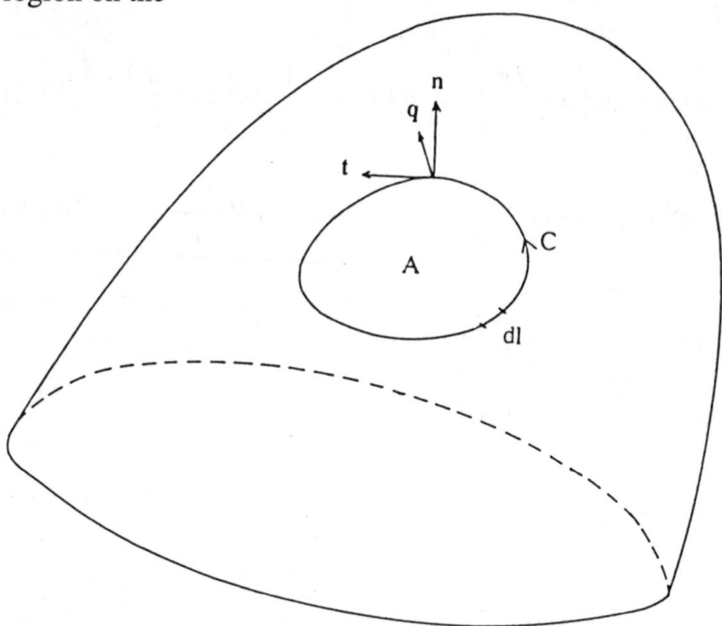

Fig A-1 Region A bounded by a closed curve C on the interface.

interface bounded by a closed curve C. Let **t** be the unit tangent vector at a given point on C, and **n** the unit vector normal to the interface. The tangent vector **t**, which defines the direction of C, and **n** form a right hand system. Let **q** = **t** x **n**, and let dl be the line element on C. The vector **q** is tangent to the interface and point outward from C. The surface tension force is pointing in the direction of **q**, reminiscent of a rubber band. Let σ be the surface tension coefficient, then the total surface tension force on C is

$$\mathbf{T} = \oint_C \sigma(\mathbf{t} \times \mathbf{n}) dl \; . \tag{A2.1}$$

Making use of the generalized Stokes theorem, (A2.1) can be rewritten as

$$\mathbf{T} = \int_A (\mathbf{n} \times \nabla) \times (\sigma\mathbf{n}) d^2\mathbf{x}$$

$$= \int_A \{-\mathbf{n}[\nabla \cdot (\sigma\mathbf{n})] + \nabla\sigma\} d^2\mathbf{x} \ . \qquad \text{(A2.2)}$$

Since A is arbitrary, take an infinitesimal region, the interfacial surface force density S_s is thus

$$S_s = -\mathbf{n}[\nabla \cdot (\sigma\mathbf{n})] + \nabla\sigma \ . \qquad \text{(A2.3)}$$

Associated with the interface is the interfacial energy E_s. If the interfacial area is A, then the interfacial energy associated with A is

$$E_s = \sigma A \ . \qquad \text{(A2.4)}$$

Now because the interfacial area is changing due to its motion, there is also corresponding change of interfacial energy. Referring to the definition of w_n in (5.1.15), the increase of interfacial energy per unit time and unit area is thus $(-\rho Q_s)$, and

$$-\rho Q_s = \sigma w_n \left(\frac{1}{R_1} + \frac{1}{R_2} \right) . \qquad \text{(A2.5)}$$

Thus, using (5.1.15), we have

$$\rho Q_s = -\frac{\sigma \frac{\partial F}{\partial t}}{|\nabla F|} \left(\frac{1}{R_1} + \frac{1}{R_2} \right) . \qquad \text{(A2.6)}$$

Books for Concurrent and Further Reading

Chandrasekhar, S., **Hydrodynamic and Hydromagnetic Stability**, Dover Publications, New York, 1981.

Drazin, P.G. and Reid, W.H., **Hydrodynamic Stability**, Cambridge University Press, 1981.

Landau, L.D. and Lifshitz, E.M., **Fluid Mechanics**, Pergamon Press, New York, 1959.

Lighthill, J., **Waves in Fluids**, Cambridge University Press, 1978.

Lin, C.C., **The Theory of Hydrodynamics Stability**, Cambridge University Press, 1955.

Whitham, G.B., **Linear and Nonlinear Waves**, John Wiley, New York, 1974.

REFERENCES

Gardner, C.S., Greene, J.M., Kruskal, M.D. & Miura, R.M., Korteweg-deVries equation and generalization, VI, Methods for exact solution, *Comm. Pure & Appl . Math.*, **27**, 97-133 (1974). [p. 148].

Gel'fand, I.M. & B.M. Levitan, On the determination of a differential equation from its spectral function, *Am. Math. Soc. Transl. Ser. 2.1*, 253-304 (1955). [p. 147].

Hasimoto, H & Ono, H., *J. Phys. Soc. Japan*, **33**, 805-811 (1972). [p. 153].

Hsieh, D.Y., Gravity Waves in Lagrangian Coordinates, in *Engineering Science, Fluid Dynamics* ed. by G.E. Yates, World Scientific Publishing Co., Singapore (1990) pp. 3-15. [p. 161].

Hsieh, D.Y., Effects of Heat and Mass Transfer on Rayleigh-Taylor Stability, *Trans. ASME*, **94D**, 156-162 (1972). [p. 234].

Hsieh, D.Y., Interfacial Stability with Mass and Heat Transfer, *Phys. Fluids*, **21**, 745-748 (1978). [p. 237].

Hsieh, D.Y., Nonlinear Rayleigh-Taylor Stability with Mass and Heat Transfer, *Phys. Fluids*, **22**, 1435-1439 (1979). [p. 237].

Hsieh, D.Y. & Ho, S.P., Rayleigh-Taylor Stability with Mass and Heat Transfer, *Phys. Fluids*, **24**, 202-208 (1981). [p. 242].

Hsieh, D.Y. & Chen, F., A Nonlinear Study of Kelvin-Helmholtz Stability, *Phys. Fluids*, **28**, 1253-1262 (1985). [p. 270].

Hsieh, D.Y., Kelvin-Helmholtz Stability and Two-Phase Flow, *Acta Math. Sci.* **9**, 189-197 (1989). [p. 282].

Ho, S.P., Linear Rayleigh-Taylor Stability of Viscous Fluids with Mass and Heat Transfer, *J. Fluid Mech.*, **101,** 111-128 (1980). [p. 234].

Landau, L.D., *Compt. Rend. (Doklady), Acad. Sci. URSS*, **44**, 139 (1944). [p. 275].

Shen, S.F., Calculated amplified oscillations in plane Poiseuille and Blasius flows, *J. Aero. Sci.*, **6**, 496-503 (1954). [p. 341].

Taylor, G.I., Stability of a viscous liquid contained between two rotating cylinders, *Phil. Trans. A*, **223**, 289-343 (1923). [p. 324].

Westervelt, P.J., Scattering of Sound by Sound, *J. Acoust. Soc. Am.*, **29**, 199-203 (1957). [p. 48].

Westervelt, P.J., Scattering of Sound by Sound, *J. Acoust. Soc. Am.*, **29**, 934-935 (1957). [p. 48].

Westervelt, P.J., Parameter Acoustic Array, *J. Acoust. Soc. Am.*, **35**, 535-537 (1963). [p. 48].

INDEX